Introduction to Fluid Dynamics

Introduction to Fluid Dynamics

Understanding Fundamental Physics

Young J. Moon
Korea University
Seoul, Korea, Rep. of

Registered Office
John Wiley & Sons, Inc., 111 River Street, Hoboken, NJ 07030, USA

Editorial Office
111 River Street, Hoboken, NJ 07030, USA

For details of our global editorial offices, customer services, and more information about Wiley products visit us at www.wiley.com.

Wiley also publishes its books in a variety of electronic formats and by print-on-demand. Some content that appears in standard print versions of this book may not be available in other formats.

Library of Congress Cataloging-in-Publication Data applied for:

ISBN: 9781119823155

Cover image: © zeta ophiuchi/Pixabay
Cover design by Wiley

Set in 9.5/12.5pt STIXTwoText by Straive, Chennai, India

SKY10035635_081722

Contents

Preface

Fluid dynamics has a long history of more than 400 years. Although it is a classical subject, fundamental physics is neither easy to understand nor simple to convey to students. This book aims to guide undergraduate and beginning graduate students in engineering and science to understand the fundamental physics of fluid dynamics. To inspire students to be curious about the physics, this book begins with questions on the nature of forces; how forces originate and get distributed in the field, how fluids react to these forces with inertia, and how fluid motions are related to hydrostatic and viscous stresses. To stimulate the readers' curiosity, an elementary level of geophysical fluid dynamics, biofluid mechanics, and aeroacoustics are introduced, and some of the problem sets are carefully selected to illustrate flow physics often encountered in nature such as the Great Red Spot on Jupiter, jet streams, flying frogs, dragon lizards (*Draco*), frogfish, whales, crab nebula in the constellation of *Taurus*, interstellar bow shock wave (*Zeta Ophiuchi*).

In this book, the fundamentals of fluid dynamics are covered in Chapters 1, 2, and 3: pressure, macroscopic balance of mass, momentum, and energy, and differential equations of motions. To provide an in-depth physical explanation of the basic concepts, each subject is dealt with the physics of fluid motion and their relevances to the conservation laws of mass, momentum, and energy. Chapter 1 focuses specifically on discussing what is the nature of pressure and how pressure varies with inertia in transient and steady flows. Chapter 2 uses the concept of fluid particles to explain how the laws of conservation of mass, momentum, and energy apply to fluids. Chapter 3 covers the fundamental physics of viscous flow, that is, the deviatoric nature of viscous stresses and strain rates. The Navier–Stokes equations are derived based on the constitutive relation between the viscous stress and the strain rate.

Chapters 4 and 5 discuss the general concepts of physics in curved motion and vortex dynamics, in which the centrifugal force coupled with viscosity governs the dynamics. To discover the complex nature of vortical flows, the Lamb acceleration vector and its divergence and curl are brought to our attention. The divergence and curl of the Lamb vector are of great importance in describing how rotating fluids do viscous shear work or dissipate energy at a rate in a shear-straining field, or how vortices change the strength while being convected, stretched, tilted, or isotropically expanded. Many curved flows in aerodynamics, geophysical fluid dynamics, and biofluid dynamics are introduced from the perspective of vortex dynamics.

In Chapters 6 and 7, the basic physical concepts covered in Chapters 1, 2, 3, 4, and 5 are practiced for various external and internal viscous flows. Chapter 6 discusses the physics of boundary layers, stability and transition, turbulence, separation, drag and lift, aerodynamics, and gliding. Chapter 7 focuses on the hydrodynamic resistance of internal viscous flows with energy supply and frictional losses. Related topics include frictional losses in pipe and duct flow, local losses in systems with flow separation and secondary flow, and valves. Biofluid mechanics share the same interests as internal viscous flows in topics such as blood vessels, branches, and stenosis (or aneurysms), but differs only in scale (e.g. mosquito food canal, fish gill).

Lastly, Chapter 8 covers the compressible flow physics of sound waves, Mach waves, shock waves, and isentropic flows in rocket nozzles. The fundamental physics of each topic is based on the speed of sound, the Doppler effect, and the volumetric dilatational rate, that is, the rate of change of volume associated with the compression or expansion work done on a fluid at a rate. The physics of compressible flows can also be observed in our planetary systems. Waves in plasma, similar to waves in fluids, are brought for discussion on the relationship between waves and forces.

This book aims to introduce students to a simple yet unique methodology by which to learn fluid dynamics through an understanding of fundamental physics, and we will all do our best to achieve it.

Seoul   *Young J. Moon*
January 2022

Acknowledgments

In preparation of this book, various textbooks and materials are referenced. Among those, I would like to express my special appreciation to *An Introduction to Fluid Dynamics* by G.K. Batchelor, *An Informal Introduction to Theoretical Fluid Mechanics* by J. Lighthill, and *Illustrated Experiments in Fluid Mechanics* by National Committee for Fluid Mechanics Films (prefaced by A.H. Shapiro). I am also grateful to the undergraduate and graduate students of the School of Mechanical Engineering and also to the graduate students in the Computational Fluid Dynamics and Acoustics Laboratory, Korea University, Seoul, where I have lectured fluid dynamics for the past 30 years. I thank Minsung Kim for his assistance in preparing this manuscript. Lastly, I would like to extend my gratitude to my dear wife, *Inmee*, and my daughters, *Min* and *Hee*, for their constant support and advices.

About the Companion Website

This book is accompanied by a companion website.

www.wiley.com/go/Moon/IntroductiontoFluidDynamics

This website includes:

- Instructor Manual

1

Pressure

1.1 Microscopic Physical Properties

An element of a fluid or solid may be viewed as a continuous space with homogeneously distributed molecules. The differences between fluids and solids lay in the molecular spacings and intermolecular forces. In a solid phase, molecules are closely and orderly arranged, having a cohesive structure. Fluids have material properties that differ vastly from those of solids. Gas molecules are spaced far apart and disorderly arranged, and the cohesiveness among the molecules is very weak. In contrast, molecules in the liquid phase have partially ordered arrangements and a cohesiveness that lies in between those of solid and gas molecules, thus exhibiting mixed properties of both states of matter. Furthermore, liquid molecules are often grouped, which constantly break and reform when liquid elements are in relative motions.[1] Many physical properties of liquids can be explained from this unique feature.

1.1.1 Continuum

Fluids are made up of billions and billions of molecules. For example, air at sea-level ($T = 15\,^{\circ}C$ and $p = 101,330\ N/m^2$) contains 3×10^7 molecules in a cube of dimensions $1\ \mu m \times 1\ \mu m \times 1\ \mu m$, i.e. a volume of $10^{-9}\ mm^3$. If our scale of interest is larger than the distance between molecules (e.g. roughly $10\ nm$), fluid properties such as density, pressure, temperature can be described by continuous functions of space and time; we then can say the medium is a *continuum*. With the continuum assumption, physical quantities can be defined at a point $\vec{x}$, representing an averaged value of the molecules surrounding that point.[2]

This continuum hypothesis is valid for a large-scale flow over an aircraft wing (e.g. chord length: $5\ m$ and span: $15\ m$). It even works for a small-scale flow over a mosquito wing (e.g. chord length: $1\ mm$ and span: $3\ mm$), since there exists roughly 0.1 million air molecules across the mosquito's wing chord. However, the continuum model fails for air at very high altitudes because the molecular spacings become too large (e.g. rarefied gas).

The validity of the continuum hypothesis is determined by the *Knudsen* number

$$\kappa_n = \lambda/l \tag{1.1}$$

where λ is the mean free path of the molecules and l is the characteristic length. For example, $\lambda = 0.1\ \mu m$ at the standard atmospheric condition. However, $\lambda = 16\ cm$ at an elevation of $100\ km$,

---

1 Batchelor [1].
2 It has been experimentally demonstrated that a constant value of the measured fluid density fluctuates, as the volume of the fluid to which instrument responds approaches zero.

Introduction to Fluid Dynamics: Understanding Fundamental Physics, First Edition. Young J. Moon.
© 2022 John Wiley & Sons, Inc. Published 2022 by John Wiley & Sons, Inc.
Companion website: www.wiley.com/go/Moon/IntroductiontoFluidDynamics

$\lambda = 50\ m$ at 160 km, and $\lambda = 20\ km$ at 300 km. If κ_n is less than 0.01, the medium can be assumed as a continuum.

1.2 Forces

In fluids, forces originate either from within fluid volumes or at external boundaries. In response to the forces, fluids move[3] with inertia; thus, forces are distributed in the flow field, changing the momentum of the fluid at a rate. It is to be noted that this interaction occurs under the laws of conservation of mass, momentum, and energy.

1.2.1 Surface Forces

Surface forces act across a surface element via direct contacts of fluid molecules, and these forces defined by per unit area are called stresses. There are two different types of stresses: pressure (or hydrostatic) and viscous stresses. Pressure represents a measure of repulsiveness of molecules against compressive forces. It acts in the direction normal and inward to the surface of a unit area, and the magnitude is direction-independent (i.e. isotropic).

Meanwhile, viscous stress represents a measure of resistance of molecules when a fluid element is strained (or deformed) at a rate against frictions. Thus, the magnitude is proportional to the viscosity[4] and the straining rate of the fluid. Due to the nature of frictional intermolecular interactions, a viscous stress vector acts on the surface at an angle, which depends on the orientation of the surface.[5] It is a direction-dependent quantity (i.e. nonisotropic).

It is worth to note that the magnitude of the surface force is proportional to the contact surface area of the fluid. For example, one way to increase the lift of an aircraft is to simply increase the surface area, i.e. a planform area of the aircraft such as wings, fuselage. On the other hand, engineers also make efforts to reduce the increased frictional drag, using surface controls such as riblets, re-laminarization of the boundary layer.[6]

1.2.2 Volumetric Body Forces

In the fields of gravity, electricity, and magnetism, a force acts through the volume of a body that carries mass or electric charge. This is called volumetric body force. The most common volumetric body forces are the gravitational body force and the Lorentz force. The centrifugal and Coriolis forces are also volumetric body forces, but these are fictitious; they are only concerned when the equation of motion is transformed from an inertial reference frame to a rotating reference frame. The Coriolis force is particularly important in geophysical fluid dynamics because it is a

---

3 To be more specific, fluids can be translated, rotated, or strained at a rate, depending on the type of forces and boundary conditions. More discussion continues in Chapter 3 *Differential Equations of Motions*.
4 Viscosity is a physical property of a fluid that represents the measure of with how much difficulty a fluid element is strained at a rate against frictional forces. It is defined as the ratio between the given viscous stress and the strain rate.
5 This angle is zero to a surface parallel to the externally applied tangential shear force vector but changes to 90 degrees to a surface rotated by 45 degrees. More discussion continues in Chapter 3 *Differential Equations of Motions* (see Section 3.3.2 *The principal axes of strain rate*).
6 The wall shear stress can be lowered by the surface control techniques. More discussion continues in Chapter 6 *External Viscous Flows*.

nonconservative body force that produces rotation; in this case, the line of action of the net force does not go through the center of mass of the fluid particles.[7]

1.3 Pressure

In a volume of space, fluid molecules in random oscillations retain a certain level of repulsiveness. If a small piece of solid is immersed in a fluid, the fluid molecules exert normal, compressive forces to the surface of the solid, and as the body's volume shrinks to a point, the normal force per unit area (or normal stress) becomes independent of direction. This omnidirectional and compressive forces per unit area is called pressure.

Gases are compressible fluids that can change volume due to normal force or heat. According to Boyle's law, if a gas is slowly compressed by normal force, pressure is inversely proportional to gas volume at a constant temperature T

$$pV = p/\rho = \text{constant}, \tag{1.2}$$

where p, ρ, and V are the pressure, density, and volume, respectively. If the gas is slowly heated up with a fixed pressure p, Charles' law states:

$$V/T = 1/(\rho T) = \text{constant} \tag{1.3}$$

With Boyle's and Charles' laws, we can now obtain the ideal gas law:

$$p = \rho R T \tag{1.4}$$

where R is the universal gas constant. The ideal gas law shows that pressure of a gas can be defined by two state variables in thermodynamic equilibrium, e.g. density and temperature.

On the contrary, liquid pressure cannot be defined by the equation of state with two thermodynamic variables. As shown in the p–V diagram (Figure 1.1), liquid volume can hardly be changed by compressive forces, but its pressure can easily be changed to suit the local force equilibrium. If an external normal force is applied to compress a liquid, an electrostatic repulsion immediately acts as a restoring force to resist the external deformation.[8] In a cup of water, for example the volume

Figure 1.1 p–V diagram and isothermal lines of liquid and vapor

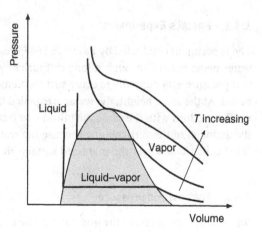

---

7 More discussion continues in Chapter 5 *Vortex Dynamics* (see Section 5.8 *Nonconservative body-force torques*).
8 In liquids, there is an optimum spacing for the molecules, which is the spacing corresponding to the minimum in the potential energy curve. Hirschfelder et al. [2].

of water remains the same, but the pressure changes vertically due to its own weight. However, the volume of the liquid more easily changes with heat, as a result, evaporation results in a large normal compressive force.

Example 1.1 *Physical illustration of pressure*
A bullet is fired into two paint cans, one of which is full of liquid and the other full of air (YouTube, *W. Lewin, MIT, Classical Physics, Lec-27*) [3]. Which one will explode and which one will not?

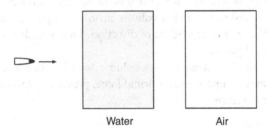

Water   Air

This experiment demonstrates three physical principles:

(i) The compressibility of the fluid: nearly incompressible water vs. compressible air.
(ii) Squeezing in the bullet into the paint can containing water causes the pressure inside the can to substantially increase (Figure 1.1). As a result, the paint can must explode because of the pressure force. The can containing air will not explode because the air locally changes the volume with an increase of pressure, forming a shock wave.
(iii) The force exerted by the bullet is transmitted through the fluid by pressure. Pascal's law states that a change in pressure at any point in an enclosed fluid at rest is transmitted undiminished to all points in the fluid.

1.4 Pressure in a Fluid at Rest

1.4.1 Pascal's Experiment

The experiment conducted by Pascal in 1646 presents many interesting truths about pressure. The water being poured through a long cylindrical vertical tube on top of a barrel applies weight so that pressure acts normally to every surface element of the water as well as to the inner wall of the barrel. At the same height, the water is expelled through the holes with equal strength since it acts in all directions with the same magnitude. In fact, the pressure in the barrel does not change with the diameter of the vertical tube because the water weight augmented by increasing the diameter is to be distributed over the increased surface area.

1.4.2 Hydrostatic Pressure

When fluids are at rest in the gravitational field, there is no other external force acting on the fluid except its own weight. In this case, the action force, i.e. gravitational body force F proportional to the fluid volume and density, is locally in static force equilibrium with the reaction force F from the other side that supports the fluid above. As a result, the fluid is in the state of compression

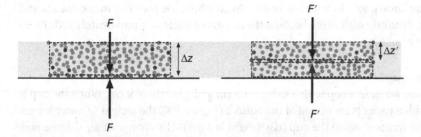

Figure 1.2 Pressure (or hydrostatic stress) determined by two forces in action and reaction; the top and bottom boundaries are exposed to the ambient pressure

Figure 1.3 Hydrostatic stresses acting on the surfaces of an infinitesimal fluid element

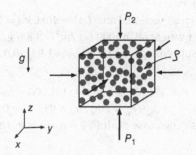

by two forces in action and reaction, and the pressure (or hydrostatic stress) at this point is defined per unit area with these two normal forces in local static equilibrium (Figure 1.2).

For an infinitesimal element of fluid, for example a static force balance can be set in the vertical direction as follows:

$$-dp \, (dx \, dy) - \rho_0 \, g \, (dx \, dy \, dz) = 0 \tag{1.5}$$

where $dp = p_2 - p_1$ is the pressure difference, the density ρ_0 is a constant for incompressible fluid, and g is the gravitational acceleration (Figure 1.3).

By integrating Eq. (1.5) from 0 to z, we can obtain a pressure distribution:

$$p(z) = p(0) - \rho_0 \, gz \tag{1.6}$$

where $p(0)$ is the pressure at the reference point, and $p(z)$ is called the hydrostatic pressure. Equation (1.5) also proves that as the volume of the fluid element approaches zero, $dp = p_2 - p_1$ must approach zero since the area of the element is one order larger.

1.4.3 Earth Surrounded by Fluids

On a grand scale, all creatures living on the Earth are surrounded by two fluids: air and water. As opposed to outer space where no matter exists, the Earth's atmosphere is filled with air molecules. At sea level, the atmospheric pressure is 101,000 Pascals (or denoted by [Pa]) because air molecules occupy the space from the troposphere (0–12 km) to the thermosphere (85–110 km).[9] Note that the atmospheric pressure can be obtained by integrating Eq. (1.5) with the equation of state, expressing the density as a function of pressure and temperature; the temperature distribution in the

---

9 The space in between is called the stratosphere (12–50 km) and mesosphere (50–85 km).

atmosphere is usually known as a function of height. Meanwhile, the pressure in the oceans and rivers varies more significantly with depth because the density of water is approximately 1000 times greater than the density of air.

Example 1.2 *Magic cup?*

Let us suppose we have water in a cup with a radius of 3 *cm* and a height of 9 *cm*. When the cup is held upside down with a paper plate placed at the bottom (Figure 1.4), the weight of water lowers the pressure of the air trapped inside the cup (e.g. height is 2 *cm*). The change of air volume may not be noticeable, but the trapped air expands by a fraction of its volume. According to Boyle's law, this very small change of air volume lowers the air pressure and creates a suction force to hold the water.

Numerics:

Cup cross-sectional area $(A) = \pi(0.03^2) = 0.00283 \ [m^2]$

Water weight (W): $1000 \ (kg/m^3) \cdot 9.8 \ (kg/s^2) \cdot 0.07 \ (m) \cdot 0.00283 \ (m^2) = 1.941 \ (N)$

Pressure change $(\Delta p) = W/A = 1.941/0.00283 = 686 \ (Pa)$

Boyle's law:

$p_o V = p' V'$; $101,000 \cdot h_a = (101,000 - 686) \cdot h_a' \ \rightarrow \ h_a'/h_a = 1.00684$

Thus, $(h_a' - h_a)/h_a = \Delta h_a/h_a = 0.0068 = 0.68\%$, and with $h_a = 2 \ cm$, $\Delta h_a = 0.136 \ mm = 136 \ \mu m$.

Density decrease of air $(\%) = \Delta \rho/\rho = h_a/\Delta h_a = 1.47\%$

1.4.4 Buoyant Force

Buoyancy is another form of gravitational effect. It is created by the hydrostatic pressure acting on the surface of an immersed matter of a different density, where the hydrostatic pressure results from the gravitational body force of the fluid. If a solid body of density ρ_s is immersed in a fluid of density ρ_f, the hydrostatic pressure acting on the solid surface varies with depth (Figure 1.5). If the hydrostatic pressure is integrated over the solid body, the forces in the horizontal direction are all cancelled off, resulting in no net force. In the vertical direction, however, there exists a net pressure force acting in the positive direction. This buoyant force B is written as follows

$$B = \sum_i dA_i(p_l - p_u)_i = \sum_i dA_i \ (\rho_f \ g \ H_i) = \sum_i (\rho_f \ dV_i) \ g = m_f \ g \qquad (1.7)$$

where i denotes a segmented vertical column of the solid body of height H_i, dA_i an incremental area projected to the horizontal plane, dV_i an incremental volume of the solid, and m_f the mass

Figure 1.4 Suction (or negative) pressure inside the cup

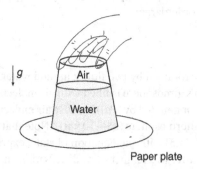

Figure 1.5 Hydrostatic stress vectors
on the surface of a solid body immersed
in a fluid; densities of solid and fluid
denoted by ρ_s and ρ_f

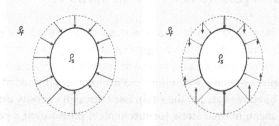

of the fluid for the volume occupied by the solid. Equation (1.7) shows that the buoyant force is
equivalent to the gravitational body force of the fluid that occupies the volume of the solid.

The net upward force acting on the submerged body is the difference between the buoyant force
and the weight of the body:

$$B - m_s g = m_f g - m_s g = (\rho_f - \rho_s) g \, V \tag{1.8}$$

It is clearly shown in Eq. (1.8) that the buoyant effect occurs due to the density difference between
the solid body and the fluid. If the solid were replaced by a fluid of the same density ρ_f, there would
be no net vertical force acting on the fluid replacing the space of solid. If $\rho_f > \rho_s$, then $B - m_s g > 0$,
and the force magnitude is proportional to the density difference and volume. There are many
examples of submergence of matter in different phases, e.g. a piece of wood or submarine in water,
a hot air balloon in the air, the chimney (or stack) effect.

Example 1.3 *Solar chimney*

EnviroMission, an Australian public company, proposed to build the first large-scale solar updraft
tower (about twice as tall as the Empire State Building) in La Paz County, Arizona. This towering
hot air chimney is designed to produce 200 MW electrical energy with turbines (Figure 1.6).

Figure 1.6 Solar chimney; two Empire State
Buildings high

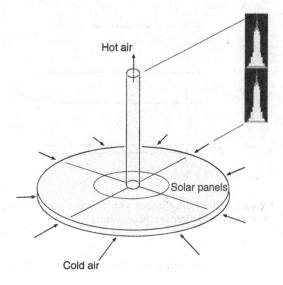

1.5 Pressure in a Fluid in Motion

We have shown so far that the pressure in a fluid at rest is determined by the local balance of action and reaction forces: gravitational body force vs. reaction force from the ground. The next question is how pressure can be defined in a fluid in motion. The answer to this question is not that simple, especially when the fluid is viscously strained at a rate. In viscous flows, viscous stresses are directly related to the strain rates through viscosity under Newton's law of viscosity. Therefore, the mean normal stress (or mechanical pressure) at a point may or may not be the same as the pressure. Nevertheless, we assume Stokes' hypothesis that mechanical pressure would be the same as pressure. Once Stokes' hypothesis is assumed, the pressure of a fluid in motion can be defined in the same manner as it is defined in a fluid at rest.[10]

In this section, we are going to demonstrate with a simple syringe-like device (Figure 1.7) that the pressure of an incompressible fluid (e.g. water) inside the device is determined by the local balance of action and reaction forces. Here, the action force is the normal force originated by the piston, and the reaction force can be the inertial resistance force of the accelerating fluid or the frictional resistance force of the shear-straining fluid. We assume that the syringe-like device is immersed in water and fixed to the ground. We also neglect any inertial effect of water being sucked into the device inlet, for the sake of simplicity.

1.5.1 Transient Inertia Force

1.5.1.1 Slug Motion

Let us suppose we have a straight tube and a force F_1 is applied to the piston to accelerate the water at rest (Figure 1.7). Then, the water in the tube is forced to be in the state of compression by forces in action and reaction: the force F_1 applied to the piston vs. the inertial resistance force of the water. In this case, the magnitude of the inertial resistance force depends not only on the

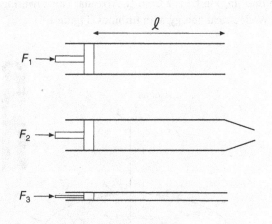

Figure 1.7 Pressure changes with different reaction forces: (i) transient inertia force, (ii) convective inertia force, and (iii) frictional resistance force

---

10 More discussion continues in Chapter 3 *Differential Equations of Motions.*

local time acceleration of the piston but also on the amount of mass of water being accelerated in time. Therefore, the pressure of the water inside the tube linearly decreases with distance and its gradient depends on the acceleration of the piston.

To quantify the pressure inside the tube, an equation of motion can be set for a water element. The inertial resistance force produced by transient acceleration of the water element dR_t is balanced with a local net pressure force:

$$dR_t = -dp \, (dy \, dz) \tag{1.9}$$

where $dp = p_2 - p_1$ is the incremental pressure difference over dx and $dydz$ is the cross-sectional area of the water element. The inertial resistance force of the water element dR_t can be written as follows:

$$dR_t = \rho \, (dx \, dy \, dz) \, (dv_p/dt) \tag{1.10}$$

where ρ is the water density, v_p is the piston speed (which is generally time-dependent), and dx is the incremental distance in the streamwise direction. Note that p_1 and p_2 are the normal stresses determined by the local balance of action and reaction forces.

Equation (1.9) can be written per unit volume as follows:

$$\rho \, \frac{dv_p}{dt} = -\frac{dp}{dx} \tag{1.11}$$

and the pressure distribution inside the tube can be obtained by integrating Eq. (1.11) with the boundary condition at the exit, $p(l) = p_\infty$[11]

$$p(x) = p_\infty + \rho \, \frac{dv_p}{dt} \, (l - x), \tag{1.12}$$

where l is the tube length.

Equation (1.12) shows that the pressure inside the tube linearly decreases with distance because the inertial resistance force is linearly proportional to the mass being accelerated in time, and the slope of pressure distribution is proportional to the magnitude of the acceleration of the piston itself. Equation (1.12) is only valid if $dv_p/dt > 0$.

[Notes] If the piston is suddenly pulled in the opposite direction, i.e. $dv_p/dt < 0$, the pressure distribution inside the tube is obtained as follows:

$$p(x) = p^* + \rho \, \frac{dv_p}{dt} \, (l - x) \tag{1.13}$$

where the pressure at the tube exit $p(l) = p^*$ is lower than p_∞, since the water outside is spatially accelerated into the tube. From the conservation law of energy (or Bernoulli's principle), the pressure at the exit can be estimated as follows:

$$p^* = p_\infty - \frac{1}{2} \, \rho \, v_p^2 \tag{1.14}$$

if we neglect the local transient acceleration of the water outside and any possible flow separations near the tube exit. Equation (1.13) shows that the pressure along the tube linearly decreases from $p(l) = p^*$ at the tube exit to $p(0) = p^* + \rho l \, (dv_p/dt)$ at the head of the piston.

---

11 Why the jet exit pressure is always the ambient pressure will be discussed in Section 1.6 *Fountain*.

Example 1.4 *Oscillation of water in a U-tube*

The water in a U-tube oscillates with the same velocity as a slug flow (Figure 1.8) [4]. Find the instantaneous pressure distribution inside the tube.

An equation of motion can be set for an infinitesimal slice of control volume $A\,ds$ in the U-tube. In this case, the gravitational body and pressure forces are balanced with an unsteady inertial force:

$$\rho\,(A\,ds)\,\frac{dv}{dt} = -dp\,(A) - \rho\,(A\,ds)\,g \tag{1.15}$$

where s is the vertical coordinate originating from the bottom and $v = ds/dt$ is the vertical velocity.

By integrating Eq. (1.15) from s to the free surface of the water, we can obtain the pressure inside the U-tube:

$$p(s) = p_{\infty} + \rho\,(g + dv/dt)\,(l - s) \tag{1.16}$$

As shown in Figure 1.9, pressure linearly increases from the free surface ($s = l$) to the bottom ($s = 0$) with a slope of $g + dv/dt$. Note that the slope changes in time with the sign and magnitude of dv/dt, while the water column is accelerating (or decelerating).

For the sake of simplicity, we assume that two vertical tubes are connected by a horizontal junction, which at both ends will have two different pressures. Due to the transient inertial effect of the fluid, pressure varies linearly across this horizontal tube:

$$p(x) = p_{r} + \rho\,(du/dt)\,(T - x) \tag{1.17}$$

where p_{r} is the pressure at the right end of the junction, T is the length, u is the horizontal velocity, $u = dx/dt$, and $|u| = |v|$. The two pressures will become the same when the two water columns level even. Note also that the gradient discontinuity of pressure at the two end points will be rounded off by the centrifugal effect, which is not included in the present model.

Figure 1.8   Oscillation of water in a U-tube

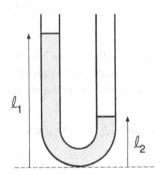

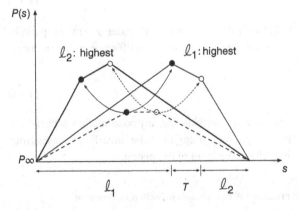

Figure 1.9   Instantaneous pressure distribution in a U-tube; thin solid (l_1: highest), dashed ($l_1 = l_2$), and thick solid (l_2: highest); filled circles (left-end junction) and hollow circles (right-end junction)

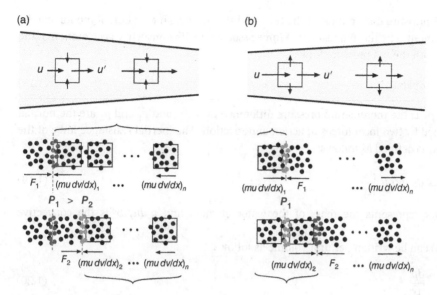

Figure 1.10 Pressure changes in the nozzle (a) and diffuser (b) with changes of fluid momentums in space; a thrust force in reaction to $m(u\,du/dx) > 0$ (nozzle) and a drag force in reaction to $m(u\,du/dx) < 0$ (diffuser)

1.5.2 Convective Inertia Force

1.5.2.1 Nozzle

If a nozzle is attached to the tube exit, a force F_2 is required to move the piston at a constant speed v_p (Figure 1.7). In this case, the water in the nozzle is forced to be in the state of compression by forces in action and reaction: the force F_2 applied to the piston vs. the inertial resistance force of the accelerating water in the nozzle.

In contrast to the transient inertial force in the previous case, the inertial resistance force is the reaction force of the fluid being accelerated in the nozzle. If we look at the water element in the nozzle, it is laterally being contracted and longitudinally being stretched (i.e. linearly strained) at a rate while it moves downstream. Over an infinitesimally small control volume fixed in space, the water is convectively accelerated in the streamwise direction,[12] and in reaction to this acceleration, the water upstream is forced backward (i.e. a principle of thrust force).

The magnitude of this inertial resistance force depends not only on the convective acceleration of the water, which is the product of the local velocity u and the straining rate of the fluid du/dx, but also on the amount of mass of water being accelerated in the nozzle. Therefore, the pressure in the nozzle is increased in the upstream direction because the amount of volume-occupying mass to be accelerated is accumulated in that direction (Figure 1.10a).

At this point, special attention should be paid to the definition of the pressure of a fluid in straining (or deforming) motion. When the water in the nozzle is linearly strained at a rate in the streamwise and transverse directions, normal viscous stresses are exerted in the direction normal to the surfaces of the water element. However, they do not build any energy density to be added to the pressure (or hydrostatic stress) since the sum of the strain rates in the x, y, and z directions is zero in incompressible fluids; hence, mechanical pressure is equal to pressure.[13]

---

12 More discussion continues in Chapter 2 *Macroscopic Balance of Mass, Momentum, and Energy*.
13 More discussion continues in Chapter 3 *Differential Equations of Motions*.

To quantify the pressure distribution in the tube and the nozzle, an equation of motion can be set for a water element. The inertial resistance force associated with convective acceleration of the fluid is balanced with the net pressure force:

$$dR_c = -dp \, (dy \, dz)$$ (1.18)

where $dp = p_2 - p_1$ is the incremental pressure difference over dx, and p_1 and p_2 are the normal stresses determined by two local forces in action and reaction. The inertial resistance force of the water element dR_c is defined as follows:

$$dR_c = \rho \, (dx \, dy \, dz) \, u \, \frac{du}{dx},$$ (1.19)

where $\rho(dx \, dy \, dz)$ represents the mass of the water element and $u \, du/dx$ is the convective acceleration.

Equation (1.18) can be written per unit volume as follows:

$$\rho \, u \, \frac{du}{dx} = -\frac{dp}{dx}$$ (1.20)

and the pressure distribution in the nozzle can be obtained by integrating Eq. (1.20) from x to the nozzle exit:

$$p(x) = p_\infty + \frac{\rho}{2} \{u_j^2 - u(x)^2\}$$ (1.21)

where u_j is the velocity at the nozzle exit. It is shown that $p(x)$ is greater than p_∞ due to the increase of kinetic energy (per unit volume) between two stations at x and the nozzle exit (i.e. production of thrust force). The pressure difference is also linearly proportional to the density, meaning that the pressure change is related to the mass in convective acceleration in the nozzle. Equation (1.21) is the well-known Bernoulli's principle.

By conservation of mass, $u_j A_j = v_p A_p$, the pressure at $x = x_e$ (denoting a nozzle entrance) is expressed as follows:

$$p(x_e) = p_\infty + \frac{\rho}{2} v_p^2 \, (\beta^2 - 1)$$ (1.22)

where $\beta = A_p/A_j \, (> 1)$ is the contraction-ratio of the nozzle.

1.5.2.2 Diffuser

In diffusers, a fluid is forced to be in the state of compression by a local balance of forces in action and reaction: an inertia force of the fluid upstream vs. a reaction force from downstream. The reaction force can be produced either by the solid fixed to the ground or by the fluid in ambient condition.

If we look at the water element in the diffuser, it is laterally stretched and longitudinally contracted at a rate. Over an infinitesimally small control volume fixed in space, the fluid is being decelerated in the streamwise direction, and in reaction to this deceleration, the fluid downstream is being forced forward (i.e. a principle of drag force). As discussed in the previous cases, the magnitude of this inertial force depends on the deceleration of the fluid as well as the mass of water being decelerated in the diffuser. Hence, the pressure is increased in the downstream direction due to an accumulation of the decelerating fluid in that direction (Figure 1.10b).

Figure 1.11 Diffuser-augmented wind turbines (DAWT)

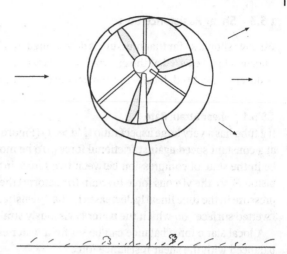

1.5.2.3 Venturi Effect

If the diffuser exit is open to the ambient fluid, the inertial force of the decelerating fluid causes the upstream pressure at the throat to be lower than the ambient pressure. This occurrence of negative pressure at the throat is called *Venturi* effect. In fact, the Venturi effect is used in many applications to readily suck the fluid in the area close to the neck.

One example is a Venturi tunnel under a car with a carefully shaped under-tray for additional downforce. The wind tamer (or skywolf wind turbine), a commercial product of "diffuser-augmented wind turbines" (DAWT) that install diffusers at the turbine blades, utilizes the Venturi effect to lower the pressure at the blades to increase the efficiency of the wind turbines (Figure 1.11).

Example 1.5 *Squid propulsion*

A squid propels by transforming itself into an instant diffuser and nozzle. In the refilling phase, external water enters the inflated tube, which we call the *mantle*, through small openings around the squid's head. The mantle forms a nozzle-diffuser-like flow passage with flaps that build a negative pressure at the throat (Figure 1.12a). In the ejection phase, the squid forcefully expels water through a small orifice by forming a *funnel*. It clamps the openings shut by contracting the mantle, as sketched in Figure 1.12b. The squid is propelled by the conservation of momentum.

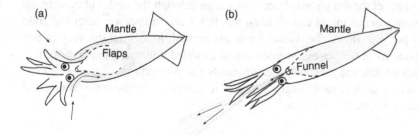

Figure 1.12 Water jet propulsion of squid by shaping an instant diffuser (a) and nozzle (b) with its body: mantle, funnel, and flaps

1.5.3 Shear Resistance Force

We have shown so far that pressure is determined by a local balance of forces in action and reaction: a normal force exerted at the external boundary (i.e. piston) vs. an inertial force of the accelerating or decelerating fluid, locally in time or convectively in space.

1.5.3.1 Shear Strain Rate

If a tube has a very large aspect ratio ($l/d \gg 1$) (Figure 1.7), a force F_3 is required to move the piston at a constant speed against frictional forces. To be more specific, the water in the tube is forced to be in the state of compression between two forces in action and reaction: the force applied to the piston F_3 vs. the viscous force to strain (or deform) the water at a rate against friction.[14] As a result, pressure in the tube linearly decreases in the downstream direction because the area of the so-called "wetted surface" on which the water is viscously strained at a rate decreases in that direction.

A local static force balance can be set for a water element in the tube. The net pressure force is balanced with the shear resistance force:

$$-dp\, A_p = dR_v, \tag{1.23}$$

where $dp = p_2 - p_1$ is the incremental pressure difference over dx and A_p is the area of the piston. The shear resistance force of the water element dR_v is defined as follows:

$$dR_v = \tau_w\, (\bar{P}\, dx), \tag{1.24}$$

where τ_w is the wall shear stress and $\bar{P}$ is the perimeter of the tube. Meanwhile, Newton's law of viscosity states that the wall shear stress is defined as the viscosity times the shear strain rate at the tube wall:

$$\tau_w = -\mu \left(\frac{\partial u}{\partial r} \right)_{r=R}. \tag{1.25}$$

A pressure distribution in the tube can finally be obtained by integrating Eq. (1.23) from x to the tube exit ($x = l$):

$$p(x) = p_\infty + \tau_w\, \bar{P}\, (l - x)/A_p \tag{1.26}$$

It is shown in Eq. (1.26) that the pressure in the tube linearly decreases as expected because the shear resistance force is proportional to the wetted area (or streamwise length). In this case, the work done on the fluid by the pressure difference force is simply used at a rate for shear-straining the fluid, not for increasing the kinetic energy of the fluid. In other words, the energy provided by the piston is dissipated at a rate into heat by viscous force work.

[Notes] The line of action of the net viscous force does not go through the center of mass of the fluid particles. Therefore, viscous forces cannot compress fluids unless they are bound by solid walls.[15] In fully developed pipe flows, the viscous forces are exerted throughout the field. In this case, transverse components of the viscous stresses are all canceled due to axisymmetry so that flows are exactly parallel, while the streamwise components are in reaction to the normal force applied at the external boundary. If then, local compressions can be set by these two action and reaction forces: pressure force vs. viscous force.[16]

---

14 We assume that the flow in the tube is entirely fully developed (e.g. a Poiseuille flow).
15 Generally speaking, viscous forces can compress fluids as long as flows are forced to be parallel or nearly parallel. More discussion continues in Chapter 3 *Differential Equations of Motions*.
16 In Couette flow, the fluid is bound by solid walls, but the viscous stresses cannot change the pressure because there is no net viscous force exerted on the fluid.

Figure 1.13 Shear stress vs. strain rate in Newtonian and non-Newtonian fluids

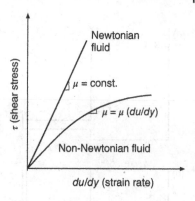

1.5.3.2 Netwon's Law of Viscosity

When a fluid element is shear strained at a rate, the fluid molecules are mutually interfered with, causing frictions. The degree of internal friction is a physical property of the fluid called *viscosity*. A more viscous fluid requires a larger viscous stress for the fluid element to yield the same straining rate (Figure 1.13).

Newton's law of viscosity states that a linear relation is held between the tangential shear stress and the shear strain rate:

$$\tau = \mu \, \frac{du}{dy}, \tag{1.27}$$

where a proportionality constant μ is the dynamic viscosity of the fluid. We call any fluid obeying this law a Newtonian fluid.

Viscosity depends on the state of the fluid. Gas molecules are more spaced-apart and active in random motion than those of liquids. Thus, viscosity of gases increases with temperature. The viscosity of gas is defined by Sutherland's law:

$$\mu(T) = \frac{(T/T_0)^{3/2} \, (T_0 + c^*)}{(T + c^*)}, \tag{1.28}$$

where T_0 is the reference temperature and c^* is the Sutherland constant.

In contrast, liquid molecules are close enough to be in the force field of others, in which a sheared motion tears off coherent groups of molecules, while reforming others. Therefore, liquid viscosity is higher than gas viscosity and inversely increases with temperature. The viscosity of liquid is defined by Andrade's equation:

$$\mu(T) = De^{B/T} \tag{1.29}$$

where D and B are case-dependent constants.

For some fluids such as water mixed with cornstarch or ketchup, the linear relation between the shear stress and the shear strain rate fails. When a demand for the shear strain rate of ketchup doubles, it requires less than twice the shear force. Meanwhile, the water mixed with cornstarch requires more than double the force for the same amount of shear straining rate. Any fluid that does not obey Newton's law of viscosity is called a non-Newtonian fluid.

1.6 Fountain

A fountain is a vertical water jet. It soars at a vertical distance h because of the hydrostatic pressure built in the reservoir (Figure 1.14a). We can now conduct an interesting experiment. If we cover the jet with a vertical pipe, the water stands at a height h (Figure 1.14b). If we remove the pipe, then the jet will develop again. Now, the question is what we can learn from this simple experiment.

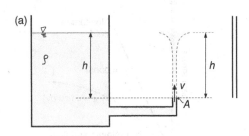

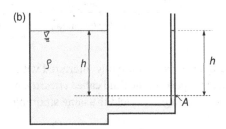

Figure 1.14 Fountain (a) and its counterpart with pipe (b)

By using the vertical pipe, the water can be laid in the state of compression by two forces in action and reaction: a gravitational body force vs. a reaction force from the ground. This is called the stack effect. Since the pressure of the water molecules acts in all directions, we need a side wall to support the hydrostatic pressure to build up. Without the pipe, the water cannot be in the state of compression by the gravitational body force. In this case, the jet withstands its weight by decelerating the water in the vertical direction; this is a natural form of energy conversion without changing pressure—converting kinetic energy into potential energy.

1.6.1 Spreading of a Vertical Jet

It is interesting to note that the fountain spreads laterally as it soars. In a vertical jet, a volumetric body force decelerates the jet, and the equation of motion can be set per unit volume as follows:

$$v \frac{dv}{dz} = -g \tag{1.30}$$

If Eq. (1.30) is integrated from z to h, we can obtain the vertical velocity:

$$v(z) = \sqrt{2g\,(h-z)} \tag{1.31}$$

It is interesting to note that the jet speed v_j ($= \sqrt{2gh}$) at $z = 0$ decays to 0 at $z = h$ at a rate of $-g/v$. The fact that the decaying rate is a function of z indicates that the fountain will greatly lose momentum as it approaches the top. This is a nonlinear effect of the inertia of the jet acting against gravity. The slower the jet becomes, the stiffer the spatial gradient of the jet speed will be, meaning that the jet dies out more quickly at lower speeds.

1.6.2 Spreading of a Horizontal Jet[†]

A horizontal jet also loses momentum at a rate by the shear forces acting on the side surfaces of the jet, and the ambient fluid is entrained in return by the jet. Then, the jet must spread laterally, enlarging its cross-sectional area of the stream tube.

---

† For advanced studies.

An equation of motion can be set in the horizontal direction as follows:

$$u\,\frac{\partial u}{\partial x} = \frac{1}{\rho}\frac{\partial \tau}{\partial r}, \tag{1.32}$$

where τ is the shear stress and r is the coordinate in the radial direction. If we further simplify the right-hand side of Eq. (1.32) as

$$\frac{1}{\rho}\frac{\partial \tau}{\partial r} \approx \nu\frac{\partial^2 u}{\partial r^2}, \tag{1.33}$$

then it reads

$$u\,\frac{du}{dx} = \nu\,f_v(x), \tag{1.34}$$

where $f_v = d^2u/dr^2$.

If we assume that $f_v = -\tau^*$, a constant with the dimension of $[m^{-1}s^{-1}]$, the jet speed is found as follows:

$$u(x) = \sqrt{2\nu\tau^* (l - x)} \tag{1.35}$$

where the jet speed at $x = 0$, $u_j = \sqrt{2\nu\tau^* l}$ decays to 0 at $x = l$ at a rate of $-\nu\tau^*/u$. Note that the jet spreading is not drastic at an immediate distance from the jet exit. Its spreading is more pronounced downstream by the same nonlinear effect of the inertia as that of the vertical jet.

[Notes] If we include the frictional shear in the vertical jet (fountain), the jet speed reads

$$v(z) = \sqrt{2\,\{g + \nu\tau^*\}(h - z)}, \tag{1.36}$$

where $\nu\tau^*$ is an effective deceleration associated with the viscous shear force. Note that the jet speed at $z = 0$, $v(z) = \sqrt{2\,(g + \nu\tau^*)\,h}$ will decay to 0 at $z = h$ at a rate of $-\{g + \nu\tau^*\}/v$.

1.6.3 Manometer

There are a variety of pressure measurement devices available today, among which a classic is the manometer. As shown in Figure 1.15, the liquid flowing in the channel fills the tube columns (i.e. manometers) by the pressure force until it reaches a static equilibrium with the barometric pressure. Hence, the height reads the hydrostatic pressure of the liquid. The liquid heights of the manometers sketched in Figure 1.15 show the pressure distribution created by the centrifugal effect and flow separation in curved channel flow.

Figure 1.15 One-column manometers

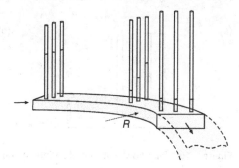

The one-column manometer represents the gage pressure, that is, the pressure relative to the atmospheric pressure. An absolute pressure is then defined as the sum of the gage pressure and the atmospheric pressure:

$$p_{abs} = p_{gage} + p_{atm}, \tag{1.37}$$

where $p_{gage} = \rho g h$ and ρ is the density of the liquid inside the column. For example, a mercury manometer of 760 *mmHg* represents the atmospheric pressure.

1.7 Manifold Diffusers

1.7.1 Internal Flow Systems

An internal flow system can be viewed as a pipe flow where the pump delivers work to the fluid, while the valve dissipates the energy through the viscosity. The flow rate in the internal system is determined by these two energy transport processes, while conserving the total energy of the system. As shown in Figure 1.16a, the pressure of the flow driven by the pump is higher than the ambient pressure because it is resisted by the valve which dissipates the mechanical energy into heat. There is also a very slight decrease of pressure in the streamwise direction due to frictional loss in the pipe.

We are now interested in investigating how a pipe with a manifold changes the pressure distribution in the internal flow system [5]. This simple manifold pipe could represent some of the branching vessels of blood in the cerebral and coronary arteries. As the flow in the pipe exits through the manifold, the pressure increases almost linearly across the manifold by the inertial force of the decelerating fluid (Figure 1.16b). This manifold pipe is another form of a diffuser. To be more specific about the pressure distribution, opening the manifold reduces system resistance, resulting in increased flow rate (or lower pressure) than the pipe without a manifold. However, discharging the fluid through the manifold continuously reduces the flow rate, while increasing the pressure.

1.7.2 Brain Aneurysm

Blood in the brain sometimes breaks out the vessel wall because of a rupture in the aneurysm. The aneurysm is often located at the stagnation point because the blood flow has to be diverted away with reduction of flow speed. Sometimes, the aneurysm appears right after the branching of the vessel (Figure 1.17b). At this spot, the flow speed is reduced with an increase of pressure (Figure 1.17a,). This case is identical to the previously discussed case of bleeding with a manifold.

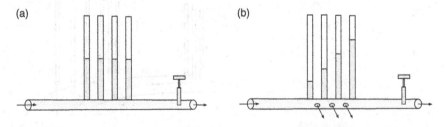

(a)   (b)

Figure 1.16   Pipe flows without bleeding (a) and with bleeding through manifold (b)

Figure 1.17 Surface pressure contours of blood flows in the cerebral vessels; normal (a), anuerysm (b)

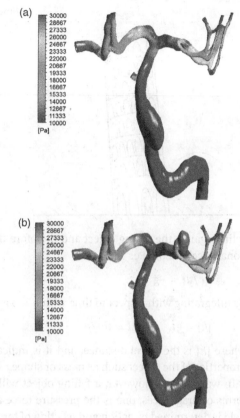

1.8 Drag Force on a Moving Body

Transport vehicles, such as aircrafts, automobiles, ships, are resisted by fluids even though they are deformable. This resistance force acts as drag on the vehicles. Generally speaking, the drag force depends on a fluid's density and viscosity, the shape of the solid body, and the rate at which the fluid is forced to change the velocity; the drag force is associated with viscous forces and pressure forces.

1.8.1 Galileo Galilei

A famous experiment which Galileo Galilei (1564–1642) may or may not have performed at the *Tower of Pisa* has provoked an issue on the drag force of a body in acceleration and also on its relevance to an existence of a matter (Figure 1.18). That is, a steel ball falls much faster than a feather in the air, while they both hit the ground at the same time in vacuum.[17] In other words, drag force does not exist without matter.

Any object falling in a vacuum has nothing to push down at the front and nothing to pull down at the back. Furthermore, there is nothing to drag by friction. As a result, there is no resistance force

---

17 Apollo 15 Commander Dave Scott demonstrated that the mass of an object does not affect the time it takes to fall, using a 1.32 *kg* aluminum geological hammer and a 0.03 *kg* falcon feather on the Moon [6].

Figure 1.18   Galileo Galile's experiment at the tower of Pisa

acting against the falling object and therefore the falling motion is only governed by the gravitational acceleration, i.e.

$$dv/dt = -g \qquad (1.38)$$

By integrating with respect to time with $v = 0$ at $t = 0$, it yields

$$|h| = gt, \quad \text{or} \quad t = |h|/g \qquad (1.39)$$

where $|h|$ is the travel distance, and it is indicated that the arrival time t is independent of any properties of the object such as mass or shape.

In water or air, however, a falling object will be resisted by the fluid. There are two different forms of drag forces; one is the pressure force and the other is the frictional force. The pressure field is determined by action and reaction of forces. For example, a falling object with mass pushes and pulls the stationary fluid around the body. Then, the fluid forced to be accelerated by the body reacts in the opposite direction but with the same magnitude of the action forces. This mechanical interaction is done with Newton's first, second, and third laws of motion. Therefore, the fluid elements under these action and reaction forces are laid in the state of compression or relief.

The motion of the falling body is thus expressed as follows:

$$dv/dt = -g + \beta(v), \qquad (1.40)$$

where $v = v(\rho, \mu, m, shape)$. The arrival time is now influenced by the physical properties of the fluid such as density and viscosity and by the mass and the shape of the falling body.

1.8.2   Body Moving at a Constant Speed

Let us consider a circular cylinder moving at a constant velocity v_o in the positive x direction in an infinite frictionless incompressible fluid at rest (Figure 1.19). The question is whether we will have the same pressure difference between the point at the nose of the moving cylinder and the point far upstream, as in the case with uniform flow passing over a stationary circular cylinder.

When the cylinder is moving at a constant speed v_0, the fluid at the centerline fore the cylinder does not move like a solid column. A negative gradient of velocity decaying from v_0 at the nose to zero far downstream is formed as the fluid escapes through the sides. The spatial profile is, however, frozen for all times because the cylinder is simply translated at the speed of v_0.

In this pure translation, the cylinder experiences two inertia forces. One is the 'transient inertia force' acting upstream to resist the body of motion. This inertia force will increase the pressure at

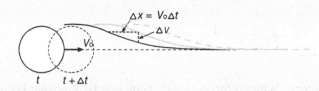

Figure 1.19 Coordinate translation (Galilean invariance); a cylinder moving at a constant speed v_0

the nose. The other is the convective inertia force acting downstream, as it is in the diffuser. This inertia force will reduce (or relieve) the pressure at the nose since the flow stream that is losing momentum on the way forces the fluid downstream by its excessive inertia; consequently, it will lower the pressure at the nose.

Now, the force balance between the pressure and inertia forces can be integrated along the stream from the stagnation point to the far field,

$$-\int_{p_{stag}}^{p_\infty} dp = \rho \int_{stag}^{\infty} (a_t + a_x)\, dx = R'_{It} + R'_{Ic} \tag{1.41}$$

where R'_{It} and R'_{Ic} correspond to the transient and convective inertia forces per unit area, respectively. As if with a slug, the former is associated with the fluid in local time acceleration:

$$R'_{It} = \rho \int_{stag}^{\infty} \frac{\partial v}{\partial t}\, dx \tag{1.42}$$

The question is how to obtain the local acceleration along the streamline. Let us suppose that the velocities are monitored in time at two points on the centerline in front of the body, x_1 and x_2 $(= x_1 + dx)$. At the time of t, velocities at these points are different, i.e. $v(x_1, t) \neq v(x_2, t)$, but the time history of the velocities at these two points will be the same, i.e. $v(x_1, t) = v(x_2, t + dt)$, where $dt = dx/v_0$, since it is purely a translation at a constant speed.[18]

Thus, the local time change of velocity can be expressed as follows:

$$\frac{\partial v}{\partial t} = -v_0 \frac{\partial v}{\partial x} \tag{1.43}$$

The flow field around the cylinder after reaching an equilibrium state (i.e. a steady-state flow field with respect to an observer moving with the body) translates in time and space. Therefore, the velocity and pressure vary in space and time, but the field is frozen; for an observer moving with the cylinder, the flow is steady. Now, Eq. (1.43) can be integrated from the stagnation point to the far field:

$$\int_{stag}^{\infty} \frac{\partial v}{\partial t}\, dx = -v_0 \int_{stag}^{\infty} \frac{\partial v}{\partial x} dx = -v_0 \int_{stag}^{\infty} dv = v_0^2 \tag{1.44}$$

The convective inertia force is rather interesting because the fluid escaping at the centerline in front of the cylinder imposes a negative pressure force to the cylinder; the pressure at the stagnation point will be lowered from the ambient pressure:

$$R'_{Ic} = \rho \int_{stag}^{\infty} v \frac{\partial v}{\partial x}\, dx = \rho \int_{stag}^{\infty} d(v^2/2) = -\rho \frac{v_0^2}{2} \tag{1.45}$$

---

18 If the body moves at a variable speed in time, then the aforementioned relations cannot be held.

Finally, the pressure at the nose has two contributions (one positive and one negative) besides the ambient pressure:

$$p_{stag} = p_\infty + \underbrace{\rho\, v_0^2}_{R_{It}} + \underbrace{\left(-\frac{1}{2}\, \rho\, v_0^2\right)}_{R_{Ic}} = p_\infty + \frac{1}{2}\, \rho\, v_0^2 \tag{1.46}$$

which proves that the pressure difference between the stagnation point and far upstream is the same for both reference frames.

1.8.3 Galilean Invariance

A Galilean transformation specifies how to transform an inertial frame to a frame of reference moving at constant velocity.

1.8.3.1 Invariance of Acceleration

By a Galilean transformation (a principle of special relativity), the velocity vectors of a material point are related between the inertia reference frame and the moving reference frame as follows:

$$\vec{v}_a = \vec{v}_b + \vec{v}_r \tag{1.47}$$

where $\vec{v}_a$ denotes the velocity vector measured on the inertial reference frame (absolute velocity vector), $\vec{v}_b$ the moving body velocity vector (constant), and $\vec{v}_r$ the velocity vector measured on the moving reference frame (relative velocity vector).

Note that velocity is a relative quantity; it only appears to be different, but the nature is the same. For example, $\vec{v}_a = d\vec{x}_p(t)/dt$, where P indicates the material point. If we take a total time derivative $dv_p(t)/dt$, then

$$\vec{a}_a = \vec{a}_r \tag{1.48}$$

since $\vec{v}_b$ is constant. The acceleration measured on the inertial reference frame equals the acceleration measured on the moving reference frame. That is, the acceleration is an invariant of the reference of frame. This is a very important concept in fluid mechanics.

Let us go back to the previous problem of a moving cylinder. If the acceleration is measured in the stationary reference frame, a_a written in the Eulerian description is the sum of the transient part, $a_t = \partial v_a/\partial t$ and the convective part, $a_x = v_a\, \partial v_a/\partial x_a$, while a_r is only $v_r\, \partial v_r/\partial x_r$. Thus, it is much simpler to handle the moving body problem with the Galilean transformation.

1.8.3.2 Galilean Invariance of Netwon's Laws

If the relative acceleration vectors in a moving reference frame are the same as the absolute acceleration vectors in the stationary (or inertia) reference frame, then the local force (or stress) field should be the same, regardless of the reference frame. It is to be noted that force is not an absolute quantity; in other words, it does not depend on an absolute position or a velocity vector. It depends rather on the relative position or velocity vectors between two interacting bodies of matter. Force is a Galilean invariant quantity:

$$\vec{F} = m\, \vec{a}_a = m\, \vec{a}_r \tag{1.49}$$

Let us go back to the previous moving cylinder problem (Figure 1.19). If a coordinate translation is taken, the fluid at the ambient pressure flows in the positive x-direction over the cylinder. Then,

Eq. (1.41) can be integrated as follows:

$$-\int_{P_\infty}^{P_{stag}} dp = \rho \int_{v_o}^{0} v_r\, dv_r$$

(1.50)

and thus

$$P_{stag} = P_\infty + \frac{1}{2} \rho v_o^2$$

(1.51)

References

1 Batchelor, G.K., *An Introduction to Fluid Dynamics*, Cambridge: Cambridge University Press, 1964.

2 Hirschfelder, J. O., C. F. Curtiss, and R.B. Bird, *Molecular Theory of Gases and Liquids*, New York: Wiley, 1954.

3 Lewin, W., *8.01x–Lect 27–Fluid Mechanics, Hydrostatics, Pascal's Principle, Atmosph. Pressure*, YouTube, 2015.

4 Sabersky, R.H., A.J. Acosta, and E.G. Hauptmann, *Fluid Flow–A First Course in Fluid Mechanics*, 2nd ed. New York: The Macmillan Company, 1971.

5 Shapiro, A.H., "Pressure Fields and Fluid Acceleration," *Illustrated Experiments in Fluid Mechanics*, National Committee for Fluid Mechanics Films, 1980.

6 https://youtu.be/KDp1tiUsZw8.

Problems

1.1 A force *F* is applied to the plate as shown below. Note that the fluid is in steady state.

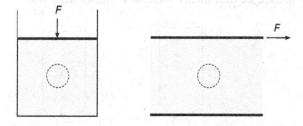

a) Sketch the stress vectors exerted on a circular fluid element.

b) Explain why pressure is a scalar quantity and how it differs from viscous stresses.

1.2 To lift up an Airbus 380 with fully loaded passengers, what pressure difference would be required between the upper and lower surfaces of the wings? Compare this required pressure difference to that of the human heart needed to pump the blood from arteries to veins, which is known to be approximately 100 *mmHg*.

Fuel: 200 *tons* (cf. fuel capacity: 323 *kL*)
850 passengers and luggage: $850 \times 100\ kg = 85\ tons$
Aircraft body: 300 *tons*
Wing span: 80 *m*
Averaged wing chord length: 10 *m*
Averaged fuselage length: 70 *m*
Averaged fuselage width: 10 *m*

1.3 Explain the physics of a rubber suction cup.

1.4 We have two bottles; one is partially filled with water and the other is empty. After connecting the two bottles, we turn them upside down, as shown below.

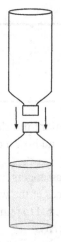 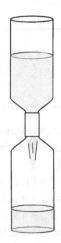

a) Explain why water does not fall easily.

b) Sketch the vertical pressure distributions for bottles with necks of different sizes.

c) Discuss (b) in terms of the function of the neck.

d) Explain why the water falls more easily if we draw a circle several times with the bottles.

1.5 Fish have swim bladders to keep them neutrally buoyant.

a) Discuss how fish float or sink with a swim bladder.

b) Discuss how submarines rise or sink.

c) Compare the buoyancy control mechanism between the two cases.

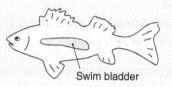

Swim bladder

1.6 In high-rise buildings, the so-called "chimney (or stack) effect" is important when the outdoor air temperature differs significantly from the indoor temperature.

a) What are the main sources that produce the stack effect in high-rise buildings?

b) Physically explain why the stack effect is proportional to the height of the building as well as to the ratio of the outdoor and indoor temperatures.

c) Discuss the forces associated with the stack effect.

d) Describe the stack effect in the heating season and in the cooling season, respectively.

1.7 When a squid ejects water from its body, a thrust force is created. This thrust force is associated with various forces acting on the squid body as well as on the water.

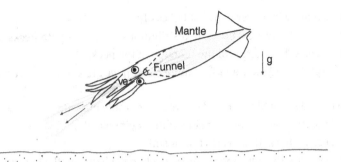

a) Describe the forces inside the funnel at the moment the water is ejected and explain how they are balanced.

b) Sketch the pressure distribution in the funnel at the moment of ejection.

c) Explain how a thrust force is created.

1.8 Effective ventilation of the interior of a helmet is created by a combination of inlet and outlet vents (with fully functional ventilation channels). Cool, fresh air enters the helmet through the vents, and damp and warm air leaves the helmet through the outlet vents for a fresh and comfortable interior. Explain how the helmet vents function aerodynamically.

1.9 A reservoir filled with water has a drainage hole at the bottom. The cross-sectional areas of the reservoir and bottom hole are A_1 and A_2, and the area contraction ratio $\beta = A_1/A_2$ is substantially large so that any unsteady effect can be neglected.

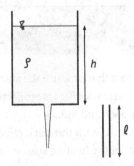

a) Find the velocity of water at the bottom drainage hole.

b) Find the cross-sectional area of the jet in terms of z, where $z = 0$ is the location of the bottom drainage hole.

c) A pipe of length l is plugged into the drainage hole, as sketched. Find the velocity of water at the bottom drainage hole.

d) Discuss how this problem compares to fountain (see Section 1.6).

1.10 Two circular disks of radius R are displaced by a vertical distance h and an airflow is blown through a vertical tube of diameter d with a volumetric flow rate Q. The disk at the lower position is allowed to vertically move and neglect any viscous effects in the system.

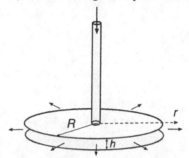

a) Find the velocity distribution $v(r)$ of the air between the disks.

b) Sketch the pressure distribution along a streamline from the entrance of the tube to the exit. Note that the ambient pressure is denoted by p_∞.

c) If the airflow is blown through the tube, what will happen to the lower disk?

1.11 In a boundary layer (see figure), viscous stresses cannot change the pressure, except at the leading-edge where the fluid flows nearly parallel. Physically explain why the net viscous force in the boundary layer cannot change the pressure.

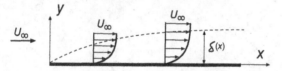

1.12 A hydraulic brake consists of a movable ram and a slightly larger cylinder, as shown in the figure. A force F is applied to move the ram at a constant speed V. The area of the cylinder and the cross-sectional area of the ram are denoted by A_c and A_r, respectively.

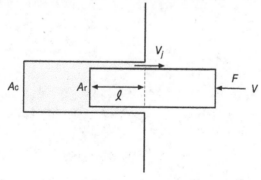

a) Sketch the instantaneous streamlines in the inertial frame of reference inside the cylinder.

b) Sketch the instantaneous streamlines in the reference frame moving with the ram.

c) Determine the pressure at the end of the cylinder (where the velocity is assumed to be zero) p_c with the given parameters, and discuss the Galilean invariance.

2

Macroscopic Balance of Mass, Momentum, and Energy

2.1 Conservation Laws of Mass, Momentum, and Energy

2.1.1 The Concept of Fluid Particles

Fluid mechanics is founded upon the conservation principles of mass, momentum, and energy. To apply these conservation laws to fluids, a lump of fluid is viewed as *a collection or aggregate of fluid elements*. In this case, an element refers to a very small fluid particle whose physical quantities such as density, velocity, pressure, correspond to an averaged value of the molecules. This concept of fluid particle is based on the fact that the dynamics of each particle is a local response to the dynamics of the whole system (e.g. the shape of a hanging chain).[1]

With this concept of fluid particles, we can express the flow field as a continuous function of flow variables such as density, pressure, velocities, while defining the forces on the surfaces or within the volume of the fluid particle. Furthermore, the conservation law of mass stating that a material derivative of mass of a fluid particle is zero at all times can be written as

$$\frac{Dm}{Dt} = 0 \qquad (2.1)$$

where $m = \rho \, dV$ is the mass of the particle.

Likewise, the conservation law of momentum (i.e. Newton's second law of motion) can be applied to a fluid particle:

$$m\frac{D\vec{v}}{Dt} = \vec{F} \qquad (2.2)$$

where the left-hand side is the rate of change of momentum of a fluid particle and $\vec{F}$ on the right represents the net force exerted on the particle.

The conservation law of energy (i.e. the first law of thermodynamics) can also be applied to a fluid particle:

$$m\frac{De_t}{Dt} = \dot{Q} + \dot{W} \qquad (2.3)$$

where $e_t = e + v^2/2$ is the specific total energy, e the specific internal energy, and v the total velocity. The right-hand side terms $\dot{Q}$ and $\dot{W}$ are the net inflow rates of heat and work to the fluid particle.

---

1 Lighthill [1].

Introduction to Fluid Dynamics: Understanding Fundamental Physics, First Edition. Young J. Moon.
© 2022 John Wiley & Sons, Inc. Published 2022 by John Wiley & Sons, Inc.
Companion website: www.wiley.com/go/Moon/IntroductiontoFluidDynamics

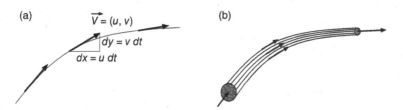

Figure 2.1 Streamline: a locus, tangent to the local velocity vectors (a); stream tube: a bundle of streamlines (b)

By dividing by the volume of the particle dV, the conservation equations applied to the fluid particle, Eqs. (2.1)–(2.3), can be cast into a general form:

$$\rho \frac{D\beta}{Dt} = s \tag{2.4}$$

where $\beta = (1, \vec{v}, e_t)$ represents the mass, momentum, and total energy of the fluid per unit mass, respectively. Note that $s = (0, \vec{f}, \dot{q} + \dot{w})$ is the source in action per unit volume.[2]

With initial and boundary conditions, Eq. (2.4) will finally be integrated over the volume of the lump to fulfill the conservation laws of the system. It is important to note that this concept of breaking down the physics to micros and reintegrating them back into macros is useful when dealing with deformable substances such as fluids.

2.1.2 Descriptions of Fluid Flows

2.1.2.1 Lagrangian vs. Eulerian

In a Lagrangian description, fluid particles are tracked in time. The density of the fluid particle i (i is the marker) denoted by $\rho_i(t)$ represents the mass of the molecules divided by the volume $dV_i(t)$, and the velocity vector $\vec{v}_i(t)$ is defined as a time derivative of the position vector $\vec{x}_i(t)$, which points to the center of mass of the particle at time t. This Lagrangian approach can be important when dealing with sprays, bubbles, particles, rarefied gases, but the main drawback is in its implementations because there are simply too many particles to track in time.

In an Eulerian description, the physical quantities are expressed as functions of space and time, i.e. $\rho(\vec{x}, t)$ and $\vec{v}(\vec{x}, t)$, as they are distributed in space at an instant. This *field concept* founded by Euler is extremely useful for continuums such as fluids. Here is given a good example of Eulerian description. As depicted in Figure 2.1, an instantaneous streamline can be defined as a locus, tangent to the velocity vectors expressed as a field function, i.e. $\vec{v} = \vec{v}(\vec{x}, t)$. The streamline is mathematically defined as

$$\frac{dx}{u} = \frac{dy}{v} = \frac{dz}{w} \tag{2.5}$$

where $\vec{v}(\vec{x}, t) = (u(\vec{x}, t), v(\vec{x}, t), w(\vec{x}, t))$.[3] The importance of the Eulerian description will be more discovered in the next sections when we obtain the rate of change of physical quantities of a fluid particle.

---

2 Equation (2.4) with $\beta = 1$, the conservation law of mass, appears to be more explicit when it is integrated over a control volume:
$$\int_{CM} \rho \frac{D(1)}{Dt} \, dV = \int_{CM} \frac{D(1)}{Dt} \, dm = \int_{CM} \frac{D(dm)}{Dt} = \frac{D}{Dt} \int_{CM} dm = 0.$$

3 A stream tube is defined as a bundle of streamlines and by definition, there can be no component of velocity that traverses the stream tube's side surfaces.

2.2 Rate of Change of a Fluid Particle

A rate of change of a physical quantity $\phi_i(t)$ of a fluid particle i is mathematically defined as follows:

$$\frac{D\phi_i(t)}{Dt} = \frac{\partial \phi}{\partial t} + \vec{v} \cdot \nabla \phi \tag{2.6}$$

where the left-hand side is referred to as the total (or material) rate of change of ϕ_i, and the two terms on the right are the local and convective rates of change of $\phi(\vec{x}, t)$. In this section, we are going to explain with graphical illustrations how the total rate of change of $\phi_i(t)$ is physically related to the local and convective rates of change of $\phi(\vec{x}, t)$.

2.2.1 Convective Rate of Change

Let us suppose we have a tube with a nozzle, and the piston is moved at a constant speed u_p by applying a force F_1. In this case, a steady-state flow is set in the tube and nozzle, with velocity and pressure distributions as sketched in Figure 2.2.[4]

For a fluid particle i moving a distance $d\vec{x} = \vec{x}_B - \vec{x}_A$ ($= \vec{v} \, dt$) from point A to point B in time dt, $\phi_i(t)$, and $\phi_i(t + dt)$ can be expressed as follows:

$$\phi_i(t) = \phi(\vec{x}_A, t) \tag{2.7}$$

$$\phi_i(t + dt) = \phi(\vec{x}_B, t + dt) = \phi(\vec{x}_B, t) \tag{2.8}$$

since in steady flow, $\phi_i(t)$ changes along the streamline, but the value remains the same at a fixed point. In other words, all the fluid particles follow the exact same time-track along the streamline.

Now, an equivalence can be set between the Lagrangian and Eulerian descriptions by subtracting Eq. (2.7) from (2.8). For example, in one-dimension,

$$\phi_i(t + dt) - \phi_i(t) = \phi(\vec{x}_B, t) - \phi(\vec{x}_A, t) = dx \cdot \left. \frac{\partial \phi}{\partial x} \right|_{x_A} \tag{2.9}$$

This shows that the amount of change of ϕ_i over dt is equivalent to a distance of travel $d\vec{x}$ ($= \vec{v} \, dt$) times the spatial gradient of $\phi(\vec{x})$.

(a)

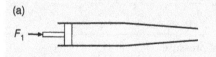

(b)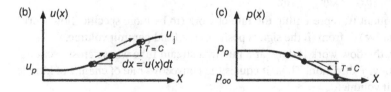

Figure 2.2 Steady flow in a tube-nozzle configuration (a); velocity and pressure distributions (b,c); solid and hollow circles mark the convecting fluid particles; convective changes of velocity and pressure over the same time interval dt (vertical arrows); C(convective) and T(total)

---

4 Sketches represent concepts only.

With $dx = u\ dt$, Eq. (2.9) becomes

$$d\phi_i(t) = u\ dt \cdot \left.\frac{\partial\phi}{\partial x}\right|_{x_A} \tag{2.10}$$

and in multidimensions, it is written as follows:

$$d\phi_i(t) = \vec{v}\ dt \cdot \nabla\phi\ |_{\vec{x}_A} \tag{2.11}$$

If we divide both sides by dt, the material derivative of ϕ can be expressed as follows:

$$\frac{D\phi_i(t)}{Dt} = \vec{v} \cdot \nabla\phi = u\ \frac{\partial\phi}{\partial x} + v\ \frac{\partial\phi}{\partial y} + w\ \frac{\partial\phi}{\partial z} \tag{2.12}$$

where the right-hand side is called the convective rate of change of ϕ.

2.2.1.1 Graphical Interpretation

A convective rate of change of velocity (or pressure) in the nozzle is illustrated with circles (Figure 2.2). The solid and hollow circles show the positions of the fluid particles traced in time (with the same time interval Δt) but at two different events. Note that the spatial gradient of velocity (or pressure) at the second movement (between the second and third circles) looks almost the same as that of the first movement (between the first and second circles). However, the amount of increase of velocity (or decrease of pressure) at the second movement is larger than that at the first movement because the distance of travel over Δt is longer (with a higher speed); thus, the convective rate of change of the velocity (or pressure) at the second movement is larger than that at the first movement.

It is clearly shown in Figure 2.2 that the difference of the velocity (or pressure) between the two circles becomes larger not only with a stiffer spatial gradient of the velocity (or pressure) but also with a higher velocity; the two circles will be farther apart because the distance of travel is longer. It is also to be noted that the magnitudes of the local velocity u and its spatial gradient du/dx in the nozzle are determined by the piston speed u_p, area contraction ratio $\beta = A_p/A_j$, and fluid density ρ [2].

2.2.1.2 Convective Rate of Change of Pressure[†]

The convective rate of change of velocity

$$\frac{D\vec{v}_i(t)}{Dt} = \vec{v} \cdot \nabla\ \vec{v} = u\ \frac{\partial\vec{v}}{\partial x} + v\ \frac{\partial\vec{v}}{\partial y} + w\ \frac{\partial\vec{v}}{\partial z} \tag{2.13}$$

represents fluid acceleration in space.

What does a convective rate of change of pressure

$$\frac{D\ p_i(t)}{Dt} = \vec{v} \cdot \nabla\ p = u\ \frac{\partial p}{\partial x} + v\ \frac{\partial p}{\partial y} + w\ \frac{\partial p}{\partial z} \tag{2.14}$$

mean then? It is a scalar quantity, representing the rate of work (to be more specific, flow work) done on (if the sign is negative) or from (if the sign is positive) the fluid per unit volume.

In a steady nozzle flow, the flow work is done at a rate in a stream tube whose cross-sectional area changes in the streamwise direction. Thus, it equals the convective rate of change of kinetic energy of the fluid per unit volume:

$$u\ \frac{\partial p}{\partial x} = -\rho u\ \frac{\partial(u^2/2)}{\partial x} \tag{2.15}$$

where the signs of both sides are switched because a fluid either attains its kinetic energy by the flow work done on the fluid per unit time (e.g. nozzle), or loses its kinetic energy by the flow work done

---

† For advanced studies.

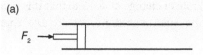

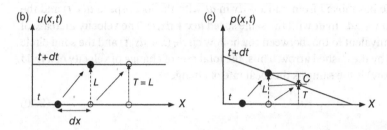

Figure 2.3 Transient slug flow in a straight tube (a); velocity and pressure distributions (b,c); total change (dashed-dotted), local time change (dashed), and convective change (dotted) over the same time interval dt; C(convective), L(local), and T(total)

on the surrounding fluid per unit time (e.g. diffuser). The convective rate of change of pressure can also occur in unsteady flows (e.g. a slug flow in transient acceleration). In this case, it will be the same as the local rate of change of kinetic energy of the fluid per unit volume (see Section 2.2.2).[5]

2.2.2 Local (Time) Rate of Change

A straight tube of length l is considered, and a force F_2 is applied to the piston to accelerate the fluid at rest. A transient slug flow is set in the tube, with velocity and pressure distributions as sketched in Figure 2.3. This case clearly shows the local rate of change of velocity and pressure. Note, however, that the local rate of change is identical to the total rate of change only for velocity. The total rate of change of pressure includes the convective rate of change (see Section 2.2.2.2 *Total rate of change of pressure*).

For a fluid particle i moving a distance $d\vec{x} = \vec{x}_B - \vec{x}_A$ ($= \vec{v}\,dt$) from point A to point B in time dt, we can express $\phi_i(t)$ and $\phi_i(t + dt)$ as follows:

$$\phi_i(t) = \phi(\vec{x}_A, t) \tag{2.16}$$

$$\phi_i(t + dt) = \phi(\vec{x}_B, t + dt) = \phi(\vec{x}_A, t + dt) \tag{2.17}$$

since the changes of ϕ, such as the velocity vector $\vec{v}$, will be the same in space.

An equivalence can be set between Lagrangian and Eulerian descriptions by subtracting Eq. (2.16) from (2.17):

$$\phi_i(t + dt) - \phi_i(t) = \phi(\vec{x}_A, t + dt) - \phi(\vec{x}_A, t) = dt \cdot \left.\frac{\partial \phi}{\partial t}\right|_{x_A} \tag{2.18}$$

This shows that the amount of change of ϕ of the fluid particle i being convected from point A to B is equivalent to the local change of $\phi(x, t)$ at point A over dt.

By dividing Eq. (2.18) by dt, the material derivative of ϕ reads

$$\frac{D\phi_i(t)}{Dt} = \left.\frac{\partial \phi}{\partial t}\right|_{x_A} \tag{2.19}$$

---

5 More discussion continues in Section 2.6.2 *Flow work*.

where the right-hand side is the local (time) rate of change of ϕ. Note that this relation only holds true for the velocity field, not for other quantities such as pressure, density, total energy (Figure 2.3).

2.2.2.1 Graphical Interpretation

The local rate of change of velocity in the tube is illustrated with the solid and hollow circles in Figure 2.3b. The solid circle has moved from x to $x + dx$ in dt with the local speed $u(x, t)$ and the hollow circle from $x + dx$ to $x + 2dx$ in dt with the same speed $u(x + dx, t)$. The velocity increase of the hollow circle in dt is equivalent to that between the hollow circle $(x + dx, t)$ and the solid circle $(x + dx, t + dt)$ represented by the dashed arrow. Thus, the total rate of change of velocity of a fluid particle (dashed-dotted arrow) is the same as the local rate of change:

$$\frac{Du}{Dt} = \frac{\partial u}{\partial t} \tag{2.20}$$

2.2.2.2 Total Rate of Change of Pressure[†]

The total rate of change of pressure of the fluid particle does not correspond to the local rate of change of pressure, as velocity does. As illustrated in Figure 2.3c, the total rate of change of pressure of the fluid particle (dashed-dotted arrow) is the local rate of change of pressure (dashed arrow) plus the convective rate of change of pressure (dotted arrow):

$$\frac{Dp}{Dt} = \frac{\partial p}{\partial t} + u \frac{\partial p}{\partial x} \tag{2.21}$$

The convective rate of change of pressure (or the rate of flow work done on the fluid per unit volume) is the same as the local rate of increase of kinetic energy of the fluid per unit volume[6]

$$u \frac{\partial p}{\partial x} = -\rho \frac{\partial (u^2/2)}{\partial t} \tag{2.22}$$

Thus, the total rate of change of pressure can be expressed as follows:

$$\frac{Dp}{Dt} = \frac{\partial p}{\partial t} - \rho \frac{\partial (u^2/2)}{\partial t} \tag{2.23}$$

During transient acceleration of a fluid particle, there must be a reduction in the local rate of change of pressure by the local rate of increase of kinetic energy of the fluid per unit volume. This is clearly shown in Figure 2.3c by the dotted arrow, which indicates a pressure-relieving effect created by the transient acceleration of the fluid particle.

2.2.3 Total (or Material) Rate of Change

If a force F_3 is applied to suddenly accelerate the fluid in the tube and the nozzle (already driven by the force F_1), the velocity and pressure distributions will have both local and convective rates of change, as illustrated in Figure 2.4.

If $\phi(\vec{x}, t)$ changes in time and space, the amount of change of ϕ of the fluid particle i over dt is the sum of the local change in time and the convective change in space.

$$d\phi_i(t) = dt \frac{\partial \phi}{\partial t} + \left(dx \frac{\partial \phi}{\partial x} + dy \frac{\partial \phi}{\partial y} + dz \frac{\partial \phi}{\partial z} \right) \tag{2.24}$$

---

† For advanced studies.
6 A fluid locally accelerated in time must accompany a negative pressure gradient in space. See Eq. (1.11) in Section 1.5.1 *Transient inertial force*, Chapter 1 *Pressure*. For this reason, the convective rate of change of pressure can be a sound source.

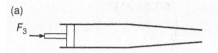

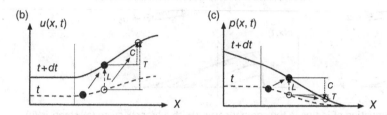

Figure 2.4 Transient flow in a tube-nozzle configuration (a); velocity and pressure distributions (b,c); total change (dashed-dotted), local time change (dashed), and convective change (dotted) over the same time interval dt; vertical dashed line: border between the tube and the nozzle; C(convective), L(local), and T(total)

where $dx = u\, dt$, $dy = v\, dt$, and $dz = w\, dt$ are the distances that the particle i at $(\vec{x}, t)$ travels in each direction over dt. By dividing Eq. (2.24) by dt, the total derivative (or a material or substantial derivative) is written as follows:

$$\frac{D\phi}{Dt} = \frac{\partial \phi}{\partial t} + \left(u\, \frac{\partial \phi}{\partial x} + v\, \frac{\partial \phi}{\partial y} + w\, \frac{\partial \phi}{\partial z} \right) \tag{2.25}$$

2.2.3.1 Graphical Interpretation

Figure 2.4 illustrates with solid and hollow circles that the rate of increase of velocity (or pressure) of the particle in the nozzle over dt is equivalent to the sum of the local and convective rates of change. In this case, the total rate of change of velocity (dashed-dotted arrow) will be greater than that of the steady-state case because the local rate of change (dashed arrow) is added to the convective rate of change (dotted arrow).

$$\frac{Du}{Dt} = \frac{\partial u}{\partial t} + u\, \frac{\partial u}{\partial x} \tag{2.26}$$

Similarly, the total rate of change of pressure will be less negative than that of the steady-state case. As shown in Figure 2.4c, the total rate of change of pressure (dashed-dotted arrow) is the local rate of change of pressure (dashed arrow) plus the convective rate of change of pressure (or the rate of flow work done on the fluid) (dotted arrow).

Thus, the convective rate of change of pressure can be expressed as the sum of the local and convective rates of increase of kinetic energy of the fluid per unit volume

$$u\, \frac{\partial p}{\partial x} = -\rho\, \frac{\partial (u^2/2)}{\partial t} - \rho u\, \frac{\partial (u^2/2)}{\partial x} \tag{2.27}$$

to which the total rate of change of pressure is finally written as follows:

$$\frac{Dp}{Dt} = \frac{\partial p}{\partial t} - \rho\, \frac{\partial (u^2/2)}{\partial t} - \rho u\, \frac{\partial (u^2/2)}{\partial x} \tag{2.28}$$

For example, when a fluid particle moves, there must be a reduction in the local rate of change of pressure by the local and convective rates of change of kinetic energy of the fluid per unit volume.

Example 2.1 *Concentration of the constituent*

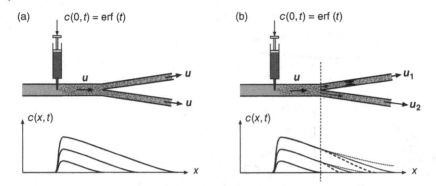

Figure 2.5 Transluminal attenuation of iodine in the coronary artery with a branch; clean (a), stenosis (b); vessel w/ stenosis (dashed) and w/o stenosis (dotted)

For a diagnostic purpose, iodine of molar concentration $c(t)$ is injected into the blood stream in a coronary artery (Figure 2.5). Show that a branch with stenosis can be detected by comparing the gradients of the iodine molar concentration along the vessel.

The conservation law of the constituent states that

$$\frac{Dc}{Dt} = \alpha \, \nabla^2 c + R \tag{2.29}$$

where α is the diffusivity coefficient and R is a source. If mass diffusion is small compared to convection and there is no source, Eq. (2.29) can be written as follows:

$$\frac{Dc}{Dt} = 0 \quad \text{or} \quad \frac{\partial c}{\partial t} = -u \, \frac{\partial c}{\partial x} \tag{2.30}$$

Equation (2.30) shows that the local rate of change of c is the same as the convective rate of change in magnitude because c is an invariant quantity. For a given injection rate (e.g. $c(t) = \mathrm{erf}(t)$), u and $\partial c/\partial x$ are inversely proportional to each other so that as the iodine particles move at a slower speed, their concentration must drop faster in space. The vessel with stenosis will show a stiffer gradient of the iodine molar concentration (dashed), whereas in the vessel in other section of the branch, the gradient will be more gradual than in the normal case (dotted) because the flow speed is faster [3].

2.3 Rate of Change of a Lump of Fluid

We introduce flux and the divergence theorem, essential concepts to extending the conservation laws of mass, momentum, and energy applied to a fluid particle to those to a lump.

2.3.1 Flux

Flux[7] represents the rate of transfer of a physical quantity per unit area. This concept of flux is useful when we count the amount of the physical quantity that has crossed the surface boundary for a given time interval; we simply multiply the flux by the total area and the elapsed time. This concept is well illustrated in Figure 2.6 by a picture of soap bubbles, which visualizes the amount of mass of the wind that has flown through a surface defined by the wire frame in a specific time interval.

---

7 In Latin, it means *flow*.

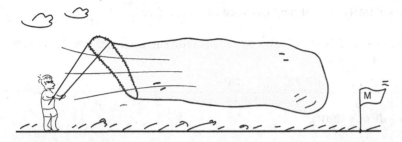

Figure 2.6 Visualization of a flux of wind through a surface of soap bubble formed by a wire

To apply with the conservation laws of mass, momentum, and energy, we should define the mass flux, momentum flux, and energy flux. A mass flux can be expressed as $\rho\,(\vec{v}\cdot\vec{n})$, where $\vec{n}$ is the outward unit vector normal to a surface and $\vec{v}\cdot\vec{n}$ is the distance of travel per unit time. The mass flux can be used to express the momentum flux, $\rho\,\vec{v}\,(\vec{v}\cdot\vec{n})$, total energy flux, $\rho\,e_t\,(\vec{v}\cdot\vec{n})$, or any flux of physical quantity that can be carried by the mass. Here is given an example. In rocket propulsion, a thrust force of the rocket is proportional to the amount of mass of the fluid particles that has flown out through the nozzle exit plane per unit time (also called a mass flow rate), and also to the particle velocity at the nozzle exit.[8]

2.3.2 Divergence Theorem

The divergence theorem, also known as Gauss's theorem, states that a volume integral of the divergence of a vector field is equal to the net outflow across the volume's boundary,

$$\int_{CV} \nabla\cdot\vec{F}\ dV = \int_{CS} \vec{F}\cdot\vec{n}\ dA \tag{2.31}$$

With $\vec{F}$ being $(\rho\beta)\,\vec{v}$, Eq. (2.31) reads

$$\int_{CV} \nabla\cdot(\rho\beta\,\vec{v}\,)\ dV = \int_{CS} (\rho\beta)\ \vec{v}\cdot\vec{n}\ dA \tag{2.32}$$

where the so-called conservative variable $\rho\beta$ represents the mass, momentum, and total energy per unit volume. Here, $\beta = (1, \vec{v}, e_t)$, and $e_t = e + v^2/2$ is the total energy of the fluid per unit mass.

If $\beta = 1$,

$$\int_{CV} \nabla\cdot(\rho\vec{v}\,)\ dV = \int_{CS} (1)\ \rho\vec{v}\cdot\vec{n}\ dA \tag{2.33}$$

represents the net outflux of mass through a closed control surface.

If $\beta = \vec{v}$,

$$\int_{CV} \nabla\cdot\{(\rho\vec{v})\,\vec{v}\,\}\ dV = \int_{CS} (\vec{v})\ \rho\vec{v}\cdot\vec{n}\ dA \tag{2.34}$$

represents the net outflux of momentum through a closed control surface in each direction.

If $\beta = e_t$,

$$\int_{CV} \nabla\cdot(\rho\,e_t\,\vec{v}\,)\ dV = \int_{CS} (e_t)\ \rho\vec{v}\cdot\vec{n}\ dA \tag{2.35}$$

---

8 Multiplication of both is the momentum flux at the nozzle exit. This interpretation is based on the conservation principle of momentum.

represents the net outflux of total energy through a closed control surface.

Example 2.2 Show that the net flux of $\vec{v} \cdot \vec{n}$ over a closed control surface represents the rate of change of volume of the control mass.

$$\int_A \vec{v} \cdot \vec{n} \ dA = \int_V \nabla \cdot \vec{v} \ dV = \int_V \frac{dV'/dV}{dt} \ dV = \frac{DV}{Dt} \tag{2.36}$$

where $dV' = dV(t + dt) - dV(t) = d(dV)$.

2.3.3 Conservation Equations in Divergence Form

A general form of the conservation equations, Eq. (2.4), can be expressed as follows:

$$\rho \frac{D\beta}{Dt} = \rho \left(\frac{\partial \beta}{\partial t} + \vec{v} \cdot \nabla \beta \right) = s \tag{2.37}$$

where the convective rate of change of β can be rearranged as

$$\rho (\vec{v} \cdot \nabla)\beta = \nabla \cdot (\rho \beta \ \vec{v}) - \beta \ \nabla \cdot (\rho \ \vec{v}) \tag{2.38}$$

By using the conservation of mass, $\nabla \cdot (\rho \vec{v})$[9] that represents a net outflux of mass at a point is written as follows:

$$\nabla \cdot (\rho \vec{v}) = -\frac{\partial \rho}{\partial t} \tag{2.39}$$

Equation (2.38) now reads

$$\rho (\vec{v} \cdot \nabla)\beta = \nabla \cdot (\rho \beta \ \vec{v}) + \beta \ \frac{\partial \rho}{\partial t} \tag{2.40}$$

If we add $\rho \, \partial \beta / \partial t$ on both sides of Eq. (2.40), it is written in a divergence form as follows:

$$\rho \left(\frac{\partial \beta}{\partial t} + \vec{v} \cdot \nabla \beta \right) = \frac{\partial (\rho \beta)}{\partial t} + \nabla \cdot (\rho \beta \ \vec{v}) \tag{2.41}$$

The conservation equation, Eq. (2.37), finally reads

$$\rho \frac{D\beta}{Dt} = \frac{\partial (\rho \beta)}{\partial t} + \nabla \cdot (\rho \beta \ \vec{v}) = s \tag{2.42}$$

where $\partial (\rho \beta)/\partial t$ is the local rate of change of mass, momentum, and total energy of the fluid per unit volume. It must be emphasized that this divergence form is crucial to extending the conservation laws defined at a point into a volume.

2.3.4 Macroscopic Balance of Mass, Momentum, and Energy

By using the divergence theorem, the conservation equation in differential form, Eq. (2.42), can be integrated over a lump of fluid (or a control mass):

$$\int_{CM} \rho \frac{D\beta}{Dt} \ dV = \int_{CV} \frac{\partial (\rho \beta)}{\partial t} \ dV + \int_{CS} (\rho \beta) \ \vec{v} \cdot \vec{n} \ dA = \int_{CV} s \ dV \tag{2.43}$$

Since the material volume moves with the fluid and the mass is an invariant quantity, it follows:

$$\frac{D}{Dt} \int_{CM} (\rho \beta) \ dV = \int_{CM} \rho \frac{D\beta}{Dt} \ dV + \int_{CM} \beta \ \overset{0}{\cancel{\frac{D(\rho \, dV)}{Dt}}} \tag{2.44}$$

---

9 See Section 3.2 *Conservation law of mass.*

Thus, Eq. (2.43) finally reads

$$\frac{D}{Dt}\int_{CM} (\rho\beta)\, dV = \int_{CV} \frac{\partial(\rho\beta)}{\partial t}\, dV + \int_{CS} (\rho\beta)\, \vec{v}\cdot\vec{n}\, dA = \int_{CM} s\, dV \tag{2.45}$$

where $\beta = (1, \vec{v}, e_t)$ are the mass, momentum, and energy of the fluid per unit mass, and $s = (0, \vec{f}, \dot{q} - \dot{w})$ are the sources in action per unit volume.

This relation, also known as the Reynolds's transport theorem, indicates that a total derivative of $\rho\beta$ of the control mass, equivalent to a sum of the local rate of change of $\rho\beta$ within the control volume and the net outflux of $\rho\beta$ (or imbalance of $\rho\beta$ flux) through the control surfaces,[10] is proportional to the sources in action [4]. Now, Eq. (2.45) can be solved with a given set of boundary and initial conditions.

[**Notes**] In control volume analysis, how to select a control volume is just a matter of choice. For instance, an unsteady problem can be viewed as a steady problem, or vice versa. Let us suppose we have two control volumes for a fluid at rest; one is stationary, and the other is enlarging in time. The conservation law of mass states that the mass in the stationary control volume is invariant, but if the control volume is enlarging in time, the mass in the control volume increases at a rate and there must be an influx of mass at the control surface to match the local rate of increase.[11] For the control volume changing shape over time, the local increase of mass, or transfer of mass at the boundary is just another way of expressing that the mass is an invariant quantity.

On the other hand, a rate of increase of momentum or energy in the control volume, or flux of these at the boundary must be evaluated with the absolute velocity, since the system is based on the inertial reference frame. Now, for the control volume changing shape over time, the conservation equations of mass, momentum, and energy can be generalized as follows:

$$\frac{\partial}{\partial t}\int_{CV(t)} \rho\beta\, dV + \int_{CS(t)} \rho\beta\, \vec{v}_{rel}\cdot\vec{n}\, dA = \int_{CV(t)} s\, dV \tag{2.46}$$

where $\beta = (1, \vec{v}, e_t)$, $\vec{v}$ is an absolute velocity vector, and $\vec{v}_{rel}$ is the velocity vector relative to the moving control surface. Some examples with a moving control volume will be presented in later sections.

2.4 Conservation of Mass ($\beta = 1$)

$$\frac{D}{Dt}\int_{CM} \rho\, dV = \frac{\partial}{\partial t}\int_{CV} \rho\, dV + \int_{CS} \rho\vec{v}\cdot\vec{n}\, dA = 0 \tag{2.47}$$

The conservation law of mass states that if there is a difference between the mass influx and outflux through the control surface (i.e. imbalance of mass flux), there must be a rate of change of mass within the control volume. In this case, the fluid density or shape of the control volume changes over time.

---

10 This is based on the fact that the fluxes of those quantities and internal stresses are shared at the common boundaries within the control volume.

11 The same logic applies to the control volume shrinking in time. The mass within the control volume decreases at a rate so that there must be an outflux of mass at the control volume; thus, both rates are the same.

Figure 2.7 Imbalance of mass, momentum, and energy fluxes of fuel in rocket combustion; Apollo 11 launched from the Kennedy Space Center, Florida (July 16, 1969)

2.4.1 Imbalance of Mass Flux

In an incompressible flow, the local rate of change of mass within the control volume is zero unless the control volume changes shape over time. If not, Eq. (2.47) becomes

$$\int_{CS} \vec{v} \cdot \vec{n}\ dA = 0 \tag{2.48}$$

indicating that the mass flux is balanced; the fluid that enters into the control volume must go out at the same rate. If the control volume enlarges in time, for example, the mass in the control volume increases at a rate, and there must be an influx of mass at the control surface to match the local rate of increase. If it shrinks in time, there must be an outflux of mass at the control surface to match the local rate of decrease.

In a compressible flow, the mass conservation law states that a net outflux of mass across the control surfaces equals the rate of decrease of mass within the control volume. In this case, the imbalance of mass flux across the control surface is created by the local rate of change of mass of the control volume, which changes shape or density over time. In a rocket engine, for instance, the combusted gas is expelled through the rocket nozzle, and the imbalance of mass flux across the control surface is balanced with a rate of reduction of mass (or density) within the rocket engine (Figure 2.7).[12]

2.5 Conservation of Momentum ($\beta = \vec{v}$)

$$\frac{D}{Dt}\int_{CM} \rho\vec{v}\ dV = \frac{\partial}{\partial t}\int_{CV} \rho\vec{v}\ dV + \int_{CS} \rho\vec{v}\ \vec{v} \cdot \vec{n}\ dA = \vec{F} \tag{2.49}$$

---

12 The fundamental physics of rocket propulsion is well explained in a video made by US Department of the Air Force, released by *Space and Missile Systems Center Los Angeles AFB*, published on Jun 5, 2014 (https://youtu.be/lnyDnruVpTw) [5].

where a net force acting on the control mass $\vec{F}$ can be expressed as follows:

$$\vec{F} = \int_{CS} \vec{n}\,[\sigma]\ dA + \int_{CV} \vec{f}_b\ dV + \vec{R} \tag{2.50}$$

It is to be noted that $\vec{n}\,[\sigma]$ is a stress vector acting on the surface of area dA with orientation specified by an outward unit normal vector $\vec{n}$, where a stress tensor $[\sigma]$ is defined by

$$[\sigma] = (-p + \lambda\,\nabla\cdot\vec{v})\,[\,I\,] + [\tau] \tag{2.51}$$

In Newtonian fluids, a constitutive relation exists between the viscous stress tensor and the strain rate tensor, to which a viscous stress tensor $[\tau]$ is defined by the viscosity multiplied by twice the strain rate of a fluid element:

$$[\tau] = 2\,\mu\,\epsilon_{ij} = \mu\left(\frac{\partial v_i}{\partial x_j} + \frac{\partial v_j}{\partial x_i}\right) \tag{2.52}$$

Note that the dilatational viscosity λ is $-2/3\,\mu$ if the Stokes hypothesis is assumed. Unlike the pressure and dilatational viscous stress that always act in the direction normal to the surface, the direction of action of $\vec{n}\,[\tau]$ depends on the orientatio n of the surface.[13]

The net force $\vec{F}$ acting on the control mass is finally written as follows:

$$\vec{F} = \int_{CS} (-p)\,\vec{n}\ dA + \int_{CS} \vec{n}\,[\tau]\ dA + \int_{CS} (\lambda\,\nabla\cdot\vec{v})\,\vec{n}\ dA + \int_{CV} \vec{f}_b\ dV + \vec{R} \tag{2.53}$$

where $\vec{f}_b$ is the volumetric body force vector per unit volume, and a reaction force from the ground $\vec{R}$ is only to be included if the control volume encloses any solid object; any solid object *in contact with a fluid* within the control volume will exert a force on the fluid.

The conservation law of momentum, Eq. (2.49), states that the rate of change of momentum of the control mass is produced in each direction by a net force vector acting on the fluid of the control mass. If there is no net force acting on the control mass, the momentum flux on the control surface can be unbalanced by the local rate of change of momentum of the control volume, which changes shape or density over time.

2.5.1 Imbalance of Momentum Flux: Steady External Flows

In steady external flows, a control volume can be selected in such a way that the pressure on the control surface becomes the ambient pressure and the reaction force from the solid body $\vec{R}$ is the only force that causes the momentum flux to be unbalanced at the control surface.

If this is the case, it can be said that a destruction or creation of momentum flux in the stream-wise direction produces a drag or thrust force on the body immersed in the fluid. Here is given an example. An aircraft in cruise is opposed to drag because the momentum influx is destructed by the aircraft; it diverts the flow in the transverse direction but cannot pull it back due to flow separations at the body surfaces. Meanwhile, the aircraft is thrust forward by the jets that create the momentum fluxes across the jet engines. Note also that generation of lift is associated with the imbalance of the momentum flux in the vertical direction, producing the so-called 'upwash' and 'downwash' of the air around the airfoil (i.e. circulation).[14]

---

13 More discussion continues in Chapter 3 *Differential Equations of Motions.*
14 The origin of lift of an airfoil is associated with the curvature of the body surface and fluid viscosity, which impose a kinetmatic condition at the trailing-edge of the airfoil, i.e. Kutta condition. More discussion continues in Chapter 4 *Curved Motions.*

2.5.1.1 Drag Force

When a uniform flow passes over a bluff body, an imbalance of momentum flux occurs in the streamwise direction due to flow separations. Because the surface divergence is too stiff, the flow accelerated on the windward side of the body cannot reverse its kinetic energy to a static pressure, as an ideal diffuser could. This separated flow behind the body, called a wake, is full of vortices, often turbulent, and recirculates with low momentum. As a result, the pressure on the leeward side of the body becomes lower than that of the windward side. This pressure difference exerts a force on the body in the streamwise direction, which we call drag force.

2.5.1.2 Thrust Force

When a rocket or a propeller engine expels a jet to ambient fluid, it creates a thrust force by conservation of momentum. The rocket engine burns the fuel which creates a high pressure and temperature in the engine chamber. Due to the differences in pressure and density between the rocket engine chamber and the ambient fluid, the highly expanded gas in the chamber is rapidly expelled through a convergent–divergent nozzle (i.e. C–D nozzles).[15] This creation of huge momentum flux produces a thrust in the streamwise direction. Similarly, turbo-engines rotate propeller blades to produce an imbalance of momentum flux. In this case, the pressure difference across the blade surface draws in the fluid and expels it out. The thrust force is equal to the net momentum flux out produced by the rotating propeller, and it is the net force exerted on the body surfaces (in the streamwise direction).

Example 2.3 *Jet deflection over a curved surface*

A jet is impinged on a concave or convex surface of a body (Figure 2.8). Explain how the drag and lift forces are exerted on the body by jet deflections.

A drag force is exerted on the concave or convex surface because the horizontal momentum flux is reduced at the control surface, i.e. $\dot{m}\,v_j(\cos\theta - 1) < 0$, where $\dot{m}\,(= \rho A_j v_j)$ is the mass flux of the jet, A_j is its cross-sectional area, and θ is the angle between the jet and the horizontal line. Likewise, a downward force (or negative lift) or lift force is exerted on the concave or convex surface with positive or negative vertical momentum flux created at the control surface, i.e. $\dot{m}\,v_j\sin\theta$.

By intuition, the drag force and the lift (negative) force exerted on the concave surface seem obvious because it blocks the flow, but those on the convex surface are not as evident. The jet in this case is attached to the convex surface all the way to the end due to the Coanda effect.[16] Thus, the jet produces a negative pressure on the convex surface with its inertia exerted in the centrifugal

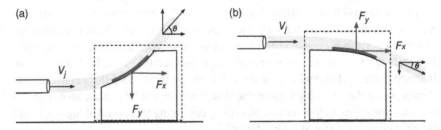

Figure 2.8 Imbalance of momentum flux across the control surface results in drag and lift forces over the concave and convex surfaces; $\theta > 0$ (a), $\theta < 0$ (b)

---

15 More discussion continues in Chapter 8 *Compressible Flows*.
16 More discussion continues in Chapter 4 *Curved Motions*.

Figure 2.9 Water jet impingement on the cart with vane

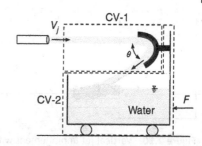

direction; the horizontal component of the resulting force exerted on the body is the drag force and the vertical component is the lift force.

Example 2.4 *Water jet impingement on the cart with vane*

A water jet of cross-sectional area A_j and speed v_j is impinged on a vane mounted on a cart with frictionless wheels. The jet turns and falls into the cart without spilling. With two control volumes as sketched in Figure 2.9, evaluate the horizontal force F required to hold the cart stationary.

If a control volume is selected to include all of the aforementioned subjects (e.g. jet, vane, and cart), the momentum influx is totally destructed by the cart and produces a drag force F on the system. The conservation law of momentum simply states

$$-F = \rho A_j(-v_j)(+v_j) \tag{2.54}$$

However, we can purposely break the control volume into two; one includes the jet and vane (CV-1), and the other the rest (CV-2). In the first control volume, the force required to support the vane F_1 is larger than F since the momentum influx is reflected back at an angle of θ, creating an additional momentum influx at the control surface:

$$F_1 = \rho A_j(-v_j)(+v_j) + \rho A_j(+v_j)(-v_j \cos\theta) \tag{2.55}$$

In the second control volume, the jet reflected at an angle of θ from the vane acts as a momentum influx but is destructed again while exerting a force on the cart in the negative direction. Thus, the force required to support the cart F_2 is balanced with the momentum influx:

$$F_2 = \rho A_j(-v_j)(-v_j \cos\theta) \tag{2.56}$$

It is to be noted that the force F_2 exactly cancels out the additional force acting on the vane created by the reflected jet. If F_1 and F_2 are summed, we end up with the same drag force F:

$$-F = F_1 + F_2 \tag{2.57}$$

Note also that if the jet completely misses the cart, the force required to hold the cart stationary will be greater than F by F_2.

Example 2.5 *Vertical jet impingement (gravitational body force)*

A vertical water jet of speed v_j is being impinged upon the plate (Figure 2.10). The area of the jet exit is A_j. What is the weight W of the plate that can be held in place at the elevation h_1?

Let us first consider a jet only. If we select a control volume to enclose the entire jet, the momentum influx of the jet at the exit is totally destructed by the gravitational body force exerted within the jet. If a solid body of mass M is put on the jet, the momentum influx of the jet is to be destructed by the weights of both jet and plate. Therefore, the jet height must be reduced from h to h_1; obviously, the heavier the plate is, the shorter the jet becomes.

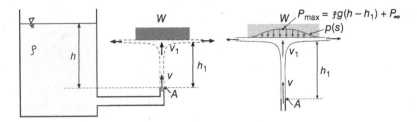

Figure 2.10 Vertical jet impingement with gravitational body force

Control Volume-I If we select a control volume to enclose the plate and the jet from the elevation h_1, where the jet is about to spread, the momentum influx of the jet at h_1 is then to be balanced with the plate's weight.

(Momentum)

$$W = \dot{m}\, v_1 \tag{2.58}$$

(Continuity)

$$\dot{m} = \rho A_j v_j \tag{2.59}$$

(Bernoulli)

$$sl_a : \quad v_j = \sqrt{2gh} \tag{2.60}$$

$$sl_b : \quad \frac{v_j^2}{2} = \frac{v_1^2}{2} + gh_1 \tag{2.61}$$

Here, sl_a is the streamline from the water free surface in the reservoir to the jet exit and sl_b is the streamline from the jet exit to the elevation h_1.

From Eqs. (2.58)–(2.61), we can express the weight of the plate W as follows:

$$W = 2\, (\rho g h) A_j \sqrt{1 - h_1/h} \tag{2.62}$$

Interpretation The weight of the plate is found to be proportional to several parameters: the potential energy of the jet per unit volume $\rho g h$, the nozzle exit area A_j, and the ratio between the two heights h and h_1, h_1/h. If $h_1/h \to 1$, $W \to 0$. This means if there were no weight, the jet would stand as high as h, as expected.

If $h_1/h \to 0$, then $W \to 2(\rho g h)A_j$. If the plate were put on the jet with no gap in between, the weight of the plate that can withstand the pressure force of the water would be expressed as $W' = (\rho g h)A_j$. If there is a gap in between, we need a plate of weight $2W'$ to block the jet. The reason is that an additional W' is needed to destroy the momentum influx of the jet from the exit.

Control volume-II Another control volume can be drawn to include the plate and the jet from the exit. In this case, the momentum flux imbalance of the jet is reacted to by the weight of the plate and the weight of the water jet up to h_1:

$$W + \int_0^{h_1} \rho g\, A(y)\, dy = \dot{m}\, v_j \tag{2.63}$$

Meanwhile, the cross-sectional area of the vertical water jet can be expressed with the continuity equation, $A(y) = A_j v_j / v(y)$, and Bernoulli's equation, $v_j^2/2 = v(y)^2/2 + gy$, as follows:

$$A(y) = A_j / \sqrt{1 - y/h} \tag{2.64}$$

From Eqs. (2.63) and (2.64), we can find the weight of the plate

$$W = 2\,(\rho g h) A_j \sqrt{1 - h_1/h} \tag{2.65}$$

which is exactly the same weight of the plate as that with the control volume-I.

Conservation Law of Energy To find v_j and v_1, the present problem may be interpreted as an imbalance of momentum flux at the control surface. Without the plate, the unbalanced jet momentum influx is reacted to by its gravitational body force acting within the jet. If we select a control volume to include the entire jet, the momentum conservation is written as:

$$-(\rho A_j v_j)\, v_j = -\int_0^h \rho g\, A(y)\, dy \tag{2.66}$$

where $A(y)dy$ is the volume of a slice of the jet. However, Eq. (2.66) cannot be solved unless $A(y)$ is known.

Instead of applying the momentum equation, we can use the conservation equation of energy (see Section 2.6):

$$\int_{CS} \rho\, e_t\, (\vec{v} \cdot \vec{n})\, dA = \int_0^h \int \rho \vec{v} \cdot (-g\vec{k})\, dA\, dy \tag{2.67}$$

where e_t is the total specific energy. Equation (2.67) now reads

$$\rho Q\, v_j^2/2 = \int_0^h \rho g Q\, dy \tag{2.68}$$

where the volumetric flow rate of the jet $Q\ (> 0)$ is defined as follows:

$$Q = A_j v_j \tag{2.69}$$

Note that Q is a constant by conservation of mass. Dividing both sides of Eq. (2.68) by Q and ρ gives us the Bernoulli equation, Eq. (2.60),

$$v_j^2/2 = \rho g h \tag{2.70}$$

where the jet exit speed can be expressed as $v_j = \sqrt{2gh}$.

If we put a plate of mass M on top of the jet, the jet column is reduced from h to h_1, where h_1 depends on the mass of the plate. In this case, the momentum flux imbalance of the jet is reacted to by the weight of the plate and that of the jet in between. Besides, there is no pressure difference force at the control surface because pressure only changes in proximity to the plate bottom surface, where the jet interacts with the plate. The conservation law of energy applied to the shrunk jet reads

$$\rho Q\, (-v_1^2/2 + v_j^2/2) = \int_0^{h_1} \rho g Q\, dy \tag{2.71}$$

where the volumetric flow rate of the jet $Q\ (> 0)$ is the same as defined in Eq. (2.69), which finally becomes the Bernoulli equation, Eq. (2.61):

$$v_j^2/2 = v_1^2/2 + g\, h_1 \tag{2.72}$$

2.5.2 Imbalance of Momentum Flux: Steady Internal Flows

In steady internal flows, it is not possible to make the pressure on the control surface equal to the ambient pressure, as in steady external flows. In this case, an imbalance of momentum flux at the control surface is attributed not only by the pressure and viscous forces acting at the control surface

(a) P_1

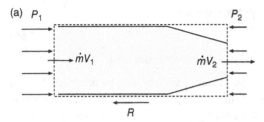

P_2

$\dot{m}V_1$

$\dot{m}V_2$

R

Figure 2.11 Control volumes taken outside (a) and inside the nozzle (b)

(b) P_1

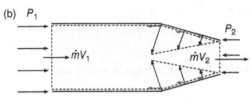

P_2

$\dot{m}V_1$

$\dot{m}V_2$

but also by the reaction force from the solid body; the volumetric body forces also attribute, if they need to be included.

Let us consider a steady flow in the nozzle where the fluid is being accelerated in the streamwise direction. The imbalance of momentum flux at the control surface is produced by the forces acting on the nozzle internal surface as well as on any solid surface upstream (not shown in the present analysis). Here, two control volumes are selected to demonstrate their equivalence (Figure 2.11).

2.5.2.1 Control Volume-I

A control volume is taken outside the nozzle (Figure 2.11a). The pressure at point 1 can be determined by an energy balance between the flow work done on the fluid at a rate by pressure forces and the rate of increase of kinetic energy of the fluid:

$$(p_1 - p_2)\, Q = \frac{1}{2}\, \rho Q \, (v_2^2 - v_1^2) \tag{2.73}$$

where $Q \,(= A_1 v_1 = A_2 v_2)$ is the volumetric flow rate and p_2 may be considered as the ambient pressure p_∞. If we divide Eq. (2.73) by Q, it yields the Bernoulli equation:

$$p_1 + \frac{1}{2}\, \rho v_1^2 = p_2 + \frac{1}{2}\, \rho v_2^2 \tag{2.74}$$

With the mass conservation,

$$\dot{m} = \rho v_1 \, A_1 = \rho v_2 \, A_2 \tag{2.75}$$

the pressure difference between points 1 and 2 can be expressed as follows:

$$p_1 - p_2 = \frac{1}{2}\, \rho v_1^2 \, (\beta^2 - 1) \tag{2.76}$$

where $\Delta p = p_1 - p_2$ is proportional to the upstream kinetic energy per unit volume and $\beta = A_1/A_2 \,(> 1)$, a nozzle area contraction ratio between points 1 and 2.

For the selected control volume, we use the momentum equation to find the reaction force R exerted on the system (or nozzle),

$$(p_1 - p_2)A_1 - R = \dot{m} \, (v_2 - v_1) \tag{2.77}$$

which yields the reaction force R,

$$R = \frac{1}{2}\, \rho v_1^2 \, A_1 \, (\beta - 1)^2 \tag{2.78}$$

[**Notes**] As shown in Eqs. (2.76) and (2.77), two forces, ΔpA_1 and R are created when $\beta > 1$. If $\beta > 1$, a net difference between these two, $\Delta pA_1 - R$, is the net force exerted on the fluid, which produces the imbalance (i.e. creation) of the momentum flux between boundaries at 1 and 2. Note that these two forces are reaction forces from the ground; p_1 is usually associated with the input power supplied from an external source.

2.5.2.2 Control Volume-II

If a control volume is taken inside the nozzle (Figure 2.11b), the momentum equation reads

$$p_1 A_1 - p_2 A_2 - \bar{p}_w (A_1 - A_2) = \dot{m} \, (v_2 - v_1) \tag{2.79}$$

where $\bar{p}_w$ is an averaged pressure on the nozzle side wall.

A static force balance on the nozzle wall shows that the reaction force R is equivalent to the force exerted inside the nozzle wall (in the streamwise direction):

$$R = (\bar{p}_w - p_2) \, (A_1 - A_2) \tag{2.80}$$

Combining Eqs. (2.79) and (2.80) with the elimination of $\bar{p}_w$ will result in the same momentum equation, Eq. (2.77), and the same reaction force R in Eq. (2.78). We can also find the averaged pressure on the nozzle side wall

$$\bar{p}_w = p_2 + \frac{1}{2} \, \rho v_1^2 \, \beta(\beta - 1) \tag{2.81}$$

Example 2.6 *Jet on the cart*

A person standing on a cart with wheels forces a piston to push out a jet through a nozzle (Figure 2.12). The cart is tied to the ground with a rope. Find the force F and the tension on the rope T.

A control volume is selected as sketched in Figure 2.12, and we neglect any unsteady effect for simplicity.

(Mass)

$$\rho_i \, A_1 v_1 = \rho_i \, A_2 v_2 = \rho_i \, Q \tag{2.82}$$

where ρ_i is the fluid density inside the tube.

(Momentum)

$$(p_1 - p_\infty)A_1 + R = \rho_i \, A_2 \, v_2^2 - \rho_i \, A_1 v_1^2 \tag{2.83}$$

where R is the reaction force from the ground.

(Bernoulli)

$$p_1 + \frac{1}{2} \, \rho_i \, v_1^2 = p_\infty + \frac{1}{2} \, \rho_i \, v_2^2 \qquad \text{(aft the piston)} \tag{2.84}$$

$$p_\infty = p' + \frac{1}{2} \, \rho_o \, v_1^2 \qquad \text{(fore the piston)} \tag{2.85}$$

where ρ_o is the fluid density outside the tube and $p' < p_\infty$, which entrains the ambient fluid through the tube entrance.

(Statics)

$$F + p'A_1 = p_1 A_1 \tag{2.86}$$

$$-F - R + T = 0 \tag{2.87}$$

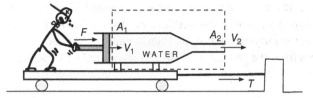

Figure 2.12 Jet on the cart

With Eqs. (2.84)–(2.86), the force applied to the piston can be expressed as follows:

$$F = (p_1 - p')A_1 = \frac{1}{2} \rho_i A_1(v_2^2 - v_1^2) + \frac{1}{2} \rho_o A_1 v_1^2$$
$$= \frac{1}{2} \rho_i A_1 v_1^2 (\beta^2 - 1 + \gamma) \tag{2.88}$$

where $\beta = A_1/A_2$ (> 1) is the nozzle area contraction ratio, $\gamma = \rho_o/\rho_i$ (< 1) is the density ratio between two fluids (e.g. air and water), and $F > 0$ indicates that a force must be exerted in the positive direction to push the piston.

The reaction force R exerted on the tube-nozzle body from the ground is written as follows:

$$R = -(p_1 - p_\infty)A_1 + \rho_i A_2 v_2^2 - \rho_i A_1 v_1^2$$
$$= -\frac{1}{2} \rho_i A_1 v_1^2 (\beta^2 - 1) + \rho_i A_1 v_1^2 (\beta - 1)$$
$$= -\frac{1}{2} \rho_i A_1 v_1^2 (\beta - 1)^2 \tag{2.89}$$

where $R < 0$ indicates that it is exerted in the negative direction.

The tensional force on the rope T then reads

$$T = F + R$$
$$= \frac{1}{2} \rho_i A_1 v_1^2 (\beta^2 - 1 + \gamma) - \frac{1}{2} \rho_i A_1 v_1^2 (\beta^2 - 2\beta + 1)$$
$$= \frac{1}{2} \rho_i A_1 v_1^2 \{2(\beta - 1) + \gamma\} \tag{2.90}$$

which is also positive, indicating $F > R$.

Finally, the forces can be nondimensionalized by $1/2 \, \rho_i A_1 v_1^2$ as follows:

$$\overline{F} = \beta^2 - 1 + \gamma \tag{2.91}$$

$$\overline{R} = -(\beta - 1)^2 \tag{2.92}$$

$$\overline{T} = 2(\beta - 1) + \gamma \tag{2.93}$$

Interpretations If $\beta = 1$, i.e. the nozzle disappears, then $\overline{R} = 0$, but $\overline{F} = \gamma$, and $\overline{T} = \gamma$. In this case, there is no reaction force exerted on the tube-nozzle body from the ground, but we still need a force $\overline{F} = \gamma$ to push the piston to move at a constant speed because the ambient fluid force of the piston must be sucked into the tube. The tensional force on the rope is also $\overline{T} = \gamma$ (Figure 2.13).

If $\gamma = 1$, it corresponds to the case of two fluids with the same density (e.g. water to water, or air to air); in this case, the inertial effect of the fluid force of the piston is comparable to that at the piston. Therefore, $F = T = 1/2 \, \rho_i A_1 v_1^2 = 1/2 \, \rho_i A_2 v_2^2$. If $\gamma = 0.001$ (e.g. air to water), the inertial effect of the force of the piston is substantially low as compared to that at the piston. Thus, almost no force is required to move the piston at a constant speed. A heavier fluid in the tube is simply

(a) 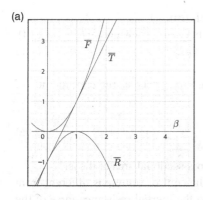   (b)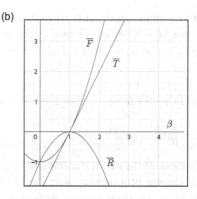

Figure 2.13 $\bar{F}, \bar{R}$, and $\bar{T}$ vs. β $(= A_1/A_2)$ $(\beta \geq 1)$; $\gamma = \rho_o/\rho_i = 1$ (a) and $\gamma = 0.001$ (b)

being translated with a very small reaction force from the piston. This results in $F = T \approx 0$. In the limit $(\gamma = 0)$, it is simply sliding with inertia so that no reaction force is exerted on any solid surfaces.

If the piston were large enough to extrude from the rear side of the tube, there would be no force required to push the piston. In this case, the γ effect will disappear. Therefore, if $\beta = 1, \bar{F} = 0, \bar{R} = 0$, and $\bar{T} = 0$. Besides, in Eq. (2.85), $p' = p_\infty$ so that Eq. (2.86) changes to $F + p_\infty A_1 = p_1 A_1$.

2.5.3 Imbalance of Momentum Flux: Unsteady Flows

In some unsteady problems, it is more preferable to use the conservation equations of mass, momentum, and energy generalized for a control volume that changes shape over time (see Section 2.3.4). To explain the basic concepts used in Eq. (2.46), we introduce a simple steady jet that horizontally expels at a speed of v_j. If we assume that no viscous forces are externally exerted on the jet, this jet will conserve its momentum.

For a control volume that enlarges in time with its surface on the right moving in the streamwise direction at a speed of v_c ($< v_j$), an imbalance of momentum flux is created at the control surface because the momentum flux is a mass-associated quantity. Since the momentum of the jet is conserved, the momentum influx, $\rho A_j v_j^2$, must be balanced with (i) the rate of increase of momentum of the jet within the control volume:

$$\frac{\partial}{\partial t} \int_{CV(t)} \rho \vec{v}\, dV = \frac{d}{dt}(\rho A_j(x - x_0)\, v_j) = \rho A_j v_c\, v_j \tag{2.94}$$

where x denotes the distance of the moving control surface and x_0 its initial position. The momentum influx should also be balanced with (ii) the momentum outflux of the jet crossing the moving control surface at x

$$\int_{CS(t)} \rho \vec{v}\ \vec{v}_r \cdot \vec{n}\, dA = \rho A_j(v_j - v_c)\, v_j \tag{2.95}$$

where $v_r = v_j - v_c$ is the relative velocity of the jet crossing the moving control surface.

If we add (i) to (ii), the sum will be the same as the momentum influx $\rho A_j v_j^2$. If $v_c = 0$, i.e. the control volume is fixed in space with $x = x_0$, the local rate of change of momentum is zero, and the momentum outflux will be the same as $\rho A_j v_j^2$. If $v_c = v_j$, i.e. the control volume is enlarging with the jet, the rate of change of momentum is the same as the momentum influx $\rho A_j v_j^2$, and the momentum outflux will be zero.

Example 2.7 *Jet impingement on the cart (transient acceleration)*

A cart with a vane of mass M_c is being accelerated by the impingement of a water jet (Figure 2.14). The jet of speed v_j is deflected at an angle of θ, and there is negligible friction between the cart and the ground. If the cart starts to move from rest, find $v_c(t)$.

If we select a control volume that enlarges over time at the same speed as the cart, the rate of change of momentum of the jet can be expressed as follows:

$$\frac{\partial}{\partial t} \int_{\text{jet}(t)} \rho \vec{v} \, dV = \frac{d}{dt}(\rho A_j \, x \, v_j) = \rho A_j v_c \, v_j \tag{2.96}$$

where x denotes the distance of the moving cart and A_j the cross-sectional area of the jet.

However, the rate of change of momentum of the jet in the control volume is different from the momentum influx of the jet at the control surface due to the inertial resistance force of the accelerating cart, i.e $M_c \, dv_c(t)/dt$. Thus, the size of the jet grows at a speed of $v_c(t)$, not v_j. Note that the rate of change of momentum of the cart is written as follows:

$$\frac{\partial}{\partial t} \int_{\text{cart}(t)} \rho \vec{v} \, dV = \frac{d}{dt}(M_c v_c) = M_c \frac{d v_c}{dt} \tag{2.97}$$

Since there is no external force acting on the control volume, these two terms are responsible for causing the momentum influx and outflux to be unbalanced at the boundaries. Note that the net momentum flux on the control surface (see Eq. (2.46)) can be written as follows:

$$\int_{\text{CS}(t)} \rho \beta \, \vec{v}_r \cdot \vec{n} \, dA = \rho A_j(v_j - v_c) \, v_x - \rho A_j v_j^2 \tag{2.98}$$

where the deflected jet passes through the control surface of area A_j at a relative speed of $v_j - v_c$. The mass within the control volume is increasing at a rate of $\rho A_j v_c$ since the cart is moving at the speed of v_c, and the mass influx is $\rho A_j v_j$ so that the mass outflux has to be $\rho A_j(v_j - v_c)$ to conserve the mass of the jet. The absolute velocity of the deflected jet in the x direction, v_x, can be obtained by adding the cart velocity to the relative speed in the x direction:

$$v_x = (v_j - v_c)\cos\theta + v_c \tag{2.99}$$

Note that the absolute velocity of the deflected jet in the x direction decreases from v_j to v_c, as the angle θ increases from 0 to $\pi/2$.

If we neglect the mass of the water jet staying on the cart, the conservation of momentum is finally written as

$$\rho A_j v_c \, v_j + M_c \frac{d v_c}{dt} + \rho A_j(v_j - v_c) \, v_x - \rho A_j v_j^2 = 0 \tag{2.100}$$

Figure 2.14 Jet impingement on the cart (transient acceleration); control volume enlarging with the moving cart

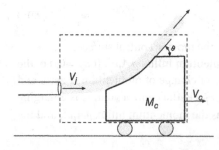

and can be rearranged as follows:

$$M_c \frac{d\,v_c}{dt} - \rho A_j(v_j - v_c)^2\,(1 - \cos\theta) = 0 \tag{2.101}$$

Equation (2.101) can be solved for v_c with an initial condition, $v_c(0) = 0$.

The solution of Eq. (2.101) reads

$$v_c(t)/v_j = 1 - \frac{1}{1 + (\rho A_j v_j/M_c)(1 - \cos\theta)\,t} \tag{2.102}$$

which shows that $v_c = 0$ at $t = 0$ and $v_c \to v_j$ as $t \to \infty$.

If we consider various resistance forces acting on the cart (e.g. friction from the ground, wind resistance, etc.), the equation of motion can be expressed as follows:

$$M_c \frac{d\,v_c}{dt} - \rho A_j(v_j - v_c)^2\,(1 - \cos\theta) = R \tag{2.103}$$

where R can be expressed as a combination of the linear and quadratic functions of v_c.

Physical Interpretations If $\theta = \pi$, the direction of the jet is totally reversed and the absolute velocity of the reflected jet becomes $v_x = -(v_j - 2\,v_c)$, which results in $v_j > 2\,v_c$. This case corresponds to the jet of two columns whose mass increases in time within the control volume. Then, the generalized conservation equation, Eq. (2.46) with $\beta = \vec{v}$ can be expressed as follows:

$$(\rho A_j v_c\,v_j + \rho A_j v_c\,v_x) + M_c \frac{d\,v_c}{dt} - \rho A_j v_x^2 - \rho A_j v_j^2 = 0 \tag{2.104}$$

where the first two terms represent the rates of change of momentum of the jet within the control volume, one for growing in the positive direction and the other for growing in the negative direction. Note that the two jets move with different absolute velocities; the one moving to the right moves with the velocity of v_j, while the other moves with the velocity of $-(v_j - 2\,v_c)$.

Equation (2.104) finally reads

$$M_c \frac{d\,v_c}{dt} - 2\,\rho A_j(v_j - v_c)^2 = 0 \tag{2.105}$$

and this is exactly the same as Eq. (2.101) for $\theta = \pi$.

[Notes] The present unsteady problem can be posed as a steady problem by coordinate transformation from an inertial reference frame to a noninertial reference frame (see Section 2.5.4). However, there is a certain class of problems that prefer to be solved with a control volume that changes shape over time (see Example 2.8).

Example 2.8 *Hydraulic brake (movable ram)*

A hydraulic brake consists of a movable ram and a slightly larger cylinder, as shown in Figure 2.15. A force F is applied to move the ram at a constant speed v. The area of the cylinder and the cross-sectional area of the ram are denoted by A_c and A_r, respectively.

(a) Determine the pressure at the end of the cylinder (where the velocity is assumed to be zero) p_c and the jet velocity v_j.
(b) Explain why p_c is lower than the pressure at the surface of the ram.
(c) Find the force F on the ram in terms of A_r, A_c, and v. Assume that the cylinder is initially full of water and that gravity effects are negligible.

(Continuity) The mass conservation reads

$$A_r\, v = (A_c - A_r)\, v_j \tag{2.106}$$

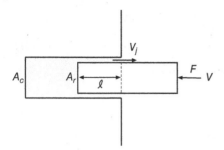

Figure 2.15 Hydraulic brake (movable ram)

(Bernoulli) With the coordinate (or Galilean) transformation,

$$p_c + \rho \frac{v^2}{2} = p_r \tag{2.107}$$

where p_c is smaller than p_r by a value of $\rho v^2/2$. This is a diffuser effect; the stationary water in the cylinder is picking up momentum in the direction opposite to the cylinder end wall at a constant rate, i.e.

$$p_c = p_r - \rho \frac{v^2}{2} = p_r - \rho \frac{v_j^2}{2} (\zeta^2 - 1) \tag{2.108}$$

where $\zeta = A_c/A_r$.

(Statics) A force balance in the ram states

$$p_r = F/A_r + p_\infty \tag{2.109}$$

(Momentum) Taking the control volume enclosing the water inside the cylinder,

$$p_c A_c - p_\infty A_g - p_r A_r = \frac{d}{dt} \{\rho l A_g \, v_j\} + \rho A_g \, v_j^2 \tag{2.110}$$

where l is the length of the jet in the gap and $A_g = A_c - A_r$ is the area of the gap. Note that the unsteady term on the right (i.e. a local rate of change of momentum of the fluid) represents transient inertial resistance of the jet through the gap, whose size increases in time as the ram intrudes into the cylinder.

From Eqs (2.106)–(2.110), the force on the ram F can be expressed as:

$$F = \frac{\rho v^2}{2} \frac{A_r A_c (A_r + A_c)}{(A_c - A_r)^2} \tag{2.111}$$

Physical Interpretation The left-hand side of Eq. (2.110) can be arranged as follows:

$$(p_c - p_\infty)A_c - (p_r - p_\infty)A_r = (p_c - p_\infty)A_c - F \tag{2.112}$$

which is the net force exerted on the fluid in the cylinder. $(p_c - p_\infty)A_c$ is the force exerted on the fluid from the inner wall of the cylinder and $(p_r - p_\infty)A_r$ is the force exerted on the fluid from the inner wall of the ram (equivalent to the force exerted on the ram outside, F). Equation (2.112) shows that this net force from the ground equals the local rate of change of the momentum plus the imbalance of the momentum flux on the control surface.

2.5.4 A Noninertial Reference Frame

A noninertial coordinate system is useful when describing the dynamics of a fluid in interaction with an accelerating and/or rotating object: e.g. the coordinates fixed to a rocket during launching or to the earth in rotation (Figure 2.16).[17]

---

17 In rocket launching, for example the accelerating fluid will act as an inertial force to the object. Similarly, low-pressure cyclones such as typhoons and hurricanes are developed in the atmosphere by the Coriolis acceleration (a virtual inertial force).

Figure 2.16 A noninertial reference frame (x', y', z') vs. an inertial reference frame (x, y, z)

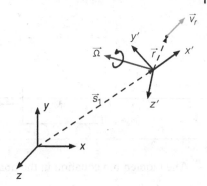

Newton's second law of motion in an inertial reference frame can be transformed into a noninertial reference frame:

$$\vec{F} = \frac{D}{Dt}\int_{sys} \rho\,\vec{v}\,dV$$

$$= \frac{D}{Dt}\int_{sys} \rho\,\vec{v}_r\,dV +$$

$$\int_{sys} \rho\left[\frac{d^2\vec{s}_1}{dt^2} + \underbrace{(2\,\vec{\Omega}\times\vec{v}_r)}_{\text{Coriolis}} + \underbrace{\vec{\Omega}\times(\vec{\Omega}\times\vec{r})}_{\text{Centripetal}} + \underbrace{\frac{d\vec{\Omega}}{dt}\times\vec{r}}_{\text{Angular}}\right]dV \qquad (2.113)$$

where $\vec{v}_r$ represents the relative velocity vector observed in the noninertial coordinate system, $\vec{s}_1$ is the position vector of the noninertial coordinate system (accelerating and/or rotating relative to the inertial coordinate system), $\vec{r}$ is the position vector from the origin of the noninertial coordinate system, and $\vec{\Omega}$ is its rotation vector.

Equation (2.113) can be rearranged as follows:

$$\vec{F} - \vec{F}_1 = \frac{\partial}{\partial t}\int_{CV} \rho\,\vec{v}_r\,dV + \int_{CS} \rho\,\vec{v}_r\,(\vec{v}_r\cdot\vec{n})\,dS \qquad (2.114)$$

where $\vec{F}_1$ can be viewed as an inertia force acting as an external force in the noninertial frame of reference:

$$\vec{F}_1 = \int_{sys} \rho\left[\frac{d^2\vec{s}_1}{dt^2} + (2\vec{\Omega}\times\vec{v}_r) + \vec{\Omega}\times(\vec{\Omega}\times\vec{r}) + \frac{d\vec{\Omega}}{dt}\times\vec{r}\right]dV \qquad (2.115)$$

Note that if the volume integral is removed and both sides are divided by dV, Eq. (2.114) becomes a differential form of the momentum equation in the noninertial frame of reference. If it is integrated along a streamline of the inviscid flow, we can obtain Bernoulli's equation in the noninertial frame of reference, which is in the same form as in the inertial frame of reference with the relative velocities used.

Example 2.9 *Jet impingement on the cart (transient acceleration): revisited*
A cart with a vane of mass M_c is being accelerated by the impingement of a water jet (Figure 2.17). The jet velocity is v_j, and there is negligible friction between the cart and the ground. If the cart starts to move from rest, find $v_c(t)$.

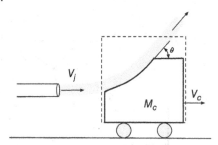

Figure 2.17 Jet impingement on the cart (transient acceleration); control volume moving with the cart

The momentum equation in the noninertial reference frame reads

$$\cancel{R_x}^{\;0} - F_I = \frac{d}{dt} \int_{cv} \rho\, v_{r_x}(t)\, dV - \dot{m}(t)\, (v_j - v_c(t))(1 - \cos\theta) \tag{2.116}$$

where $\dot{m}(t) = \rho A(v_j - v_c(t))$.

It is obvious that the relative velocity of the cart is zero, but the relative velocity (x-direction) of the fluid within the accelerating control volume, $v_{r_x}(t) \approx (v_j - v_c(t))\cos\theta$ changes in time from $v_{r_x}(0) = v_j \cos\theta$ to $v_{r_x}(\infty) \to 0$. However, we assume that the mass of the jet within the control volume is smaller than the mass of the cart, and we can then approximate the unsteady terms as follows:

$$\frac{d}{dt} \int_{CV} \rho\, v_{r_x}\, dV = -\frac{d}{dt} \int_{jet} \rho\, v_c(t) \cos\theta\, dV \approx 0 \tag{2.117}$$

and

$$F_I = \int_{sys} \rho\, \frac{dv_c}{dt}\, dV = (M_c + M_j) \frac{dv_c}{dt} \approx M_c \frac{dv_c}{dt} \tag{2.118}$$

Finally, Eq. (2.116) reads

$$M_c \frac{dv_c}{dt} = \rho A\, (v_j - v_c)^2 (1 - \cos\theta) \tag{2.119}$$

which shows that the transient inertia force of the cart, $M_c\,(dv_c/dt)$ acts as a resistance force against the net momentum influx of the jet. This ordinary differential equation for v_c has a general solution,

$$\frac{v_c(t)}{v_j} = \frac{\beta}{\beta + 1/t} \tag{2.120}$$

where $\beta = \rho A(1 - \cos\theta)v_j/M_c$ and $v_c(0) = 0$ changes in time to $v_c \to v_j$ as $t \to \infty$.

Example 2.10 *Squid propulsion*

Let us consider a squid attempting to propel from a standstill. It has a mass of m_0 and expels water at a rate of $\dot{m}_e$ (Figure 2.18). Find the condition of mass flow rate at the funnel exit for a squid of $m_0 = 1.2\ kg$ to take off vertically. We assume that the funnel exit is a circle of diameter $d = 1\ cm$ and the density of water is $1000\ kg/m^3$. Neglect any drag and buoyant forces.

The equation of motion can be set by applying the momentum equation to a control volume moving at a speed of v

$$-m(t)\, g - m(t)\, dv/dt = -\dot{m}_e\, v_e \tag{2.121}$$

where $m(t) = m_0 - \dot{m}_e\, t$ and $v_e = \dot{m}_e/(\rho_w A_e)$. Equation (2.121) can be arranged as follows:

$$m(t)\, (g + dv/dt) = \dot{m}_e^2/(\rho_w A_e) \tag{2.122}$$

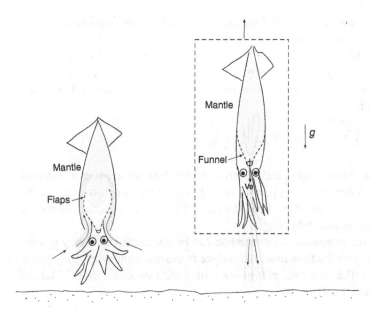

Figure 2.18 Squid propulsion; control volume moving with the squid

or

$$dv/dt = \frac{\dot{m}_e^2}{\rho_w A_e} \frac{1}{(m_0 - \dot{m}_e t)} - g \tag{2.123}$$

It is interesting to note that the right hand of Eq. (2.123) must be positive at $t = 0$ for a squid to take off vertically, that is, the mass flow rate at the exit must be greater than a threshold value:

$$\dot{m}_e > \sqrt{\rho_w A_e m_0 g} \tag{2.124}$$

For the given values of A_e and m_0, $\dot{m}_e$ must be larger than 0.961 kg/s to take off vertically in the positive direction. Otherwise, it will sink down for a certain time interval and then will move vertically up afterwards.

If we integrate Eq. (2.123) with an initial condition $v(t = 0) = 0$, the velocity is expressed as a function of t

$$v(t) = \frac{\dot{m}_e}{\rho_w A_e} \ln\left(\frac{1}{1 - \dot{m}_e/m_0 t}\right) - g t \tag{2.125}$$

where this solution is valid up to $t = m_0/\dot{m}_e$.

2.6 Conservation of Energy ($\beta = e_t$)

$$\frac{D}{Dt} \int_{CM} \rho e_t \, dV = \frac{\partial}{\partial t} \int_{CV} \rho e_t \, dV + \int_{CS} \rho e_t \, \vec{v} \cdot \vec{n} \, dA = \dot{Q} + \dot{W} \tag{2.126}$$

where $e_t = e + v^2/2$ is the specific total energy (e is the specific internal energy), and $\dot{Q}$ and $\dot{W}$ are the net inflow rates (or influx) of heat and work from the surroundings to the fluid, respectively.

The conservation law of energy states that a rate of change of the total energy within the control mass equals the net inflow rate of heat and work from the surroundings to the control mass. Here, the rate of change of the total energy is the sum of the local rate of change of total energy within the control volume and the imbalance of total energy flux at the control surface. If there is no net inflow rate of heat and work to the control mass, the total energy flux at the control surface can be unbalanced by the local rate of change of total energy within the control volume, which changes shape or density over time. Note also that the reaction force does no work on the fluid since the solid body is fixed to the ground.

2.6.1 Works

As stated in Section 1.2 *Forces*, forces originate either from within fluid volumes or at external boundaries. In response to the forces, fluids move with inertia; thus, forces are distributed in the flow field while changing the momentum of the fluid at a rate. This interaction occurs under the conservation laws of mass, momentum, and energy.

It is interesting to note that the momentum conservation can be interpreted as energy conservation. While being distributed over the flow field, the surface forces and volumetric body forces locally do works on the fluid particle at a rate, as it moves with a local velocity vector $\vec{v}$. The net force acting on the fluid particle $\vec{F}$ reads

$$\vec{F} = \vec{F}_S + \vec{F}_B = \int \vec{n}\,[\sigma]\ dA + \vec{f}_b\ dV \tag{2.127}$$

where $\vec{n}\,[\sigma]$ is the stress vector acting on the surface of area dA with orientation specified by an outward unit normal vector $\vec{n}$ and $\vec{f}_b = -\rho g\,\vec{k}$ is the gravitational body force per unit volume. Here, $[\sigma] = (-p + \lambda\,\nabla\cdot\vec{v})\,[\,|\,] + [\tau]$ is the stress tensor, λ is the dilatational viscosity, and $[\tau]$ is the viscous stress tensor. The stress vector $\vec{n}\,[\sigma] = \vec{t} = (t_x, t_y, t_z)$ exerted on the surface of unit area specified by a unit outward normal vector $\vec{n} = (n_x, n_y, n_z)$ is

$$t_x = n_x\sigma_{xx} + n_x\sigma_{yz} + n_x\sigma_{zx}$$

$$t_y = n_x\sigma_{xy} + n_x\sigma_{yy} + n_x\sigma_{zy}$$

$$t_z = n_x\sigma_{zx} + n_x\sigma_{zy} + n_x\sigma_{zz}$$

The net inflow rate of work done on the control mass can be obtained by integrating each work done at a rate by a net force $\vec{F}$ acting on the fluid particle

$$\dot{W} = \dot{W}_S + \dot{W}_B = \int_{CS} \vec{n}\,[\sigma]\cdot\vec{v}\ dA + \int_{CV} \vec{f}_b\cdot\vec{v}\ dV \tag{2.128}$$

With the divergence theorem, the net inflow rate of surface force work done on the control mass can be expressed as follows:

$$\dot{W}_S = \int_{CS} \vec{n}\,[\sigma]\cdot\vec{v}\ dA = \int_{CS} \vec{v}\,[\sigma]\cdot\vec{n}\ dA$$

$$= \int_{CS} \left(-p\,\vec{v} + \vec{v}\,[\tau] + (\lambda\,\nabla\cdot\vec{v})\,\vec{v}\right)\cdot\vec{n}\ dA$$

$$= \int_{CV} \nabla\cdot\left(-p\,\vec{v} + \vec{v}\,[\tau] + (\lambda\,\nabla\cdot\vec{v})\,\vec{v}\right)\ dV$$

$$= \left[-\int_{CV} \vec{v} \cdot \nabla p \ dV + \int_{CV} \vec{v} \cdot \nabla \ [\tau] \ dV + \int_{CV} \vec{v} \cdot \nabla (\lambda \ \nabla \cdot \vec{v}) \ dV \right]_{\textcircled{1}}$$

$$+ \left[-\int_{CV} p \ \nabla \cdot \vec{v} \ dV + \int_{CV} [\tau] \cdot [\nabla \vec{v}] \ dV + \int_{CV} \lambda \ (\nabla \cdot \vec{v})^2 \ dV \right]_{\textcircled{2}}$$

$$(2.129)$$

where ① and ② denote works related to the mechanical and thermal energies, respectively.[18]

For incompressible fluids, Eq. (2.129) is simplified to

$$\dot{W}_S = -\int_{CV} \vec{v} \cdot \nabla p \ dV + \int_{CV} \vec{v} \cdot \nabla \ [\tau] \ dV + \int_{CV} [\tau] \cdot [\nabla \vec{v}] \ dV \qquad (2.130)$$

where the first term on the right hand (called flow work) represents the rate of work done on the fluid by the pressure difference. Note that it equals $\int_{cs} (-p \ \vec{v}) \cdot \vec{n} \ dA$.

2.6.2 Flow Work

In conservation of momentum, the pressure difference across the control surface is considered as force. It creates a momentum flux imbalance at the control surface, considering the reaction force from the ground. In conservation of energy, the pressure difference is viewed as flow work transferred at a rate to or from the fluid, as the fluid moves through the control surface with a certain speed and direction [6].

It is to be noted that the flow work done on the fluid at a rate can increase the kinetic energy at a rate in time or space, or be dissipated into heat at a rate while deforming (or straining) the fluid against frictions. In this section, we are going to illustrate three different types of energy conversion with the flow work, employing the tube-piston configurations as presented in Figure 2.19.

2.6.2.1 Local Rate of Change of Kinetic Energy (Case 1)

Let us suppose we have a straight tube with a piston (Figure 2.19a). If the piston is suddenly pushed by a force F_1, the fluid at rest in the tube (assumed to be incompressible) is locally accelerated like a slug. As a result, the pressure is linearly distributed in the tube.

Figure 2.19 Flow works done on the fluid at a rate, balancing with the local rate of change of kinetic energy (a), imbalance of kinetic energy flux (b), and imbalance of internal energy flux (c), respectively

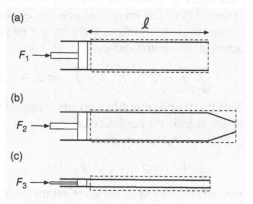

---

18 It is to be noted that $\vec{n} \ [\sigma]$ is a stress vector acting on the surface of area dA specified by an outward unit normal vector $\vec{n}$ and its dot-product with the velocity vector $\vec{v} \cdot (\vec{n} \ [\sigma]) \ dA$ is the rate of work done by the surface force vector projected on the plane of a velocity vector. This equals $(\vec{v} \ [\sigma]) \cdot \vec{n} \ dA$ because the stress tensor $[\sigma]$ is a symmetric tensor.

For the control volume as selected, the conservation law of energy states that the kinetic energy of the fluid within the control volume is locally increased at a rate by the force F_1 doing work on the fluid at a rate

$$\frac{\partial}{\partial t} \int_{CV} \rho (v^2/2) \, dV = F_1 \, v_p \tag{2.131}$$

where the external force F_1 applied by the piston does work on the fluid at a rate at the control surface:

$$F_1 \, v_p = \int_{CS} (-p \, \vec{v}) \cdot \vec{n} \, dA = (p_p - p_\infty) \, A_p v_p \tag{2.132}$$

In this case, there is no imbalance of kinetic energy flux at the control surface because there is no change of velocity in space (i.e. slug flow). A volumetric flow rate $Q = A_p v_p$ produced by the piston moving at a speed v_p is also shared by the fluid in the tube. Note that $F_p \, v_p$ equals the sum of the rate of each flow work done on the fluid particle with pressure differences distributed in the flow field, i.e. $\int_{CV} \nabla \cdot (-p \, \vec{v}) \, dV$.

Equations (2.131) and (2.132) can now be written as follows:

$$\rho Q \int_0^l \frac{\partial v_p}{\partial t} \, dx = (p_p - p_\infty) \, Q \tag{2.133}$$

where l is the tube length. If we divide both sides by the volumetric flow rate Q, it reads

$$\rho \int_0^l \frac{\partial v_p}{\partial t} \, dx = p_p - p_\infty \tag{2.134}$$

which is, indeed, the Bernoulli equation for unsteady flow. Since the local acceleration is uniformly constant, Eq. (2.134) returns to Eq. (1.12) with $x = 0$:

$$p_p = p_\infty + \rho \frac{\partial v_p}{\partial t} l \tag{2.135}$$

2.6.2.2 Imbalance of Kinetic Energy Flux (Case 2)

If a nozzle is attached to the end of the tube, a force F_2 is required to move the piston at a constant speed v_p. In this case, the fluid in the nozzle is forced to accelerate to a speed v_j at the exit (Figure 2.19b). The energy conservation equation states that an imbalance of kinetic energy flux across the control surface equals the net rate of flow work done on the fluid by pressure forces acting on the control surface:

$$\int_{CS} \rho (v^2/2) \, \vec{v} \cdot \vec{n} \, dA = \int_{CS} (-p \, \vec{v}) \cdot \vec{n} \, dA \tag{2.136}$$

Note that the pressure difference at the control surface is created by the force F_2 acting on the fluid, as discussed in the previous case.

Equation (2.136) now reads

$$\frac{1}{2} \rho (v_j^2 - v_p^2) \, Q = (p_p - p_\infty) \, Q \tag{2.137}$$

where both sides share the same volumetric flow rate $Q = A_p v_p = A_j v_j$ (A_p and A_j are the piston and jet cross-sectional areas, respectively), and this is the same solution as obtained by Eq. (1.22) with $x = x_e$,

$$p_p - p_\infty = \frac{1}{2} \rho (v_j^2 - v_p^2) = \frac{1}{2} \rho v_p^2 \, (\beta^2 - 1) \tag{2.138}$$

where $\beta = A_p/A_j$ (> 1) is the nozzle area contraction ratio.

[Notes] Let us suppose we have two fluids of different densities, density ρ_1 (left) and density ρ_2 (right) in the tube, which are immiscible during motion. Then, the imbalance of the total energy flux across the control surface is written as follows:

$$\int_{CS} \rho \, e_t \; \vec{v} \cdot \vec{n} \, dA = \frac{1}{2} \left(\rho_2 \, v_j^2 - \rho_1 \, v_p^2 \right) Q \tag{2.139}$$

where two fluids share the same volumetric flow rate, $Q = A_p v_p = A_j v_j$, but not the mass flow rate. In this case, the two fluids carry kinetic energy with different amounts of mass per unit time, i.e. $\dot{m}_1 = \rho_1 \, Q$ and $\dot{m}_2 = \rho_2 \, Q$.

The final energy balance equation reads

$$(p_p - p_\infty) \, Q = \frac{1}{2} \left(\rho_2 \, v_j^2 - \rho_1 \, v_p^2 \right) Q \tag{2.140}$$

and can be expressed as

$$p_p - p_\infty = \frac{1}{2} \left(\rho_2 \, v_j^2 - \rho_1 \, v_p^2 \right) = \frac{1}{2} \, \rho_1 \, v_p^2 \left(\alpha \, \beta^2 - 1 \right) \tag{2.141}$$

where $\alpha = \rho_2 / \rho_1$ is the density ratio. This equation returns to Eq. (2.138), if $\alpha = 1$.

If the fluid on the right is heavier than that on the left (i.e. $\alpha > 1$) (e.g. air to water), p_p in Eq. (2.141) is higher than that in Eq. (2.138). In other words, a larger force is required to push the piston while keeping the same piston speed v_p. If we use a lighter fluid on the right (i.e. $\alpha < 1$) (e.g. water to air), then it will show the opposite results. This density effect is directly related to fluid inertia; in other words, pressure mostly changes in the nozzle due to inertial resistance of the fluid.

2.6.2.3 Imbalance of Internal Energy Flux (Case 3)

If the tube has an aspect ratio greater than, let us say, 50, a force F_3 is required to move the piston at a constant speed v_p against frictions (Figure 2.19c). If there is no heat transfer, the conservation law of energy states that an imbalance of internal energy flux at the control surface is created by force F_3 doing work on the fluid at a rate

$$\int_{CS} \rho e \; \vec{v} \cdot \vec{n} \, dA = \int_{CS} (-p \, \vec{v}) \cdot \vec{n} \, dA \tag{2.142}$$

assuming that the flow is fully developed so that there exists no acceleration of the fluid within the tube.

Since $\Delta e = c_v \, \Delta T$, Eq. (2.142) reads

$$\rho c_v \, (T_j - T_p) \, Q = (p_p - p_\infty) \, Q \tag{2.143}$$

where both sides shares the same volumetric flow rate $Q = A_p v_p = A_j v_j$. The temperature at the tube exit can then be expressed as follows:

$$T_j = T_p + \frac{p_p - p_\infty}{\rho \, c_v} = T_p + \frac{\tau_w \, (\overline{P} \, l / A_p)}{\rho \, c_v} \tag{2.144}$$

Based on the fact that there is no viscous work done on the fluid at the control surface, it follows that

$$\int_{CS} \vec{v} \cdot \vec{n} \, [\tau] \, dA = \int_{CS} \vec{v} \, [\tau] \cdot \vec{n} \, dA = 0 \tag{2.145}$$

With the divergence theorem, Eq. (2.145) becomes

$$\int_{CV} \nabla \cdot \vec{v} \, [\tau] \, dV = 0 \tag{2.146}$$

or

$$-\int_{CV} \vec{v} \cdot \nabla \, [\tau] \, dV = \int_{CV} [\tau] \cdot [\nabla \vec{v}] \, dV > 0 \tag{2.147}$$

which shows that all the works done on the fluid particles at a rate by the net viscous forces must be viscously dissipated into heat (or internal energy) at the same rate. The right-hand side of Eq. (2.147) is often referred to as the rate of viscous dissipation Φ.

2.6.3 Viscous Force Work

2.6.3.1 Local Rate of Change of Internal Energy

Let us suppose we have a channel filled with a fluid of density ρ and viscosity μ, and one of the plates is moving at a constant speed U_0, while the other is fixed to the ground (Figure 2.20). After reaching steady state, the fluid between the plates is uniformly shear strained (or rotated and angularly strained at the same rate) by the viscous force tangentially applied to the plate.

In this simple shear flow (or Couette flow), the conservation law of energy states that if there is no heat transferred through the control surface, the net inflow rate of viscous force work done on the fluid at the control surface creates a local rate of change of internal energy of the fluid:

$$\frac{\partial}{\partial t} \int_{CV} \rho e \, dV = \int_{CS} \vec{n} \, [\tau] \cdot \vec{v} \, dA \tag{2.148}$$

For a volume of fluid defined by the channel height h and a unit width and length, Eq. (2.148) reads

$$h \frac{\partial(\rho e)}{\partial t} = h \, c_v \frac{\partial(\rho T)}{\partial t} = U_0 \, \tau_w \tag{2.149}$$

where the right-hand side is the power input to the fluid (per unit area) by viscous force work and the left-hand side is the local rate of change of internal energy of the fluid (per unit area).

In this case, the viscous work done on the fluid at a rate $U_0 \, \tau_w$ is dissipated into heat at the same rate via viscosity:

$$\int_{CS} \vec{v} \cdot \vec{n} \, [\tau] \, dA = \overset{0}{\cancel{\int_{CV} \vec{v} \cdot \nabla [\tau] \, dV}} + \int_{CV} [\tau] \cdot [\nabla \vec{v}] \, dV \tag{2.150}$$

where the first term on the right-hand side is zero because there is no net viscous force acting on the fluid within the control volume. The left-hand side is the power delivered to the fluid by the viscous forces exerted on the control surface:

$$\text{Input power} = \int_{CS} \vec{v} \cdot (\vec{n} \, [\tau]) \, dA = U_0 \, \tau_w \, (1 \times 1) = \frac{\mu}{h} \, U_0^2 \tag{2.151}$$

where $\tau_w = \mu \, du/dy = \mu \, U_0/h$.

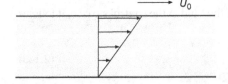

$\longrightarrow U_0$

Figure 2.20 Simple shear flow (Couette flow) driven by the top surface moving at a constant speed of U_o

Meanwhile, the second term on the right hand in Eq. (2.150) is the rate of viscous dissipation denoted by Φ and can be obtained as follows:

$$\Phi = \int_0^h [\tau] \cdot [\nabla \vec{v}] \ dy \, (1 \times 1) = \int_0^h 2 \mu \, \epsilon_{ij} \, \epsilon_{ij} \, dy$$

$$= 2\mu \cdot (\epsilon_{11} \, \epsilon_{11} + \epsilon_{12} \, \epsilon_{12} + \epsilon_{21} \, \epsilon_{21} + \epsilon_{22} \, \epsilon_{22}) \cdot h$$

$$= 2\mu \cdot \left\{ 2 \cdot \frac{1}{4}(du/dy)^2 \right\} \cdot h = \frac{\mu}{h} \, U_o^2 \tag{2.152}$$

where $\tau_{ij} = 2 \, \mu \, \epsilon_{ij}$ and $\epsilon_{ij} = \epsilon_{ji}$.[19] Thus, it proves that the two terms on the left- and right-hand sides of Eq. (2.150) yield the same results.

It is to be noted that steady Couette flow is a special case where the velocity distribution is independent of viscosity. This implies that there is no net viscous force present in the field and vortices (or rotation rates) are uniform throughout the field. Furthermore, the energy supplied at a rate by the wall moving at a constant speed is dissipated at the same rate by viscously straining the fluid.

2.6.4 Gravitational Body Force Work

2.6.4.1 Imbalance of Kinetic Energy Flux

Let us consider a reservoir filled with water of density ρ up to height H. The water in the reservoir is being ejected at a constant speed of v_j through a small gate close to the ground (Figure 2.21).

For a control volume selected to enclose the reservoir, the energy conservation law states that an imbalance of kinetic energy flux at the control surface is created solely by the gravitational body force work done on the fluid at a rate.

$$\int_{CS} \rho \, (v^2/2) \ \vec{v} \cdot \vec{n} \ dA = \int_{CV} \vec{v} \cdot (-g \, \vec{k}) \, \rho \, dV \tag{2.153}$$

There is no flow work done on the fluid, since

$$\int_{CS} (-p \, \vec{v}) \cdot \vec{n} \ dA = -p_\infty v_1 A_1 + p_\infty v_j A_j = 0 \tag{2.154}$$

where the inlet at the free surface and the outlet at the jet exit share the same volumetric flow rate $Q = v_1 A_1 = v_j A_j$ by the conservation of mass.

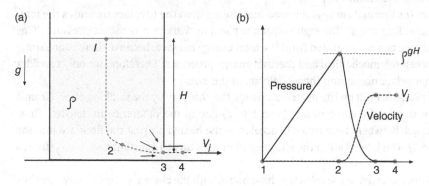

Figure 2.21 Water jet through a small gate via gravitational body force work done at a rate in the reservoir (a); pressure and velocity distribution along the steamline (b)

---

19 More discussion continues in Chapter 3 *Differential Equations of Motions*.

The left-hand side of Eq. (2.153) reads

$$\int_{CS} \rho(v^2/2) \; \vec{v} \cdot \vec{n} \; dA = \rho\left(v_j^2/2 - v_1^2/2 \overset{0}{\diagup}\right) Q \tag{2.155}$$

It is to be noted that the downfall speed of the free surface can be expressed as $v_1 = \beta \, v_j$, where $\beta = A_j/A_t$ is assumed to be a very small but finite number. The kinetic energy density ratio between the two stations will then become β^2 so that the rate of kinetic energy transferred at the free surface can be neglected.

The right-hand side of Eq. (2.153) is written as follows:

$$\int_{CV} \vec{v} \cdot (-g \, \vec{k}) \, \rho \, dV = \rho g \int_0^H Q \, dz = \rho g H \, Q \tag{2.156}$$

where the volumetric flow rate Q is defined as

$$Q = \int_{CV} \vec{v} \cdot (-\vec{k}) \, dA = \text{constant} \tag{2.157}$$

and $\rho g H$ is the potential energy of the water per unit volume (or potential energy density). Combining Eq. (2.155) with Eq. (2.156) yields the water jet speed

$$v_j = \sqrt{2gH} \tag{2.158}$$

[**Notes**] This example clearly shows how the water in the reservoir transfers its energy (i.e. a volumetric body force work done by gravity) to kinetic energy. This process of energy transfer is done at a rate with the mass of water flowing at a volumetric flow rate of Q. One simple way to increase the rate of energy transfer is to increase the exit area A_j and the reservoir height H, not the diameter of the reservoir D. Note also that the jet speed could have been obtained by the conservation law of momentum if we knew the exact pressure distribution along the inner wall of the reservoir or the reaction force exerted on the reservoir from the ground. Therefore, the conservation law of energy is much more useful since the height of the reservoir is the only information necessary to obtain the jet speed.

Example 2.11   *Chimney (or stack) effect*
Consider a house with a fireplace and a chimney of height H (Figure 2.22). Find the velocity of the heated air at the exit of the chimney v_j.

Note that the heat (or thermal energy) released into the air from the fireplace increases the temperature, while expanding the air through volume expansion work at a constant pressure. This thermal energy balance can be separated from the total energy balance, because there is no strong coupled effect between the mechanical and thermal energy processes. Therefore, we only consider the mechanical energy balance to find the jet velocity at the exit.

The hydrostatic pressure built on the heated air inside the chimney by the stacking effect (from 3 to 4) is lower than that in the ambient air (from 1 to 2) due to the difference in density. Thus, the pressure difference between these two will accelerate the heated air past the fireplace into the chimney (from 2 to 3). The heated air in the chimney then rises up at a constant speed of v_j (from 3 to 4).

For a control volume selected to include the whole house with the chimney, the conservation law of energy states that an imbalance of kinetic energy flux at the control surface $\Delta \dot{K}$ is created by the net rate of work done on the fluid $\Delta \dot{W}$

$$\Delta \dot{K} = \Delta \dot{W} \tag{2.159}$$

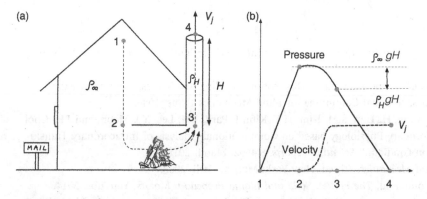

Figure 2.22 Air jet from the chimney via gravitational body force works done at a rate outside and inside the chimney (a); pressure and velocity distribution along the steamline (b)

where the net kinetic energy outflux across the control surface is

$$\Delta \dot{K} = \int_{CS} \rho \, (v^2/2) \; \vec{v} \cdot \vec{n} \, dA = \left(\rho_h \, v_j^2 / 2 - \cancel{\rho_\infty v_1^2 / 2}^{\,0} \right) Q \tag{2.160}$$

and $Q = A_1 v_1 = A_j v_j$ is the volumetric flow rate shared by the cold and hot air streams.

The net influx of work to the system can be expressed as follows:

$$\Delta \dot{W} = \Delta \dot{W}_S + \Delta \dot{W}_B \tag{2.161}$$

where the net rate of flow work reads

$$\Delta \dot{W}_S = (p_\infty^* A_1) \, v_1 - (p_\infty^* A_j) \, v_j = 0 \tag{2.162}$$

because p_∞^* is the pressure at the same elevation H for the ambient air and for the heated air leaving the chimney as well.

The net gravitational body force work $\Delta \dot{W}_B$ is done on the fluid at a rate by weight difference between the ambient and heated air streams that flow at a volumetric flow rate Q

$$\Delta \dot{W}_B = \int_\infty \vec{v} \cdot (-g \, \vec{k}) \, \rho_\infty \, dV + \int_h \vec{v} \cdot (-g \, \vec{k}) \, \rho_h \, dV$$

$$= (\rho_\infty - \rho_h) \, gHQ \tag{2.163}$$

where the volume of the heated air leaving the chimney is replaced by the same amount of volume of the ambient air for the same time interval, i.e.

$$Q = \int_\infty \underbrace{\vec{v} \cdot (-\vec{k})}_{>0} \, dA = -\int_h \underbrace{\vec{v} \cdot (-\vec{k})}_{<0} \, dA = A_j v_j = \text{const.} \tag{2.164}$$

From Eq. (2.159), we obtain the chimney jet exit speed

$$v_j = \sqrt{2 \, g \, H \left(\frac{\rho_\infty}{\rho_h} - 1 \right)} \tag{2.165}$$

Note that the kinetic energy of the heated air at the chimney exit is created at a rate by the gravitational body force works done on two columns of air with different densities (i.e. difference of potential energy densities). Therefore, v_j is proportional to the square root of the density ratio ρ_∞/ρ_h as well as to the height of the chimney H. This creation of jet speed is called the chimney effect (or stack effect).

References

1 Lighthill, J., *An Informal Introduction to Theoretical Fluid Mechanics*, Oxford: Oxford University Press, 1988.

2 Lumley, J.L., "Eulerian and Lagrangian Descriptions in Fluid Mechanics," *Illustrated Experiments in Fluid Mechanics*, National Committee for Fluid Mechanics Films, 1980.

3 Bae, Y.G., S.T. Hwang, H. Han, S.M. Kim, H.Y. Kim, I. Park, J.M. Lee, Y.J. Moon, and J.H. Choi, "Non-invasive Coronary Physiology based on Computational Analysis of Intracoronary Transluminal Attenuation Gradient," *Scientific Reports*, 8:4692, March 2018.

4 White, F.M., *Fluid Mechanics*, 6th ed. New York: McGraw-Hill, 2008.

5 Space Systems Command, *The Rocket: Solid and Liquid Propellant Motors*, YouTube. 2014.

6 Sabersky, R.H., A.J. Acosta, and E.G. Hauptmann, *Fluid Flow–A First Course in Fluid Mechanics*, 2nd ed. New York: The Macmillan Company, 1971.

7 Indowild, *Giant Frogfish Swimming – Crazy Jet Propulsion System*, YouTube. 2016.

Problems

2.1 The dynamics of a hanging chain often compares to the dynamics of fluids (see below).

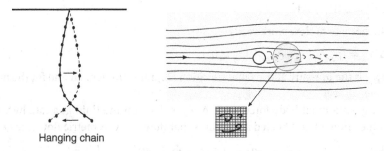

Hanging chain

a) Explain how the concept of fluid particles enables the conservation laws of mass, momentum, and energy applied to fluids.

b) Explain how the concept of fluid particles can represent the flow field as a continuous function of flow variables such as density, pressure, velocity.

2.2 Iodine is being injected into a blood stream of an artery vessel. We have two vessels A and B as sketched below, and the volumetric inflow rate of the blood Q_0 is kept the same for the two cases. In vessel A, the iodine is being injected at a constant volumetric flow rate Q_{id} such that $c(0, t) = C_0$, while in vessel B, it is being injected at a rate of $C(t) = B_0$ erf, where an error function is defined as:

$$\text{erf}(t) = \frac{2}{\sqrt{\pi}} \int_0^t e^{-z^2} \, dz$$

and B_0 is a constant. It is assumed that the iodine concentration $c(x, t)$ is subject to the law of conservation of mass, i.e. $Dc/Dt = 0$, neglecting molecular diffusion during convection with the blood.

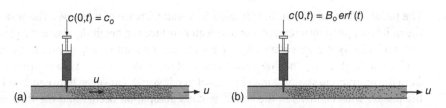

a) For vessel A, sketch the iodine concentration distribution $c(x,t)$ at three instants, t_1, t_2, and t_3 where $t_1 < t_2 < t_3$.
b) For vessel A, sketch two distributions of $c(x,t)$ at three instants, t_1, t_2, and t_3, for Q_{id} and Q'_{id} where $Q'_{id} > Q_{id}$.
c) For vessel A, sketch two distributions of $c(x,t)$ at three instants, t_1, t_2, and t_3, for Q_0 and Q'_0 where $Q'_0 > Q_0$.
d) For vessel B, sketch the iodine concentration distributions $c(x,t)$ at three instants, t_1, t_2, and t_3.
e) For vessel B, sketch two distributions of $c(x,t_1)$ for B_0 and B'_0 where $B'_0 > B_0$.
f) For vessel B, sketch two distributions of $c(x,t_1)$ for Q_0 and Q'_0 where $Q'_0 > Q_0$.
g) Explain the results of the problems (a)–(f) with the conservation law of the constituent, i.e. $Dc/Dt = 0$.

2.3 A kettle is partially filled with water of density ρ and volume V_0. Let us suppose water from the kettle is being poured into a cup at a volumetric flow rate of Q.
a) Discuss the imbalance of mass flux and the local rate of change of mass in the kettle for the control volume that encloses the kettle.
b) Discuss the imbalance of mass flux and the local rate of change of mass in the cup for the control volume that encloses the cup.
c) Discuss the imbalance of mass flux and the local rate of change of mass in the cup for the control volume that encloses the kettle and cup.

2.4 Squids create propulsion by coupling the local rate of change of mass of water with conservation of momentum.

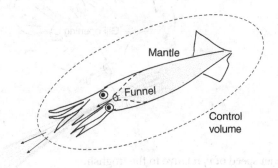

a) Describe the imbalance of momentum flux for the control volume that encloses the squid's body, including the mantle and funnel (see figure).
b) Explain how squids can increase the thrust of the jet.

2.5 The paddle wheel of a sawmill is located in a water stream of width W. The water driving the mill is supplied by a reservoir whose water surface is a height h_1 above the ground. The water of velocity V is accelerated by a passage under a sluice gate, approaching the paddle wheel at a depth of h_2. The flow downstream of the paddle wheel has a depth h_3. The flow from the reservoir to station 2 may be considered inviscid. However, between 2 and 3, where the paddle wheel churns in the water, viscous dissipation occurs and the flow cannot be considered inviscid. The viscous shear force on the stream bed is assumed negligibly small.

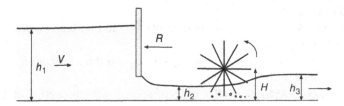

a) Express the velocities v_2 and v_3 at stations 2 and 3, in terms of g, h_1, and h_2. (Hint: use the Bernoulli equation along the streamline between stations 1 and 2.).

b) Express the force R necessary to hold the plate in place, in terms of ρ, g, W, h_1, and h_2.

c) Physically interpret the exit velocity of the sluice gate v_2, comparing with the reference case (i.e. reservoir).

d) Determine the horizontal force F exerted by the flowing stream on the paddle wheel.

e) Simplify the horizontal force F for the reference case.

2.6 A frogfish walks, hops, and swims with strokes of the caudal fin. It also jet-propels by gulping water and forcing it out through its gill openings of a cross-sectional area A_g at a rate of β [7]. Assume that the frogfish of mass m_0 gulps water (mass m_w and density ρ) and that the outflow rate of the mass β is constant.

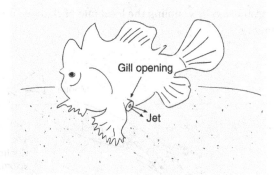

Gill opening

Jet

a) Find the jet speed of v_j relative to the frogfish.

b) Find the acceleration and speed of the frogfish.

c) Find the propulsive thrust exerted on the frogfish and show that it is constant over time.

2.7 A plate of mass M is withheld between two opposing jets, each of area A, velocity v_0, and density ρ. The plate is suddenly given a vertical velocity v_p. Determine the subsequent motion of the plate. Consider the flow at any instant as quasi-steady and neglect friction. We assume that the plate takes motion only along the y-direction.

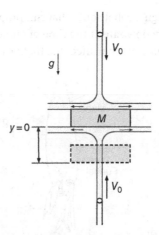

a) Derive the equation of motion of the plate by applying the conservation law of momentum.

b) Find $v(t)$ and $y(t)$, defining a new parameter $\beta = 4\, \rho A v_0 / M$.

c) Draw $v(t)$ and $y(t)$ against t for $v_p = (1, 0, -1, -2, -3)$ with $g/\beta = 1$ and $\beta = 1$.

d) Discuss the solutions as $t \to \infty$.

2.8 A blue whale swimming at a speed of v_0 gulps water to feed on a mass of krill. The instantaneous velocity of the slowing whale is $v(t)$ and the density of the water-krill mixture is assumed to be ρ_m. The mass of the whale is m_0 and the cross-sectional area of the whale's mouth is A_w.

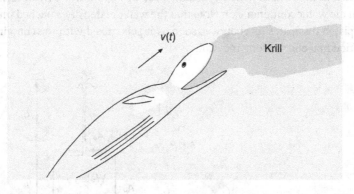

a) Express the inflow rate of mass of the water-krill mixture into the whale, in terms of the given parameters.

b) Prove the conservation of momentum for the whale feeding on a mass of krill, i.e. $m(t)\, v(t) = m_0\, v_0$, where $m(t)$ and $v(t)$ are the instantaneous mass and velocity of the whale.

c) Find the instantaneous velocity and mass of the whale $v(t)$ and $m(t)$, in terms of the given parameters. Discuss how significantly opening the mouth causes the whale to slow down.

d) Calculate the drag force exerted on the whale.

2.9   A student wants to design a robot squid that can propel in water. In squid propulsion (see Example 2.10), the mass flow rate at the time of launching, i.e. $\dot{m}_e^* = \sqrt{\rho_w A_e m_0 \, g}$, is one of the most critical parameters that affect the flight dynamics of the squid.

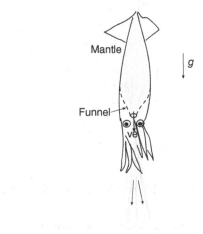

a) Exercise different values of $\dot{m}_e^*$ and plot $v(t)$ for each.
b) Derive $h(t)$ and plot for different values of $\dot{m}_e^*$.
c) Discuss the results.

2.10   As shown in the figure, a water container has two ports: at the bottom, flow is coming in through an area A_p while at the top, flow is expelled through an area A_e. A person is pushing a piston at the bottom with a constant speed of V_p to wet the fly hovering at a distance h_f from the water container exit. Note that the water of density ρ pushed into the container by the piston dissipates its energy as soon as it gets mixed with the container water. Assume no frictional effects along the wall.

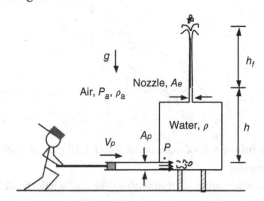

a) Express V_p in terms of given parameters.
b) Determine the pressure in the region close to the sudden enlargement and physically explain the answer.
c) Find the force required to push the piston at a constant speed of v_p.
d) Find the power required by the person to wet the fly hovering at h_f.

2.11 Two streams of an incompressible fluid are flowing concentrically as shown below, each occupying a flow area A, but the central stream's velocity is twice that of the outside stream. When they are mixed at constant area, the final velocity is uniform.

Concentric tubes
of equal areas

a) Express the conservation of mass.
b) Find the pressure difference between two stations 1 and 2.
c) Assuming that the wall shear stress is zero, find the rate of dissipation of mechanical energy due to the mixing process.

2.12 Let us revisit the reservoir discussed in Section 2.6.4. If the diameter D and height H of the container are marginally larger compared to the diameter of the exit pipe d, an unsteady effect should be considered to properly handle the problem.

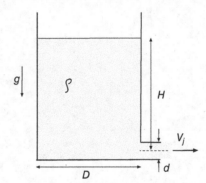

a) For the control volume that changes shape over time with the water free surface, express the mass conservation equation.
b) Express the energy conservation equation in the form of an ordinary differential equation.

3

Differential Equations of Motions

3.1 Fluid Motions

Fluid motions, often superimposed or mixed, can be grouped into solid body motion and deformable body motion. Depending on the type of forces and boundary conditions, fluids can be translated or rotated at a rate like solids, or can be linearly or angularly strained (or deformed)[1] at a rate (Figure 3.1).

The definition of fluid motion is directly associated with the laws of conservation of mass, momentum, and energy. For example, the volumetric dilatation rate $\nabla \cdot \vec{v}$ is essential in defining the continuity equation, and twice the rotation rate vector cross-producted with the velocity vector[2] equals a gradient vector of Bernoulli's function across the streamlines in effectively inviscid flows. In viscous flows, strain rate tensor $[\nabla \vec{v}]$ is related to viscous stress tensor $[\tau]$ through viscosity. In this chapter, we first define the rates of strain and rotation of a fluid element and use the definitions to derive the conservation equations of mass, momentum (i.e. the Navier–Stokes equations), and energy.

3.1.1 Linear Strain Rate

A linear strain ε is defined by the ratio of stretching (or contraction):

$$\varepsilon = \frac{l' - l}{l}, \tag{3.1}$$

where l is the initial length and l' is the length after stretching (or contraction). In fluids, straining is a continuous process and, thus, is defined at a rate basis:

$$\varepsilon = \dot{\varepsilon} = \lim_{\Delta \to 0} \frac{(l' - l)/l}{\Delta t} \tag{3.2}$$

A linear strain rate of a fluid element can be derived by tracing the size of the fluid element at time t, $dx(t)$, $dy(t)$, and $dz(t)$ in time (Figure 3.2). For example, the linear strain rate in the x direction is defined as follows:

$$\epsilon_{xx} = \frac{(dx(t + dt) - dx(t))/dx(t)}{dt} \tag{3.3}$$

---

1 To be more specific, fluids are stretched and contracted at a rate along the principal axes. Principal axes are the axes at which the rate of stretching or contraction is the maximum. More discussion continues in Section 3.3.3. The principal axes of strain rate.

2 It is called the Lamb acceleration vector, $\vec{L} = 2\,\vec{\Omega} \times \vec{v}$. More discussion on the Lamb acceleration vector continues in Chapter 4 *Curved Motions*.

Introduction to Fluid Dynamics: Understanding Fundamental Physics, First Edition. Young J. Moon.
© 2022 John Wiley & Sons, Inc. Published 2022 by John Wiley & Sons, Inc.
Companion website: www.wiley.com/go/Moon/IntroductiontoFluidDynamics

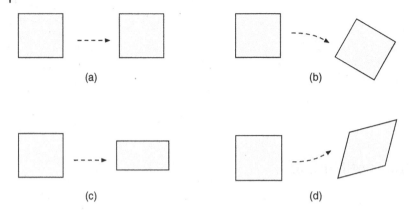

Figure 3.1 Fluid motions; solid body motions (a,b): translation (a), rotation (b); deformable body motions (c,d): linear straining (c), angular straining (d)

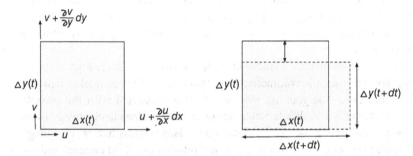

Figure 3.2 Linear strain rate

or can be written in terms of velocities:

$$\epsilon_{xx} = \frac{(u(x+dx,y,z,t) - u(x,y,z,t))\ dt\ /\ dx(t)}{dt} \tag{3.4}$$

If $u(x+dx,y,z,t)$ is expanded with Taylor's series, it becomes

$$\epsilon_{xx} = \left\{ u(x,y,z,t) + \frac{\partial u(x,y,z,t)}{\partial x}\ dx - u(x,y,z,t) \right\} \bigg/ dx = \frac{\partial u}{\partial x} \tag{3.5}$$

Similarly, the linear strain rate in the y direction reads

$$\epsilon_{yy} = \left\{ v(x,y,z,t) + \frac{\partial v(x,y,z,t)}{\partial y}\ dy - v(x,y,z,t) \right\} \bigg/ dy = \frac{\partial v}{\partial y} \tag{3.6}$$

and the linear strain rate in the z direction reads

$$\epsilon_{zz} = \left\{ w(x,y,z,t) + \frac{\partial w(x,y,z,t)}{\partial z}\ dz - w(x,y,z,t) \right\} \bigg/ dz = \frac{\partial w}{\partial z} \tag{3.7}$$

3.1.2 Angular Strain Rate and Rotation Rate

The angular strain rate and rotation rate of a fluid element (Figure 3.3) are defined by measuring the turning angle $d\theta$ of the vertical side of the element over dt in the clockwise direction:

$$d\theta_{cw} \simeq \tan d\dot{\theta} \approx \left[\frac{\{(u + \partial u/\partial y\ dy) - u\}\ dt}{dy} \right] \bigg/ dt = \frac{\partial u}{\partial y} \tag{3.8}$$

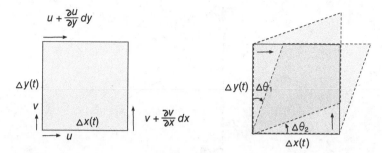

Figure 3.3 Angular strain rate and rotation rate

and the turning angle $d\theta$ of the horizontal side over dt in the counterclockwise direction:

$$d\dot{\theta}_{ccw} \simeq \tan d\dot{\theta} \approx \left[\frac{\{(v + \partial v/\partial x \; dx) - v\} \; dt}{dx}\right]\Big/ dt = \frac{\partial v}{\partial x} \tag{3.9}$$

Here, an angular strain rate is defined by the rate of angular distortion of the fluid element, taking an average of the two turning rates of the vertical and horizontal sides of the rectangle ($dx \times dy$):

$$\epsilon_{xy} = \frac{1}{2}\left(\frac{\partial u}{\partial y} + \frac{\partial v}{\partial x}\right) \tag{3.10}$$

Note that the straining rate in three-dimensional space can generally be written as follows:

$$\epsilon_{ij} = \frac{1}{2}\left(\frac{\partial v_i}{\partial x_j} + \frac{\partial v_j}{\partial x_i}\right) \tag{3.11}$$

where the diagonal terms represent the linear straining rates in the x, y, and z directions, respectively, while the off-diagonal terms represent the angular straining rates, which are symmetric.

The rotation rate of fluid can be defined similarly by considering the sign convention of rotation, i.e. positive for counterclockwise turning and negative for clockwise:

$$\Omega_z = \frac{1}{2}\left(-\frac{\partial u}{\partial y} + \frac{\partial v}{\partial x}\right) \tag{3.12}$$

The rotational rate is often expressed in terms of vorticity, where the vorticity is defined as follows:

$$\omega_z = 2\,\Omega_z = \left(-\frac{\partial u}{\partial y} + \frac{\partial v}{\partial x}\right) \tag{3.13}$$

In general, the vorticity vector is mathematically defined as follows:

$$\vec{\omega} = \nabla \times \vec{v} = 2\,\vec{\Omega} \tag{3.14}$$

Example 3.1 *Angular strain and rotation rates in streamline coordinates*

Express the angular strain and rotation rates in streamline (or natural) coordinates s and n, following the convention shown in Figure 3.4.

The turning rate of $O - B$ to $O - B'$ and that of $O - A$ to $O - A'$ are defined by the speed differences in the radial and streamwise directions, respectively, i.e. Δv_{O-B} and Δv_{O-A}:

$$\Delta v_{O-B} \approx \left\{v + \frac{\partial v}{\partial n}\; dn\right\} - v = \frac{\partial v}{\partial n}\; dn \tag{3.15}$$

$$\Delta v_{O-A} \approx \left(v + \frac{\partial v}{\partial s}\; ds\right)\sin d\theta_s - 0 \approx v\; d\theta_s \tag{3.16}$$

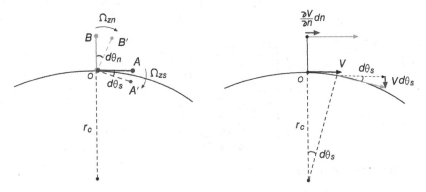

Figure 3.4 Angular strain and rotation rates of a fluid element in the s and n coordinates

where $v + (\partial v/\partial s)\, ds$ is the Taylor's series representation of the tangential speed of O at $t + dt$ and its sine of $d\theta_s$ is the vertical speed created by the motion on the curved streamline. Note that $d\theta_s > 0$ and $d\theta_n > 0$.

Thus, the turning rates of $O - B$ and $O - A$ are

$$\Omega_{zn} = -\frac{d\theta_n}{dt} \approx -\frac{\tan d\theta_n}{dt} = -\frac{\{(\partial v/\partial n)\, dn \cdot dt\}\, /dn}{dt} = -\frac{\partial v}{\partial n} \tag{3.17}$$

$$\Omega_{zs} = -\frac{d\theta_s}{dt} \approx -\frac{\tan d\theta_s}{dt} = -\frac{\{v\, d\theta_s \cdot dt\}\, /ds}{dt} = -\frac{v}{r_c} \tag{3.18}$$

where $ds = r_c\, d\theta_s$ and z is the coordinate perpendicular to the s and n coordinates.

The rotation rate is then

$$\Omega_z = \frac{1}{2}(\Omega_{zn} + \Omega_{zs}) = -\frac{1}{2}\left(\frac{\partial v}{\partial n} + \frac{v}{r_c}\right) \tag{3.19}$$

and the vorticity about the z-axis is

$$\omega_z = 2\,\Omega_z = -\left(\frac{\partial v}{\partial n} + \frac{v}{r_c}\right) \tag{3.20}$$

while the angular strain rate is

$$\epsilon_{\theta n} = \epsilon_{n\theta} = \frac{1}{2}(-\Omega_{zn} + \Omega_{zs}) = \frac{1}{2}\left(\frac{\partial v}{\partial n} - \frac{v}{r_c}\right) \tag{3.21}$$

3.2 Conservation Law of Mass

3.2.1 Continuity Equation

The conservation law of mass states that a material derivative of mass of a fluid element or particle is zero for all times:

$$\frac{D(dm)}{Dt} = 0 \tag{3.22}$$

since it is an invariant quantity.

With $dm = \rho\, dV$, Eq. (3.22) can be rearranged as follows:

$$\frac{1}{\rho}\frac{D\rho}{Dt} + \frac{1}{dV}\frac{D\,dV}{Dt} = 0 \tag{3.23}$$

showing that the total rate of change of density of the fluid particle per unit density is the same as the negative total rate of change of volume of the fluid particle per unit volume.

Since an incremental change of dV over dt can be expressed as follows:

$$
\begin{aligned}
D(dV) &= dV(t + dt) - dV(t) \\
&= \left(dx + \frac{\partial u}{\partial x}\,dx\,dt\right)\left(dy + \frac{\partial v}{\partial y}\,dy\,dt\right)\left(dz + \frac{\partial w}{\partial z}\,dz\,dt\right) - dx\,dy\,dz \\
&= \left(\frac{\partial u}{\partial x} + \frac{\partial v}{\partial y} + \frac{\partial w}{\partial z}\right)dx\,dy\,dz\,dt + \dots
\end{aligned}
\tag{3.24}
$$

the total rate of change of volume of the fluid particle per unit volume, or volumetric dilatation rate $\dot{D}$, is written as the sum of the linear strain rate in each direction (i.e. isotropic straining via compressibility of the fluid):

$$\dot{D} = \frac{1}{dV}\frac{D(dV)}{Dt} = \nabla \cdot \vec{v} \tag{3.25}$$

Note that it is a relative measure of volume change per unit time,[3] and can be related to the compressibility coefficient (or bulk modulus).[4]

Now, with Eqs. (3.23) and (3.25), the mass conservation law is finally written per unit volume as follows:

$$\frac{\partial \rho}{\partial t} + \nabla \cdot (\rho\, \vec{v}) = 0 \tag{3.26}$$

which is also referred to as the continuity equation. For incompressible flows, the volumetric dilatation rate is zero and the continuity equation reads

$$\nabla \cdot \vec{v} = 0 \tag{3.27}$$

3.2.2 Solenoidal Vector

3.2.2.1 Solenoidal Properties

A velocity vector that satisfies the divergence-free condition:

$$\nabla \cdot \vec{v} = 0 \tag{3.28}$$

is called *solenoidal*.[5] The solenoidal vector is better described when it is integrated over a stream tube. Using the divergence theorem (or Gauss theorem), it is shown that

$$\int_{CV} \nabla \cdot \vec{v}\, dV = \oint_{CS} \vec{v} \cdot \vec{n}\, dA = 0 \tag{3.29}$$

and since there is no normal component of velocity at the side surface of the stream tube, it further reads

$$\int_{\text{in}} \vec{v} \cdot \vec{n}\, dA + \int_{\text{out}} \vec{v} \cdot \vec{n}\, dA = 0 \tag{3.30}$$

---

3 If $\dot{D} = -0.1\ (1/s)$, it means that 10% of the original volume has been shrunk per unit time, or that the volume has been compressed to be 90% of the original volume per unit time. Similarly, if $\dot{D} = 0.15\ (1/s)$, it means that 15% of the original volume has been expanded per unit time, or that the volume has been expanded to become 115% per unit time.

4 More discussion continues in Chapter 8 *Compressible Flows*.

5 In Greek, *solen* means a pipe or channel.

i.e. the volumetric flow rate at the inlet surface of the stream tube must be the same as that at the outlet surface.

If we multiply the density on both sides, it shows that the mass flow rate is the same at any cross section of the stream tube. This is due to the fact that the solenoidal velocity vector, $\vec{\nabla} \cdot \vec{v} = 0$ is the same as the continuity equation in incompressible flows. *Solenoidal velocity vector* physically means *a flow of mass of an incompressible fluid*, and further implies that the stream (or streamline) cannot simply end in open space but must either end on a solid surface or form a closed loop.[6]

[**Natural coordinates**] A natural coordinate (s, n) is employed to describe the divergence-free condition of velocity vector:

$$\underbrace{\frac{\partial v}{\partial s}}_{A} + \underbrace{v \frac{\partial \theta}{\partial n}}_{B} = 0 \tag{3.31}$$

where θ is the angle between two streamlines. Here, A corresponds to the rate of stretching or contraction and B the rate of *confluence* or *diffluence*.

Multiplying an infinitesimal volume $dV = ds \, dA$ and a time interval dt to Eq. (3.31) reads

$$\underbrace{\frac{\partial v}{\partial s} \cdot ds \, dA \, dt}_{A'} + \overbrace{v \underbrace{\frac{\partial \theta}{\partial n}}^{dn'/dn} \cdot ds \, dA \, dt}^{} \;\;_{B'} = 0 \tag{3.32}$$

where A′ represents the volume increase (or decrease) over dt by the speed difference $\Delta v = v(s + ds) - v(s)$, due to the change in cross-sectional area along the stream tube (i.e. confluence or diffluence of the stream tube).

Here, B′ represents the volume decrease (or increase) over dt by the confluence (or diffluence) of the stream tube, noting that $d\theta \cdot ds \, (= dn')$ is the normal distance decreased by the confluence. Thus, $dn'/dn \, (= \varphi)$ is the contraction ratio (if $\varphi = 0$ means no confluence of the stream tube) and $\varphi \, (v \, dt \cdot dA)$ indicates how much the volume is reduced by the confluence of the stream tube. The law of mass conservation for incompressible flow states that A′ and B′ must be equal in magnitude but opposite in sign.

3.3 Surface Force Vectors

In fluids, forces originate either from within fluid volumes or at external boundaries. In response to the forces, fluids move with inertia; thus, forces are distributed in the flow field while changing the momentum of the fluid at a rate. Unlike the volumetric body forces, surface forces act across a surface element by direct contacts of fluid molecules (i.e. molecular interactions), and by per unit area, these are defined as stresses. The surface forces and stresses acting on a surface element specified by an outward unit normal vector $\vec{n}$ are defined by Cauchy's stress theorem. Cauchy's stress principle is discussed in more detail in the next section.

---

6 A vorticity vector, $\vec{\omega} = \nabla \times \vec{v}$ also has the solenoidal properties. More discussion continues in Chapter 4 *Curved Motions*.

3.3.1 Stress Vector

A surface force vector $\vec{F}$ is defined as follows:

$$\vec{F} = \vec{t}\,(\vec{x}, t, \vec{n})\,dA \tag{3.33}$$

where $\vec{t}\,(\vec{x}, t, \vec{n})$ is a stress (or traction) vector at $\vec{x}$ and t, acting on the surface of area dA specified by an outward unit normal vector $\vec{n}$ (Figure 3.5). Note also that, by action and reaction, the surface force exerted across the surface on the other side (i.e. with the outward normal vector $-\vec{n}$) is of the same magnitude but in the opposite direction and expressed as follows:

$$-\vec{F} = \vec{t}\,(\vec{x}, t, -\vec{n})\,dA \tag{3.34}$$

3.3.1.1 On a Tetrahedral Element

A stress vector $\vec{t}\,(\vec{x}, t, \vec{n})$ can be defined with the stress vectors expressed in the Cartesian coordinate system and Newton's second law of motion applied to a tetrahedral element of fluid (Figure 3.6). Here is given an example. A resultant or net force vector $\vec{R}$ acting on the mass of the tetrahedral element can be expressed as the sum of the forces acting on the surfaces ($dA, dA_1, dA_2,$ and dA_3) that bound the tetrahedral element:

$$\vec{R} = \vec{t}\,(\vec{n})\,dA + \vec{t}\,(-\vec{i})\,dA_1 + \vec{t}\,(-\vec{j})\,dA_2 + \vec{t}\,(-\vec{k})\,dA_3 \tag{3.35}$$

where the stress vectors are defined on the surfaces with outward unit normal vectors of $\vec{n}, -\vec{i}, -\vec{j}, -\vec{k}$.

Since dA_1, dA_2, dA_3 are the projection area of dA in the direction of i, j, k, they are expressed as follows:

$$dA_1 = (\vec{i} \cdot \vec{n})\,dA, \quad dA_2 = (\vec{j} \cdot \vec{n})\,dA, \quad dA_3 = (\vec{k} \cdot \vec{n})\,dA \tag{3.36}$$

Figure 3.5 A surface force vector $\vec{F}$ acting on the surface of area dA specified by an outward normal vector $\vec{n}$ at $(\vec{x}, t)$

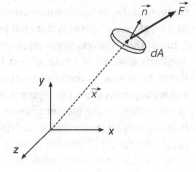

Figure 3.6 Stress vectors acting on a tetrahedral element of fluid

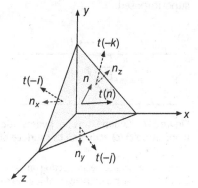

where $\vec{n} = (n_x, n_y, n_z)$ and $n_x = (\vec{i} \cdot \vec{n})$, $n_y = (\vec{j} \cdot \vec{n})$, $n_z = (\vec{k} \cdot \vec{n})$ are the directional cosines. Equation (3.35) is then written as follows:

$$\vec{R} = \{\vec{t}\,(\vec{n}) - \vec{t}\,(\vec{i})\,n_x - \vec{t}\,(\vec{j})\,n_y - \vec{t}\,(\vec{k})\,n_z\}\,dA \tag{3.37}$$

where $\vec{t}\,(\vec{i}) = -\vec{t}\,(-\vec{i})$, $\vec{t}\,(\vec{j}) = -\vec{t}\,(-\vec{j})$, $\vec{t}\,(\vec{k}) = -\vec{t}\,(-\vec{k})$.

Now, Newton's second law of motion, $\vec{R} = (\rho\,dV)\,\vec{a}$, states that if dV of the tetrahedral element approaches zero, the mass of an element and the volumetric body forces also approach zero as dV, while the surface forces approach zero as dA. From this, we can conclude that

$$\vec{t}\,(\vec{n}) = \vec{t}\,(\vec{i})\,n_x + \vec{t}\,(\vec{j})\,n_y + \vec{t}\,(\vec{k})\,n_z \tag{3.38}$$

where $\vec{t}\,(\vec{n})$, a stress vector acting on a surface with normal $\vec{n}$, can generally be expressed as the sum of $\vec{t}\,(\vec{i}), \vec{t}\,(\vec{j}), \vec{t}\,(\vec{k})$ acting on three orthogonal surfaces with each multiplied by the directional cosine vectors, n_x, n_y, n_z. The definition of $\vec{t}\,(\vec{i})$, $\vec{t}\,(\vec{j})$, $\vec{t}\,(\vec{k})$ will be discussed in the following sections.

3.3.2 Hydrostatic and Viscous Stress Vectors

Let us suppose a fluid of viscosity μ and density ρ is shear-driven in a channel by forcing one of the plates to move at a constant speed U (Figure 3.7a). After reaching steady state, the fluid between the plates is simultaneously rotated and angularly strained at the same rate; therefore, the flow in the channel is parallel to the plates.

In this simple shear flow (or Couette flow), two different types of stresses are present: hydrostatic (or pressure) stress and viscous stress. Hydrostatic stress represents a measure of repulsiveness of molecules against compressive forces. It always acts normal and inward to the surface specified with an outward unit normal vector $\vec{n}$. Thus, it is isotropic and its magnitude is omnidirectional. Figure 3.7b shows the isotropic hydrostatic stress vectors acting on the surface of a circular fluid element. If the circular element shrinks to a point, the hydrostatic stress can be defined at a point by a single scalar quantity denoted by pressure p.

In contrast, viscous stress represents a measure of hardness of a fluid element to be linearly or angularly strained at a rate against frictional forces. Though the flow in the channel is unidirectional, i.e. $u = u(y)$, the viscous stress vectors acting on the surface of a circular fluid element are nonisotropic. As shown in Figure 3.7c, the viscous stress vectors can be tangential, normal, or both to the surface of the element.[7] Note that the viscous stress vector acts on the surface at an angle that depends on the orientation of the surface specified with an outward unit normal vector $\vec{n}$. This deviatoric nature of viscous stress is due to the fact that the fluid molecules are frictionally interfered against each other, while two fluid motions (i.e. rotation and angular straining) are being superimposed.

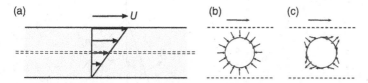

Figure 3.7 Couette flow (a); hydrostatic stress (or pressure) vectors (isotropic) (b) and viscous stress vectors (nonisotropic) (c) acting on the surface of a circular element

---

7 The viscous stress vectors acting on the circular fluid element (see Figure 3.7c) are constructed by applying Eq. (3.38) to the small triangles along the periphery of the circular element.

3.3.3 The Principal Axes of Strain Rate

In Couette flow, the horizontal viscous forces applied to the plates stretch and contract the circular fluid element at the maximum rate along the axes at $\pm45°$ from the *shear-plane*, which is the plane parallel to the externally applied tangential viscous forces. The tangential viscous stresses acting on the horizontal and vertical planes transform into normal tensile and compressive viscous stresses at $\pm45°$ from the shear-plane. At angles between 0 and $\pm45°$, the viscous stress vector is split into tangential and normal components and the magnitude ratio between the two depends on the projection area. Note that the principal axes of strain rate are the axes where the tangential shear strain rate disappears or the linear strain rate is the maximum [1].

The photo presented in Figure 3.8 shows a glacier that flows very slowly at a typical speed of 25 *cm* per day. What we seem from this photo is that the ice particles in the glacier are undertaking tremendous stresses. Due to its enormous weight, the ice particles are in great compression with extremely high hydrostatic stresses (or pressure). At the same time, they are viscously strained at a rate by the glacier that moves relatively faster at the free surface and in the middle than at the bottom or sides. This photo clearly shows one clear evidence of viscous stresses, *crevasses*, created by viscous stresses, which act normally on the principal axes of the strain rates. Note that the crevasses created by viscous stresses are always inclined 45° to the flow direction of the glacier, regardless of whether the glacier flows north, south, east, or west.

Example 3.2   *A stress vector on an inclined surface*

In two-dimensional space, a stress vector acting on an inclined surface of a fluid element can be constructed by a pair of stress vectors that act on two orthogonal surfaces; a triangle is constructed with these three surfaces (Figure 3.9).

Figure 3.9a shows four normal hydrostatic stress (pressure) vectors acting on the inclined surfaces of a fluid element in the nozzle flow. The normal stress vector acting on the inclined surface can be constructed with two normal stress vectors that act on two orthogonal surfaces. Figure 3.9b shows two pairs of tangential and normal viscous stress vectors acting on the surfaces of two fluid elements in different orientations in the Couette flow. The tangential viscous stress vectors disappear on the surface planes at $\pm45°$, where the tangential stresses are transformed into tensile and compressive normal stresses. These normal stress vectors can be constructed by two tangential viscous stress vectors acting on two orthogonal surfaces.

Figure 3.8   Crevasses created by viscous stresses acting along the principal axes of strain rate in a glacial flow; photographed in Kenai Fjords National Park, Alaska, USA (July 9, 2017)

(a)

(b)

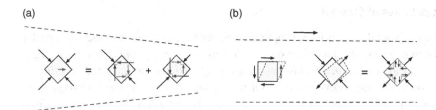

Figure 3.9 Hydrostatic stress vectors on the surfaces of a fluid element in an inviscid nozzle flow (a); viscous stress vectors on the surfaces of two shear-straining fluid elements in different orientations in a simple shear flow (b)

3.3.4 Deviatoric Nature of Viscous Stress

The characteristics of viscous flows can be represented by the followings: (i) nonisotropic viscous stresses and (ii) the principal axes of strain rate. Figure 3.10 shows the isotropic and nonisotropic natures of hydrostatic stress (or pressure) and viscous stress. Note that degree to which the Couette flow rotates is indistinguishable from the hydrostatic stress vectors (Figure 3.10a). However, the viscous stress vectors clearly show the rotation of the Couette flow (Figure 3.10b).

Furthermore, the viscous stress vectors acting on a circular fluid element show the same pattern; to be more specific, the principle axes of strain rate are only rotated by the same inclination angle. In fact, the viscous stress vectors at a different inclination angle can be constructed by rotating the shear-plane and the Cartesian stress tensor from σ_{ij} defined in x, y coordinates to σ'_{ij} in x', y' coordinates, where $(x', y') = [R] (x, y)^T$ and $[R]$ is the rotation matrix:

$$[R] = \begin{bmatrix} \cos\theta & -\sin\theta \\ \sin\theta & \cos\theta \end{bmatrix} \tag{3.39}$$

This stress transformation clearly shows that the viscous stress vectors are nonisotropic (or deviatoric). If we look at the circular element in the absolute reference frame, the viscous stresses acting on the surfaces around it are in all different directions, i.e. nonisotropic, although the nature of the shear is the same – sliding one of the plates, while the other is fixed. In contrast, the hydrostatic stress vectors always act normal and inward to the surfaces around the circular element, i.e. isotropic; thus, the hydrostatic stress is defined by a single scalar quantity, i.e. pressure (Figure 3.10a).

(a)

(b)

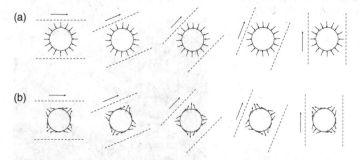

Figure 3.10 Hydrostatic stress vectors (a) and viscous stress vectors (b) acting on the surfaces of a circular fluid element; the Couette flow inclined at 0, 22.5, 45, 67.5, and 90° (from left to right)

Figure 3.11 Normal and tangential stresses on the six surfaces of a cubic fluid element

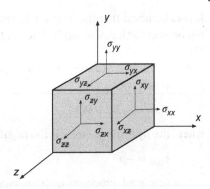

3.3.5 Stress Tensor

Let us imagine a circular fluid element in simple shear flow shrink to a point and ask ourselves what will happen to those hydrostatic and viscous stress vectors. The former is rather simple because the hydrostatic stress vectors are all pointing inward and normal to the point with its magnitude denoted by a scalar quantity, i.e. pressure. On the other hand, the viscous stress vectors are all pointing at different angles, which also change if the orientation of the Couette flow changes. Thus, a more generalized concept is needed to define a viscous stress vector acting on the surface specified by an outward unit normal vector $\vec{n}$.

We first need to define a second-order Cartesian stress tensor (or Cauchy's stress tensor) $[\sigma]$ acting on the three mutually orthogonal surfaces of an infinitesimal cubic element at a point (x, y, z):

$$[\sigma] = \begin{bmatrix} \sigma_{xx} & \sigma_{xy} & \sigma_{xz} \\ \sigma_{yx} & \sigma_{yy} & \sigma_{yz} \\ \sigma_{zx} & \sigma_{zy} & \sigma_{zz} \end{bmatrix} \tag{3.40}$$

where σ_{ij} represents one normal and two tangential stress vectors that act on each orthogonal surface; the first index i denotes the surface element on which the stress vector is acting, and the second index j is the direction of the stress vector in action (Figure 3.11).

With the second-order stress tensor, we can define $\vec{t}\,(\vec{i})$, $\vec{t}\,(\vec{j})$, $\vec{t}\,(\vec{k})$

$$\vec{t}\,(\vec{i}) = \sigma_{xx}\,\vec{i} + \sigma_{xy}\,\vec{j} + \sigma_{xz}\,\vec{k}$$

$$\vec{t}\,(\vec{j}) = \sigma_{yx}\,\vec{i} + \sigma_{yy}\,\vec{j} + \sigma_{yz}\,\vec{k} \tag{3.41}$$

$$\vec{t}\,(\vec{k}) = \sigma_{zx}\,\vec{i} + \sigma_{zy}\,\vec{j} + \sigma_{zz}\,\vec{k}$$

and also a viscous stress vector acting on the surface specified with an outward unit normal vector $\vec{n}$ as in Eq. (3.38):

$$\vec{t}\,(\vec{n}) = \vec{n}\,[\sigma] \qquad \text{or} \qquad t_i = \sigma_{ji}\, n_j \tag{3.42}$$

where $\vec{t}\,(\vec{n}) = (t_x(\vec{n}), t_y(\vec{n}), t_z(\vec{n}))$ reads

$$t_x(\vec{n}) = \sigma_{xx}\, n_x + \sigma_{yx}\, n_y + \sigma_{zx}\, n_z$$

$$t_y(\vec{n}) = \sigma_{xy}\, n_x + \sigma_{yy}\, n_y + \sigma_{zy}\, n_z \tag{3.43}$$

$$t_z(\vec{n}) = \sigma_{xz}\, n_x + \sigma_{yz}\, n_y + \sigma_{zx}\, n_z$$

It is to be noted that the magnitude-shares of the three components of viscous stress vectors acting on each orthogonal surface depend on the local direction of the shear-plane, as displayed in Figure 3.6.

As discussed in Section 3.3.2, the second-order stress tensor σ_{ij} defined at a point can be decomposed as follows:

$$\sigma_{ij} = \sigma_{\text{hyd}}\, \delta_{ij} + D_{ij} \tag{3.44}$$

where the hydrostatic stress is the negative of pressure:

$$\sigma_{\text{hyd}} = -p \tag{3.45}$$

which acts in all directions with the same magnitude; in other words, it is a direction-independent scalar quantity (i.e. energy density).

The deviatoric viscous stress D_{ij} represents the degree of fluid distortion via straining and can be expressed as follows:

$$D_{ij} = \tau_{ij} + \lambda\, (\nabla \cdot \vec{v})\, \delta_{ij} \tag{3.46}$$

where τ_{ij} is the viscous stress tensor, and λ is a second or dilatational viscosity related to volumetric dilatation rate of the fluid. Note that the definition of the second-order viscous stress tensor τ_{ij} is based on the second-order strain rate tensor ϵ_{ij}; in fact, they are related through viscosity under Newton's viscosity law, i.e. $\tau_{ij} = 2\,\mu\,\epsilon_{ij}$. How the deviatoric stress tensor is expressed as Eq. (3.46) will be explained in the following sections.

3.4 Constitutive Relations between the Stress and the Strain Rate

It is important to note that the viscous stresses and strain rates are direction-dependent on the surface of the fluid (i.e. nonisotropic). In Couette flow, a squared fluid element is simply sheared into a parallelogram at a rate by tangential viscous stresses τ_{xy} and τ_{yx}. In this case, the squared element is rotated and angularly strained at the same rate (Figure 3.12a). If the Couette flow is inclined by 30°, sliding the inclined plate imposes not only tangential viscous stresses (τ_{xy} and τ_{yx}) but also normal

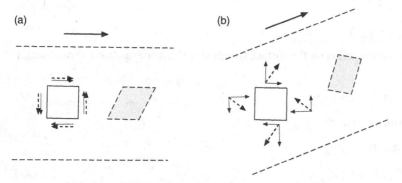

(a)

(b)

Figure 3.12 Viscous stress vectors (dashed) and their tangential and normal components (solid) acting on the surfaces of a fluid element in the Couette flow inclined at 0 (a) and 30° (b)

viscous stresses (τ_{xx} and τ_{yy}) to the surfaces of the same squared element (Figure 3.12b). Thus, the element is shear deformed into a parallelogram in different form; besides rotation and angular straining, the normal components also linearly strain the squared element at a rate. Meanwhile, if the squared element in the horizontal Couette flow is rotated 30°, we will have the same result of viscous stresses and strain rates of the nonrotated element in the Couette flow inclined at the same angle. This simple example of Couette flow well illustrates the nature of viscous stresses and strain rates.

3.4.1 Tangential and Normal Viscous Stresses

3.4.1.1 Tangential Viscous Stresses

It is important to note that the tangential viscous stress is directly related to the sum of the turning rates of two mutually orthogonal surfaces of the element. For example, in two dimensions,

$$\dot{\alpha} + \dot{\beta} = \frac{\partial u}{\partial y} + \frac{\partial v}{\partial x} = 2\,\epsilon_{xy} \tag{3.47}$$

where $\dot{\alpha}$ represents the clockwise turning rate of the vertical side of the fluid element by a local force balance between two tangential viscous stresses (denoted by τ_{yx}) acting on the horizontal surface elements (e.g. top and bottom) in the opposite direction, whereas $\dot{\beta}$ represents the counterclockwise turning rate of the horizontal side of the fluid element by a local force balance between two tangential viscous stresses (denoted by τ_{xy}) acting on the vertical surface elements (e.g. right and left) in the opposite direction (Figure 3.13).

Then, the tangential viscous stress can be defined as follows:

$$\tau_{yx} = \tau_{xy} = \mu\left(\frac{\partial u}{\partial y} + \frac{\partial v}{\partial x}\right) = 2\,\mu\,\epsilon_{xy} \tag{3.48}$$

where τ_{yx} acting in the x-direction on the surface facing the y-direction cannot be independent of τ_{xy} acting in the y-direction on the surface facing the x-direction, because both turning rates are the result of frictional interference of fluid molecules. To be more specific, τ_{yx} and τ_{xy} are the resultant shear stresses via mutual interactions of the fluid molecules while the fluid is being angularly strained and rotated at the same time, but not necessarily at the same rates.[8]

Figure 3.13 Tangential viscous stresses τ_{yx} and τ_{xy} at (x, y); $\tau_{xy} = \tau_{yx}$ (by symmetry)

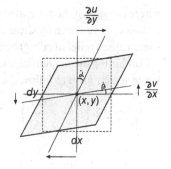

---

8 Tangential viscous stress can be present when fluids angularly deform but not rotate, e.g. free (or irrotational) vortex. Note, however, that no net viscous force is exerted in the flow field.

This constitutive relation between the strain rates and the stresses can be generalized to the surfaces of a cubic element, and this symmetry is also true for other directions, i.e.

$$\tau_{yz} = \tau_{zy} = \mu \left(\frac{\partial v}{\partial z} + \frac{\partial w}{\partial y} \right) \tag{3.49}$$

$$\tau_{zx} = \tau_{xz} = \mu \left(\frac{\partial u}{\partial z} + \frac{\partial w}{\partial x} \right) \tag{3.50}$$

It is to be noted again that the tangential viscous stress vectors act as a pair with the same magnitude on two surfaces of the element that share a common boundary. Thus, they are symmetric, i.e. $\tau_{ij} = \tau_{ji}$, since the tangential viscous force acting on one surface of the element is viscously resisted by the tangential viscous force acting on the other surface; in other words, they are the coupled forces per unit area and work as a pair on two surfaces of the element that share a common boundary. On the six surfaces of a cube of the fluid element, there are three pairs of tangential viscous stress vectors.

[**Notes**] If there is a difference in magnitude among these three counterfacing pairs of tangential viscous stress vectors, a net tangential viscous force is produced which will tangentially accelerate the fluid while, via viscous diffusion, the tangential momentum of the fluid will be transported in the transverse direction until the rotational rate of the fluid becomes uniform (e.g. Couette flow). In a fully developed Poiseuille flow, the lateral transport of tangential momentum also stops at the centerline and a steady-state flow field is established because tangential viscous forces will be completely balanced by the pressure difference force. In external viscous flows, this lateral transport of tangential momentum is continuously convected downstream so that there will exist a boundary that separates the flow into two: one is under viscous stresses and the other is out of reach. This is the boundary layer flow at steady state.

3.4.1.2 Normal Viscous Stresses

As stated before, linearly straining a fluid element at a rate causes frictional resistance in the direction normal to two mutually facing surfaces that bound the element (Figure 3.14). The magnitude of the stress vector is proportional to the viscosity μ and two times the linear strain rate of the element because it is a mutual interaction between two surfaces:

$$\tau_{xx} = 2\mu \frac{\partial u}{\partial x}, \quad \tau_{yy} = 2\mu \frac{\partial v}{\partial y}, \quad \tau_{zz} = 2\mu \frac{\partial w}{\partial z} \tag{3.51}$$

where the sign of the normal viscous stress vector is determined by that of the linear strain rate. For instance, contraction by shear causes additional compression besides pressure, while stretching by shear causes relief of compression.

Combining the tangential and normal viscous stresses, the viscous stresses can generally be written in tensor form as follows:

$$\tau_{ij} = \mu \left(\frac{\partial v_i}{\partial x_j} + \frac{\partial v_j}{\partial x_i} \right) = 2\mu\, \epsilon_{ij} \tag{3.52}$$

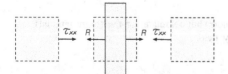

Figure 3.14 Normal viscous stresses τ_{xx} exerted on two mutually facing surfaces of a linearly straining fluid element at a rate by the frictional resistance, $R = 2\mu\, du/dx$

Example 3.3 *Couette flow inclined at an angle*

A Couette flow inclined at an angle θ is generally expressed by the velocity vector $\vec{v} = (u, v)$

$$u = V \cos \theta, \quad v = V \sin \theta$$

where

$$V = a \, (-x \sin \theta + y \, \cos \theta)$$

and

$$|x| = |y| \quad (-1 \leq x \leq 0, \quad 0 \leq y \leq 1)$$

Note that $a = |\vec{\omega}|$ is the magnitude of the vorticity vector $\vec{\omega} = \nabla \times \vec{v}$.

(a) Find the viscous stress tensor τ_{ij} in the Couette flow at an angle θ.
(b) Show that the magnitude of the resultant or net stress vector is the same at any inclination angle.
(c) Draw τ_{xx} and τ_{xy} vs. θ, and discuss what we can learn from these two graphs.

The linear and angular strain rates of the fluid are

$$\epsilon_{xx} = \frac{\partial u}{\partial x} = -a \, \sin \theta \cos \theta$$

$$\epsilon_{yy} = \frac{\partial v}{\partial y} = a \, \sin \theta \cos \theta$$

$$\epsilon_{xy} = \frac{1}{2} \left(\frac{\partial u}{\partial y} + \frac{\partial v}{\partial x} \right) = \frac{1}{2} \, a \, (\cos^2 \theta - \sin^2 \theta)$$

and thus, the normal and tangential viscous stress vectors are written as follows:

$$\tau_{xx} = 2 \, \mu \, \epsilon_{xx} = -2 \, \mu \, a \, \sin \theta \cos \theta$$

$$\tau_{yy} = 2 \, \mu \, \epsilon_{yy} = 2 \, \mu \, a \, \sin \theta \cos \theta$$

$$\tau_{xy} = \tau_{yx} = 2 \, \mu \, \epsilon_{xy} = \mu \, a \, (\cos^2 \theta - \sin^2 \theta)$$

If $\theta = 0$ (horizontal case), the normal stresses disappear and only the tangential stresses exist:

$$\tau_{xx} = 0, \quad \tau_{yy} = 0, \quad \tau_{xy} = \tau_{yx} = \mu a$$

If $\theta = \pi/6$, both the normal and tangential stresses exist:

$$\tau_{xx} = -\frac{\sqrt{3}}{2} \, \mu a, \quad \tau_{yy} = \frac{\sqrt{3}}{2} \, \mu a, \quad \tau_{xy} = \tau_{yx} = \frac{1}{2} \, \mu a$$

If $\theta = \pi/4$, only the normal stresses exist, and the tangential stresses disappear (Figure 3.15):

$$\tau_{xx} = -\mu a, \quad \tau_{yy} = \mu a, \quad \tau_{xy} = \tau_{yx} = 0$$

If $\theta = \pi/3$, both the normal and tangential stresses exist again:

$$\tau_{xx} = -\frac{\sqrt{3}}{2} \, \mu a, \quad \tau_{yy} = \frac{\sqrt{3}}{2} \, \mu a, \quad \tau_{xy} = \tau_{yx} = -\frac{1}{2} \, \mu a$$

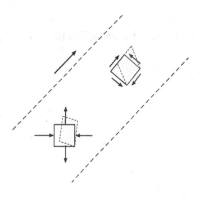

Figure 3.15 Normal and tangential viscous stress vectors acting on the surfaces of two fluid elements in different orientations; the Couette flow inclined at 45°

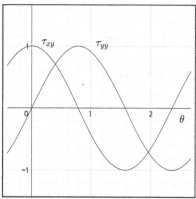

Figure 3.16 τ_{xy} and τ_{yy} vs. θ; $\mu a = 1$

If $\theta = \pi/2$ (vertical case), the normal stresses disappear and only the tangential stresses exist:

$$\tau_{xx} = 0, \quad \tau_{yy} = 0, \quad \tau_{xy} = \tau_{yx} = -\mu a$$

It is to be noted that the magnitude of the resultant stress vector is the same at any inclination angle. As shown in Figure 3.16, τ_{xy} and τ_{yx} switch sign at 45° from positive to negative and their magnitude is proportional to $\cos 2\theta$; it drops slowly at very low angles but almost linearly afterwards until the angles become close to 90°. However, τ_{xx} and τ_{yy} change their magnitudes almost linearly from 0 to 45° but becomes plateau around angles near 45°.

3.4.2 Mechanical Pressure

3.4.2.1 Incompressible Fluids

As stated earlier, contraction by shear causes additional compression besides pressure, while stretching causes relief of compression. Thus, the normal stresses σ_{xx}, σ_{yy}, and σ_{zz} can be written as follows:

$$\sigma_{xx} = -p + \tau_{xx} = -p + 2\mu \, \frac{\partial u}{\partial x} \tag{3.53}$$

$$\sigma_{yy} = -p + \tau_{yy} = -p + 2\mu \, \frac{\partial v}{\partial y} \tag{3.54}$$

$$\sigma_{zz} = -p + \tau_{zz} = -p + 2\mu \, \frac{\partial w}{\partial z} \tag{3.55}$$

At this point, the normal viscous stress vectors defined by Eqs. (3.53)–(3.55) are not direction-independent like pressure is. They represent tensile and compressive stress vectors acting in three mutually orthogonal directions, respectively; thus, the normal viscous stresses, are called deviatoric stress vectors.

We now need to define an averaged normal stress, naming its minus the *mechanical pressure $\bar{p}$* as follows:

$$\bar{p} = -\frac{1}{3}(\sigma_{xx} + \sigma_{yy} + \sigma_{zz}) = p - \frac{2}{3}\,\mu\,(\nabla \cdot \vec{v}) \tag{3.56}$$

where the mechanical pressure represents only the translational mode of energy of the molecules. In contrast, the thermodynamic pressure represents the vibrational, rotational, and translation modes of energy of the molecules (thus, temperature, density (or volume), and pressure are related to each other by an equation of state).

If the fluid is incompressible, i.e. volumetric dilatation rate is zero, then the mechanical pressure will be the same as the thermodynamic pressure:

$$\bar{p} = p \tag{3.57}$$

which implies that the pressure in an incompressible flow is actually the mechanical pressure. Note that the absolute pressure in an incompressible fluid is indeterminate; only its gradient is determined from the equation of motion; in other words, a pressure difference force is balanced with the inertial or frictional resistance forces, as discussed in Chapter 1 *Pressure*.

3.4.2.2 Compressible Fluids

If a fluid is compressible, the difference between the pressure and the mechanical pressure reads as follows:

$$\bar{p} - p = -\frac{2}{3}\,\mu\,(\nabla \cdot \vec{v}) \tag{3.58}$$

where the magnitude of the difference is proportional to constant times the viscosity and the volumetric dilatation rate. In compressible flows, this term is usually quite small, though the volumetric dilatation rate is not zero.

If the fluid's volume changes at a very fast rate (i.e. rapid isotropic compression or expansion), internal frictions act as a normal stress, and the magnitude of this normal stress will be proportional to some kind of frictional measure (called a second or dilatational viscosity, λ) and the volumetric dilatation rate. Then, the normal stress in each direction can be written as follows:

$$\sigma_{xx} = -p + 2\mu\,\frac{\partial u}{\partial x} + \lambda\,(\nabla \cdot \vec{v}) \tag{3.59}$$

$$\sigma_{yy} = -p + 2\mu\,\frac{\partial v}{\partial y} + \lambda\,(\nabla \cdot \vec{v}) \tag{3.60}$$

$$\sigma_{zz} = -p + 2\mu\,\frac{\partial w}{\partial z} + \lambda\,(\nabla \cdot \vec{v}) \tag{3.61}$$

If we sum Eqs. (3.59)–(3.61) and take an average, the mechanical pressure is related as follows:

$$\bar{p} = p - \left(\lambda + \frac{2}{3}\,\mu\right)(\nabla \cdot \vec{v}) \tag{3.62}$$

and the difference between the thermodynamic pressure and the mechanical pressure is expressed as follows:

$$\bar{p} - p = -\beta\,(\nabla \cdot \vec{v}) \tag{3.63}$$

where the bulk viscosity β is defined as follows:

$$\beta = \lambda + \frac{2}{3}\mu \tag{3.64}$$

Note that β depends on the molecular composition of the fluid.

3.4.2.3 Stokes Hypothesis

The bulk viscosity β, a nonnegative value asserted by the second law of thermodynamics, is associated with the dissipation mechanism during a change of volume at a finite rate, e.g. a damping of volumetric vibrations. Here is given an example. When a fluid particle crosses a shock wave from supersonic to subsonic, a rapid reduction of volume in a very short distance would not occur unless mechanical energy is viscously dissipated; the translational mode of energy of the molecules are converted into the vibrational mode of energy. In this case, the mechanical pressure is larger than the thermodynamic pressure because $\nabla \cdot \vec{v} < 0$.

In monatomic gases, the bulk viscosity β is zero so that the thermodynamic and mechanical pressures are the same, since monatomic gas molecules only have a translational mode of energy. However, these two pressures are different (i) in multiatomic gases or liquids that have bulk viscosities 10 or 100 times larger than μ, or (ii) when fluids undergo a nonequilibrium thermodynamic process. Though the bulk viscosity β is not zero, the mechanical pressure is considered the same as the thermodynamic pressure in many practical problems because the volumetric dilatation rate is usually quite small. This assumption is called Stokes hypothesis; the bulk viscosity is zero, $\beta = 0$, or

$$\lambda = -\frac{2}{3}\mu \tag{3.65}$$

so that $\bar{p} = p$. The Stokes hypothesis has been proven true for dilute monatomic gases by kinetic theory, but it is generally not true for multiatomic gases or liquids [2].

The normal stresses are then written with the bulk viscosity $\beta = 0$ as follows:

$$\sigma_{xx} = -p + \tau_{xx} - \frac{2}{3}\mu\,(\nabla \cdot \vec{v}) \tag{3.66}$$

$$\sigma_{yy} = -p + \tau_{yy} - \frac{2}{3}\mu\,(\nabla \cdot \vec{v}) \tag{3.67}$$

$$\sigma_{zz} = -p + \tau_{zz} - \frac{2}{3}\mu\,(\nabla \cdot \vec{v}) \tag{3.68}$$

[Notes] The stress tensor can be expressed in matrix form as follows:

$$[\sigma] = (-p + \lambda\,\nabla \cdot \vec{v})[I] + [\tau] \tag{3.69}$$

or in tensor form as follows:

$$\sigma_{ij} = \left(-p + \lambda\,\frac{\partial v_i}{\partial x_j}\right)\delta_{ij} + \tau_{ij} \tag{3.70}$$

where the viscous stress tensor τ_{ij} is defined by

$$\tau_{ij} = \mu\left(\frac{\partial v_i}{\partial x_j} + \frac{\partial v_j}{\partial x_i}\right) \tag{3.71}$$

and λ is the second or dilatational viscosity and $\lambda = -2/3\,\mu$, assuming Stokes hypothesis.

3.5   Conservation Law of Momentum

3.5.1   The Navier–Stokes Equations

Newton's second law of motion applied to a fluid particle of mass $m = \rho\, dV$ states:

$$m\,\frac{D\vec{v}}{Dt} = \vec{F} \tag{3.72}$$

where the net force acting on the fluid particle $\vec{F}$ is the sum of the surface force and the volumetric body force:

$$\vec{F} = \vec{F}_s + \vec{F}_b \tag{3.73}$$

With the divergence theorem, the net surface force $\vec{F}_s$ is defined as follows:

$$\vec{F}_s = \int \vec{t}\,(\vec{n})\ dA = \int \vec{n}\,[\sigma]\ dA = \nabla \cdot [\sigma]\ dV \tag{3.74}$$

where $\vec{n}\,[\sigma]$ $(= \sigma_{ji}\, n_j)$ is the stress vector $\vec{t}\,(\vec{n})$ exerted across the surface of area dA specified by an outward unit normal vector $\vec{n}$, $\nabla \cdot [\sigma]$ $(= \partial\sigma_{ij}/\partial x_j)$ is the net surface force vector exerted on the fluid particle of a unit volume, and the stress tensor $[\sigma]$ is

$$[\sigma] = (-p + \lambda\,\nabla \cdot \vec{v})[I] + [\tau] \tag{3.75}$$

A net volumetric body force is defined as follows:

$$\vec{F}_b = \rho\,\vec{a}_f\ dV \tag{3.76}$$

where $\vec{a}_f$ is the acceleration vector; for example $\vec{a}_f = \vec{g}$ is the gravitational acceleration vector.

The equation of motion, Eqs. (3.72)–(3.76), can be written per unit volume as follows:

$$\rho\left(\frac{\partial \vec{v}}{\partial t} + \vec{v} \cdot \nabla\,\vec{v}\right) = \nabla \cdot [\sigma] + \vec{f}_b$$

$$= -\nabla p + \nabla(\lambda\,\nabla \cdot \vec{v}) + \nabla \cdot [\tau] + \vec{f}_b \tag{3.77}$$

where $\lambda = \beta - 2/3\,\mu$ is the dilatational viscosity and β is the bulk viscosity. If the Stokes hypothesis is used, i.e. $\beta = 0$, then $\lambda = -2/3\,\mu$.

3.5.2   In Tensor Form

The Navier–Stokes equations can be written in a tensor form as follows:

$$\rho\,\frac{Dv_i}{Dt} = \frac{\partial\sigma_{ij}}{\partial x_j} + f_{bi} \tag{3.78}$$

where f_i is the volumetric body force per unit volume and the surface stress tensor σ_{ij} is

$$\sigma_{ij} = -p\,\delta_{ij} + D_{ij} \tag{3.79}$$

The deviatoric stress tensor D_{ij} is defined as follows:

$$D_{ij} = \tau_{ij} + \lambda\,(\nabla \cdot \vec{v})\,\delta_{ij} \tag{3.80}$$

where the shear stress tensor $\tau_{ij} = 2\,\mu\,\epsilon_{ij}$ and the angular strain rate tensor ϵ_{ij} is defined by

$$\epsilon_{ij} = \frac{1}{2}\left(\frac{\partial v_i}{\partial x_j} + \frac{\partial v_j}{\partial x_i}\right) \tag{3.81}$$

With Stokes hypothesis, Eq. (3.80) gives us

$$D_{ij} = 2\mu\left(\epsilon_{ij} - \frac{1}{3}\,\nabla\cdot\vec{v}\,\delta_{ij}\right) \tag{3.82}$$

Note that D_{ij} in Eq. (3.80) makes zero contribution to the mean normal stress (i.e. $\bar{p} = p$),

$$D_{ii} = (2\mu + 3\lambda)\,(\nabla\cdot\vec{v}) = 0 \tag{3.83}$$

if the dilatational viscosity is

$$\lambda = -\frac{2}{3}\,\mu \tag{3.84}$$

The Navier–Stokes equations are finally written as follows:

$$\rho\frac{Dv_i}{Dt} = -\frac{\partial p}{\partial x_i} + \frac{\partial}{\partial x_j}\left\{2\mu\left(\epsilon_{ij} - \frac{1}{3}\frac{\partial v_i}{\partial x_j}\,\delta_{ij}\right)\right\} + f_{bi} \tag{3.85}$$

and if the fluid is incompressible and the viscosity is assumed constant, the Navier–Stokes equations read

$$\rho\frac{Dv_i}{Dt} = -\frac{\partial p}{\partial x_i} + \mu\frac{\partial^2 v_i}{\partial x_j \partial x_j} + f_{bi} \tag{3.86}$$

3.6 Conservation Law of Energy

3.6.1 The First Law of Thermodynamics for a Fluid Particle

The first law of thermodynamics states that the rate of change of total energy stored in a fluid particle of mass $m = \rho\,dV$ equals the net inflow rate (or influx) of energy from its surroundings to the fluid particle via heat and work:

$$m\frac{De_t}{Dt} = \dot{Q} + \dot{W} \tag{3.87}$$

where $e_t = e + v^2/2$ is the specific total energy.

With the divergence theorem, the net inflow rate of heat from the surroundings to the fluid particle is expressed as follows:

$$\dot{Q} = \int -\vec{q}\cdot\vec{n}\,dA = \int k\,\nabla T\cdot\vec{n}\,dA = \nabla\cdot(k\,\nabla T)\,dV \tag{3.88}$$

where $\vec{q} = -k\,\nabla T$ is the heat flux vector and k is the thermal conductivity of the fluid. Note that $k\,\nabla T\cdot\vec{n}$ is the influx of heat through the surface of area dA specified by an outward unit normal vector $\vec{n}$ (per unit area) and $\nabla\cdot(k\,\nabla T)$ is the net influx of heat from the surroundings to the fluid particle (per unit volume).

Meanwhile, the net inflow rate of work done on the fluid particle is written as the sum of the net inflow rates of the surface force work and the volumetric body force work:

$$\dot{W} = \int \dot{w}_s\,dA + \dot{w}_b\,dV \tag{3.89}$$

With the divergence theorem, the net inflow rate of surface force work done on the particle is written as follows:

$$\int \dot{w}_s \, dA = \int \vec{n} \, [\sigma] \cdot \vec{v} \, dA = \int \vec{v} \, [\sigma] \cdot \vec{n} \, dA = \nabla \cdot \vec{v} \, [\sigma] \, dV \tag{3.90}$$

where $\vec{n} \, [\sigma] \cdot \vec{v} \, (= \sigma_{ji} \, n_j \, u_i)$ is the influx of surface force work done on the surface of area dA specified by an outward unit normal vector $\vec{n}$ (per unit area) and $\nabla \cdot \vec{v} \, [\sigma] \, (= \partial(u_i \, \sigma_{ij})/\partial x_j)$ is the net influx of surface force work done on the particle (per unit volume).

The net inflow rate of volumetric body force work done on the fluid particle of volume dV can generally be expressed as follows:

$$\dot{w}_b \, dV = \rho \, \vec{a}_f \cdot \vec{v} \, dV \tag{3.91}$$

where $\vec{a}_f$ is the acceleration vector.

3.6.2 Mechanical and Thermal Energy Equations

The total energy equation can be obtained in differential form by writing Eq. (3.87) per unit volume:

$$\rho \frac{De_t}{Dt} = \nabla \cdot \vec{v} \, [\sigma] + \nabla \cdot (k \, \nabla T) + \dot{w}_b$$

$$= \nabla \cdot \left(-p \, \vec{v} + \vec{v} \, [\tau] + (\lambda \, \nabla \cdot \vec{v}) \, \vec{v} + k \, \nabla T \right) + \dot{w}_b$$

$$= \frac{\partial}{\partial x_j} \left(-p \, u_j + u_i \tau_{ij} + \lambda \left(\frac{\partial v_i}{\partial x_i} \right) u_j + k \, \frac{\partial T}{\partial x_j} \right) + \dot{w}_b \tag{3.92}$$

If we inner-product the momentum equation (Eq. (3.77)) with a velocity vector and subtract the inner-producted equation from Eq. (3.92), the total energy equation can be split into the mechanical energy equation:

$$\rho \frac{D(v^2/2)}{Dt} = \underbrace{-\vec{v} \cdot \nabla p}_{①A} + \underbrace{\vec{v} \cdot (\nabla \, [\tau])}_{②A} + \underbrace{\vec{v} \cdot \nabla(\lambda \, \nabla \cdot \vec{v})}_{③A} + \underbrace{\dot{w}_b}_{④} \tag{3.93}$$

and the thermal energy equation:

$$\rho \frac{De}{Dt} = \underbrace{-p \, \nabla \cdot \vec{v}}_{①B} + \underbrace{[\tau] \cdot [\nabla \, \vec{v}]}_{②B} + \underbrace{\lambda \, (\nabla \cdot \vec{v})^2}_{③B} + \underbrace{\nabla \cdot (k \, \nabla T)}_{⑤} \tag{3.94}$$

It is to be noted that the terms $①_A$, $②_A$, $③_A$, $④$ change the kinetic energy at a rate by accelerating or decelerating the fluid, while the terms $①_B$, $②_B$, $③_B$, $⑤$ change the internal energy at a rate by altering the fluid volume in a reversible process, deforming (or straining) the fluid against frictions (dissipating the energy into heat), or exchanging heat.

3.6.3 Deformation Work and Viscous Dissipation

In the total energy conservation equation, the net inflow rate of surface force work done on the fluid particle of a unit volume (denoted by ⑤) is the sum of the rate of deformation work (denoted by ⑥) and the rate of mechanical work[9] (denoted by ⑥), both done on the fluid particle of a unit volume:

---

9 This is the work done by a net difference of stress vectors.

$$\nabla \cdot \vec{v} \, [\sigma] = \underbrace{\frac{\partial(v_i \, \sigma_{ij})}{\partial x_j}}_{\text{(s)}} = \underbrace{\sigma_{ij} \frac{\partial v_i}{\partial x_j}}_{\text{(d)}} + \underbrace{u_i \frac{\partial \sigma_{ij}}{\partial x_j}}_{\text{(k)}} \tag{3.95}$$

To more physically explain the nature of deformation work, the total energy equation Eq. (3.95) can be split into the mechanical energy equation:

$$\rho \frac{D(v^2/2)}{Dt} = \underbrace{\frac{\partial(v_i \, \sigma_{ij})}{\partial x_j}}_{\text{(s)}} - \underbrace{\sigma_{ij} \frac{\partial v_i}{\partial x_j}}_{\text{(d)}} + \dot{w}_b \tag{3.96}$$

and the thermal energy equation:

$$\rho \frac{De}{Dt} = \underbrace{\sigma_{ij} \frac{\partial v_i}{\partial x_j}}_{\text{(d)}} + \nabla \cdot (k \, \nabla T) \tag{3.97}$$

The two split equations show that the kinetic energy viscously dissipated at a rate in the mechanical energy equation is converted at a rate into the internal energy in the thermal energy equation by the rate of deformation work ⓓ. This term acts as a sink for the kinetic energy in the flow.

To be more specific, the rate of deformation work ⓓ can be written as follows:

$$\sigma_{ij} \frac{\partial v_i}{\partial x_j} = \left(\left(-p + \lambda \, \frac{\partial v_k}{\partial x_k} \right) \delta_{ij} + \tau_{ij} \right) \frac{\partial v_i}{\partial x_j}$$

$$= \underbrace{-p \, \frac{\partial v_j}{\partial x_j}}_{\text{(e)}} + \underbrace{\tau_{ij} \frac{\partial v_i}{\partial x_j} + \lambda \left(\frac{\partial v_j}{\partial x_j} \right)^2}_{\text{(v)}} \tag{3.98}$$

where the first term ⓔ on the right-hand side is the rate of work done in a reversible process via expansion or contraction of volume (or compressibility effect). This term can be positive or negative, depending on the sign of the volumetric dilatation rate; for instance, it increases or decreases the internal energy by volume of the contraction or expansion, respectively.

The second and third terms ⓥ that occur in an irreversible process represent the rate of dissipation of mechanical energy through viscosity. It is always positive because

$$\tau_{ij} \frac{\partial v_i}{\partial x_j} = 2 \, \mu \, \epsilon_{ij} \, \frac{\partial v_i}{\partial x_j} = 2 \, \mu \, \epsilon_{ij} \, (\epsilon_{ij} - \omega_{ij}) = 2 \, \mu \, \epsilon_{ij} \, \epsilon_{ij} > 0 \tag{3.99}$$

since $\epsilon_{ij} \, \omega_{ij} = 0$,[10] and

$$\lambda \left(\frac{\partial v_j}{\partial x_j} \right)^2 > 0 \tag{3.100}$$

The positiveness indicates that these terms unidirectionally dissipates the mechanical energy into the internal energy at a rate, while shear deforming (or straining) the fluid at a rate against frictions [3].

---

10 $\epsilon_{ij} \, \omega_{ij} = \epsilon_{ji} \, \omega_{ji} = -\epsilon_{ij} \, \omega_{ij}$.

3.6.4 Total Energy Equation in Conservative Form

It is important to note that the net inflow rate of surface force work is done on the fluid particle of a unit volume, the term ⑤ as expressed in divergence form, only transports or redistributes the kinetic energy without creating or losing it. To better explain this transport term, we multiply the continuity equation by the specific total energy e_t and add it to the total energy equation. Then, Eq. (3.92) can be cast into a conservative form[11]

$$\frac{\partial(\rho\, e_t)}{\partial t} + \nabla \cdot (\rho\, e_t \vec{v}) = \nabla \cdot (\vec{v}\, [\sigma]) + \dot{w}_b + \nabla \cdot (k\, \nabla T) \tag{3.101}$$

If the flow is assumed steady and $\dot{w}_b$ is not of importance, it can be integrated over a control volume:

$$\int_{CV} \nabla \cdot (\rho\, e_t \vec{v})\, dV = \int_{CV} \nabla \cdot (\vec{v}\, [\sigma] + k\, \nabla T)\, dV \tag{3.102}$$

If we apply the divergence theorem and an adiabatic condition is assumed, it reads

$$\int_{CS} (\rho\, e_t \vec{v}) \cdot \vec{n}\, dA = \int_{CS} (\vec{v}\, [\sigma]) \cdot \vec{n}\, dA$$

$$= \int_{CS} \left(-p\, \vec{v} + \vec{v}\, [\tau] + \underbrace{(\lambda \nabla \cdot \vec{v})\, \vec{v}}_{=0} \right) \cdot \vec{n}\, dA \tag{3.103}$$

where the terms related to viscous stresses are zero at the control surfaces because $\vec{v}\, [\tau] = 0$ and the volumetric dilatation rate term is usually negligible.

If we put all the terms into one surface integral, it shows

$$\int_{CS} (\rho\, e_t + p)\, \vec{v} \cdot \vec{n}\, dA = 0 \tag{3.104}$$

Equation (3.104) indicates that the scalar terms in the bracket, called *transport terms*, are the divergence fluxes that change their values from one point to the other (or *redistribute* the values), while obeying the conservation principle of their sum; the total cannot be created or destroyed if there is no net influx of these terms at the boundaries.

In ducts of gradually varying area, for example the viscous stress-related terms are zero at the boundaries. Along the streamline, Eq. (3.104) can thus be written as follows:

$$e + \frac{p}{\rho} + \frac{v^2}{2} = h + \frac{v^2}{2} = h_t = \text{constant} \tag{3.105}$$

where h is the enthalpy. This shows that the stagnation enthalpy h_t is conserved along the streamline. Note also that this equation is valid for inviscid or viscous flows, as long as the duct walls are adiabatically treated.

In ideal gases, Eq. (3.105) also shows

$$c_p T + \frac{v^2}{2} = \frac{\gamma}{\gamma - 1} \frac{p}{\rho} + \frac{v^2}{2} = \text{constant} \tag{3.106}$$

which implies that if the fluid is expanded, it accelerates as its density, pressure, and temperature are decreased. If the flow is compressed, the opposite will occur. This equation is called the Bernoulli equation for compressible flows.[12]

---

11 This equation is useful when we deal with compressible flow physics, for instance, sound waves, linear acoustics, duct flows.

12 More discussion continues in Chapter 8 *Compressible Flows*.

3.6.5 Total Rate of Change of Pressure[†]

For ideal gases, the thermal energy equation Eq. (3.94) can be expressed with mass conservation and $\rho\, e = c_v\, \rho\, T = p/(\gamma - 1)$ as

$$\frac{Dp}{Dt} = \frac{\partial p}{\partial t} + \vec{v} \cdot \nabla\, p$$

$$= -\gamma\, \left\{ p\, \nabla \cdot \vec{v} \right\} + (\gamma - 1) \left\{ [\tau] \cdot [\nabla\, \vec{v}\,] + \lambda\, (\nabla \cdot \vec{v}\,)^2 + \nabla \cdot (k\, \nabla T) \right\} \tag{3.107}$$

where $\gamma = c_p/c_v$ is the heat-specific ratio.

Equation (3.107) shows that the total rate of change of pressure can be changed by the terms on the right-hand side. Among those, the first term is related to elastic potential energy stored in compressed or expanded fluid, e.g. sound wave generation through a reversible process. Meanwhile, the last three terms are responsible for pressure changes caused by viscous energy dissipation and heat conduction in irreversible processes (e.g. turbulent flow noise, combustion noise). In turbulent shear flows, unsteady fluctuations of turbulent eddies will produce fluctuations of pressure with ②$_B$, generating aerodynamic noise, e.g. jet noise, turbulent pipe flow noise. It is to be noted that this total rate of change of pressure must be differentiated from that associated with the local and convective rates of change of kinetic energy of the fluid per unit volume, while the fluid is accelerating or decelerating in incompressible flows.[13]

3.7 A Concentric Tube with Two Rotating Surfaces

Let us consider a concentric tube with the inner and outer tubes rotating at two different speeds; the gap between the two tubes is also an important parameter as well. A general solution of the Navier–Stokes equations illustrates various important flow physics related to angular straining and rotation of the fluid in the gap between two circular surfaces (one is convex and the other is concave) – how their rates are portioned or distributed in the gap, and when or how the total energy of the fluid is dissipated into heat or conserved throughout the gap. Note that the general solution is valid for two extreme cases – (i) an infinite domain with the inner tube rotating at a rate (e.g. a free vortex) and (ii) a finite domain with only the outer tube rotating at a rate (without inner tube) (e.g. a forced vortex).

3.7.1 A General Solution of the Navier–Stokes Equations

Let us consider a fluid inside a concentric rotating tube, where the tubes rotate with two different angular speeds; the inner and outer tubes rotate at Ω_i and Ω_0, respectively. The Navier–Stokes equations in polar coordinates can be reduced to

$$\frac{dp}{dr} = \rho \frac{v^2}{r} \tag{3.108}$$

$$\frac{dv}{dt} = v \left(\frac{\partial^2 v}{\partial r^2} + \frac{1}{r} \frac{\partial v}{\partial r} - \frac{v}{r^2} \right) \tag{3.109}$$

---

† For advanced studies.
13 See Chapter 2 *Macroscopic Balance of Mass, Momentum, and Energy.*

Figure 3.17 Outer concave surface in clockwise rotation

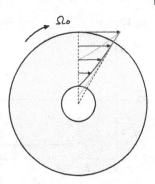

where v is the circumferential component of velocity. The first equation shows that the radial variation of pressure simply supplies the force necessary to keep the fluid elements moving in a circular path. The second equation represents the relation of the rate of increase of angular momentum of a cylindrical shell of fluid under the action of torques exerted by friction at its inner and outer faces.[14]

Steady motions with circular streamlines must be maintained by the motion of rigid boundaries at r_i and r_0 which rotate steadily with angular velocities Ω_i and Ω_0. The solution with $dv/dt \to 0$ can be expressed as follows:

$$v(r) = \frac{1}{r}\left(\frac{\Omega_i - \Omega_0}{r_i^{-2} - r_0^{-2}}\right) + r\left(\frac{\Omega_i r_i^2 - \Omega_0 r_0^2}{r_i^2 - r_0^2}\right) \tag{3.110}$$

where Ω_i and Ω_0 represent the angular speeds of cylindrical surfaces at $r = r_i$ and r_0, respectively. The viscosity does not appear in the solution because the net frictional torque on every cylindrical shell of fluid is zero; in this respect, steady flow with circular streamlines is the circular analogue of flow between parallel rigid planes which are in relative sliding motion.

3.7.2 Rotation of the Outer Tube

For $\Omega_0 < 0$ and $\Omega_i = 0$ (Figure 3.17), the velocity distribution is given by

$$v(r) = -\frac{\Omega_0\, r_i}{1 - 1/\beta^2}\left(\frac{r}{r_i} - \frac{r_i}{r}\right) \tag{3.111}$$

and its nondimensional form reads

$$v^*(r^*) = \frac{1}{1 - 1/\beta^2}\left(r^* - \frac{1}{r^*}\right) \tag{3.112}$$

where $v^* = v/(-\Omega_0\, r_i)$, $\beta = r_0/r_i$, and $r^* = r/r_i$.

With the following expressions:

$$\frac{dv^*}{dr^*} = \frac{1}{1 - 1/\beta^2}\left(1 + \frac{1}{r^{*2}}\right) \tag{3.113}$$

$$\frac{v^*}{r^*} = \frac{1}{1 - 1/\beta^2}\left(1 - \frac{1}{r^{*2}}\right) \tag{3.114}$$

---

14 Batchelor [4].

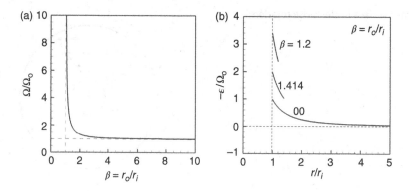

Figure 3.18 Rates of rotation and angular straining (nondimensionalized by Ω_0 (< 0))

the nondimensionalized rates of rotation and angular strain can be written as follows:

$$\Omega^* = -\frac{1}{2}\left(\frac{dv^*}{dr^*} + \frac{v^*}{r^*}\right) = -\frac{1}{1 - 1/\beta^2} \tag{3.115}$$

$$\epsilon^* = \frac{1}{2}\left(\frac{dv^*}{dr^*} - \frac{v^*}{r^*}\right) = \frac{1}{1 - 1/\beta^2}\frac{1}{r^{*2}} = -\frac{\Omega^*}{r^{*2}} \tag{3.116}$$

where $\Omega^* = \Omega/|\Omega_0|$, $\epsilon^* = \epsilon/|\Omega_0|$, and $r^* > 1$.

It is interesting to note that a presence of the inner convex surface enforces viscous shears to the fluid, even with a very small r_i. In addition, the gap size β greatly affects the rates of rotation and angular strain (Figure 3.18a). As β becomes less than $\sqrt{2}$, for instance the rotational rate drastically increases from $2\,\Omega_0$ but is uniform throughout the fluid. However, the magnitude of the angular strain rate depends on the radius in the gap. The angular strain rates at the inner and outer surfaces, respectively, read

$$\epsilon^*(r^* = 1) = \epsilon_i = \Omega^* \tag{3.117}$$

$$\epsilon^*(r^* = \beta) = \epsilon_0 = \Omega^*/\beta^2 \tag{3.118}$$

which show, regardless of β, that the angular strain rate at the inner surface is always greater than that at the outer surface by Ω_0, i.e. $\epsilon_i^* - \epsilon_0^* = 1$. Even when the gap size approaches infinity, i.e. $\beta \to \infty$, the angular straining rate starts to grow from 0 as $r^* < 4$ and becomes Ω_0 at $r^* = 1$. If the gap size becomes smaller, ϵ^* varies from 1 to 2 for $\beta = \sqrt{2}$, or from 2.273 to 3.273 for $\beta = 1.2$ (Figure 3.18b).

If there is no inner surface, i.e. $r_i \to 0$, Eqs. (3.115) and (3.116) show that the fluid shear driven by the outer concave surface is rotated at a rate Ω_0, as if it is a solid body. As a result, $\Omega^* \to 1$ and $\epsilon^* \to 0$, everywhere. This is a special case solution in which there is no shear stress present in the fluid; the entire fluid simply rotates at a constant rate as if it is a frozen fluid. However, the total mechanical energy is not conserved throughout the field, since the energy has been dissipated by viscous stresses exerted on the fluid in transient period. The viscous stresses have rotated and angularly strained the fluid at a rate at the interface between the fluid in rotation and the fluid at rest.

3.7.3 Rotation of the Inner Tube

For $\Omega_i < 0$ and $\Omega_0 = 0$ (Figure 3.19), the velocity distribution is given by

$$v(r) = -\frac{\Omega_i\,r_i}{(\beta - 1/\beta)}\left(\frac{r_0}{r} - \frac{r}{r_0}\right) \tag{3.119}$$

Figure 3.19 Inner convex surface in clockwise rotation

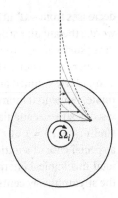

and its nondimensional form reads

$$v^*(r^*) = \frac{1}{(\beta - 1/\beta)}\left(\frac{\beta}{r^*} - \frac{r^*}{\beta}\right) \tag{3.120}$$

where $v^* = v/(-\Omega_i\, r_i)$, $\beta = r_0/r_i$, and $r^* = r/r_i$.

With the following expressions:

$$\frac{dv^*}{dr^*} = \frac{1}{(\beta - 1/\beta)}\left(-\frac{\beta}{r^{*2}} - \frac{1}{\beta}\right) \tag{3.121}$$

$$\frac{v}{r} = \frac{\Omega_i\, r_i}{(\beta - 1/\beta)}\left(\frac{\beta}{r^{*2}} - \frac{1}{\beta}\right) \tag{3.122}$$

the rates of rotation and angular strain are written in normalized form as follows:

$$\Omega^* = -\frac{1}{2}\left(\frac{dv^*}{dr^*} + \frac{v^*}{r^*}\right) = \frac{1}{\beta^2 - 1} \tag{3.123}$$

$$\epsilon^* = \frac{1}{2}\left(\frac{dv^*}{dr^*} - \frac{v^*}{r^*}\right) = -\left(\frac{\beta^2}{\beta^2 - 1}\right)\frac{1}{r^{*2}} = -\beta^2\frac{\Omega^*}{r^{*2}} \tag{3.124}$$

where $\Omega^* = \Omega/|\Omega_i|$, $\epsilon^* = \epsilon/|\Omega_i|$, and $r^* > 1$.

A presence of the outer concave surface enforces a nonslip condition at the outer surface so that the fluid is forced to rotate at a rate in the opposite direction. As shown in Figure 3.20a, the rotational rate gradually grows as β decreases and starts to significantly increase from $-\Omega_i$ when β becomes smaller than $\sqrt{2}$. Note again that the rotational rate is uniform within the gap. The angular straining rate also similarly behaves as in the previous case. There exists a unique difference of angular strain rate between the two curved surfaces. Regardless of β, the angular strain rate

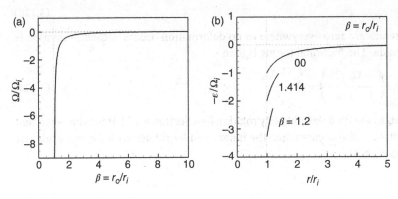

Figure 3.20 Rates of rotation and angular straining (nondimensionalized by Ω_i (< 0))

decreases from $-\Omega^*$ at the inner surface and the difference is always $\Omega^* = 1$ in magnitude. In other words, the angular strain rate at the outer surface is always Ω_i lower in magnitude than that at the inner surface, regardless of the gap size.

If there is no outer surface (i.e. $r_0 \to \infty$ or $\beta \to \infty$), the fluid driven by the inner convex surface is angularly strained at a rate but no rotation occurs, i.e. $\Omega^* = 0$. As soon as the inner tube starts to rotate, tangential momentum is being transported via frictional intermolecular interactions, while viscous stresses angularly strain the fluid at a rate but cannot rotate it. If we look at the angular strain rate, $\epsilon^* = 1$ at $r^* = 1$, then gradually decays to zero at the far field (Figure 3.20b). This is a special case of solution where the fluid is in circular motion like a free-vortex. In this case, the total mechanical energy is conserved throughout the field and pressure is fully recovered across the streamline by centrifugal forces since there is no net viscous force acting on the fluid.

[**Notes**] In summary, when the outer tube rotates, the angular strain rate at the inner convex surface is the same as the rotational rate of the fluid inside. However, at the outer concave surface, it is $1/\beta^2$ times the rotational rate. Hence, $\epsilon_i^* = -\Omega^*$ but $\epsilon_0^* = -\Omega^*/\beta^2$, where $\Omega^* = \Omega/\Omega_0$. On the contrary, when the inner tube rotates, the angular strain rate at the outer concave surface is the same as the rotational rate of the fluid inside. However, at the inner convex surface, it is β^2 times the rotational rate. Hence, $\epsilon_i^* = -\Omega^* \beta^2$ but $\epsilon_0^* = -\Omega^*$, where $\Omega^* = \Omega/\Omega_i$. With reference to the Couette flow ($\epsilon^* = -\Omega^*$), the flow driven by the outer concave surface shows unfavorable angular deformations near the concave surface, whereas the flow driven by the inner convex surface shows favorable angular deformations near the convex surface.

3.7.4 Frictional Torque with Two Rotating Surfaces

The frictional torque exerted on a cylindrical surface of radius r is written as follows:

$$\tau \cdot A_{wet} \cdot r = \mu\,(2\,\epsilon) \cdot (2\pi\,r)(1) \cdot r = -4\pi\mu \left(\frac{\Omega_i - \Omega_0}{r_i^{-2} - r_0^{-2}} \right) = -4\pi\mu \left(\frac{(\Omega_i - \Omega_0)\,r_i^2}{1 - 1/\beta^2} \right) \quad (3.125)$$

where $\epsilon = (dv/dr - v/r)/2$ is the angular strain rate of the fluid in polar coordinates, and the torque is independent of r, as expected. It is interesting to note that the viscous torque is a function of $\Omega_i - \Omega_0$, $\beta = r_0/r_i$, μ, and r_i, not r.

3.7.4.1 Forced Vortex

The velocity distributions for two special cases can be obtained. If we let $\Omega_i = 0$ and $r_i = 0$ (so that $\beta \to \infty$), then the solution represents a fluid in solid body rotation:

$$v = \Omega_0\, r \quad (3.126)$$

where the tangential stresses are zero everywhere as no deformation occurs, although rotation is uniformly distributed inside. The frictional torque is then

$$\tau \cdot A_{wet} \cdot r = -4\pi\mu \left(\frac{-\Omega_0\, r_i^2}{1 - 1/\beta^2} \right) = 0 \quad (3.127)$$

This solution corresponds to the fluid in solid body rotation (see Section 4.1.2). It is to be noted that the flow of solid body rotation also occurs when the inner cylinder rotates with the same angular speed as the outer cylinder, i.e. $\Omega_0 = \Omega_i$.

3.7.4.2 Free Vortex

The other extreme case is the flow in an infinite domain outside a single rotating cylinder. Letting $\Omega_0 = 0$ and $r_0 \to \infty$ (so that $\beta \to \infty$), the velocity of this irrotational, free-vortex flow reads

$$v = \frac{r_i^2 \, \Omega_i}{r} = \frac{\Gamma_i}{2\pi r} \tag{3.128}$$

where $\Gamma_i = 2\pi \, r_i^2 \, \Omega_i$. The frictional torque is then

$$\tau \cdot A_{\text{wet}} \cdot r = -4\pi\mu \left(\frac{\Omega_i \, r_i^2}{1 - 1/\beta^2} \right) = -4\pi\mu \left(\Omega_i \, r_i^2 \right) \tag{3.129}$$

and the power supplied by the inner rotating cylinder is

$$\tau \cdot A_{\text{wet}} \cdot r \cdot \Omega_i = -4\pi\mu \, \Omega_i^2 \, r_i^2 \tag{3.130}$$

In this case, the power supplied by the inner rotating cylinder is dissipated at a rate to create a free vortex in the infinite domain.

References

1 Lumley, J.L., "Deformation of Continuous Media," *Illustrated Experiments in Fluid Mechanics*, National Committee for Fluid Mechanics Films, 1980.
2 Schlichting, H., *Boundary Layer Theory*, 4th ed. New York: McGraw-Hill, 1979.
3 Kundu, P.K., *Fluid Mechanics*, San Diego, CA: Academic Press, 1990.
4 Batchelor, G.K., *An Introduction to Fluid Dynamics*, Cambridge University Press, 1967.
5 Shercliff, J.A., "Magetohydrodynamics," *Illustrated Experiments in Fluid Mechanics*, National Committee for Fluid Mechanics Films, 1980.
6 NASA Goddard, *Insights on Comet Tails Are Blowing in the Solar Wind*, YouTube., 2020.

Problems

3.1 A vector $\vec{\phi}$ that satisfies $\nabla \cdot \vec{\phi} = 0$ is solenoidal.
   a) What does the solenoidal vector $\vec{\phi} = \vec{v}$ (velocity vector) that satisfies $\nabla \cdot \vec{v} = 0$ physically mean?
   b) What does the solenoidal vector $\vec{\phi} = \vec{\omega} = \nabla \times \vec{v}$ (vorticity vector) that satisfies $\vec{\omega} = 0$ physically mean?

3.2 In Couette flow, rotation rate is constant and particles released at the same time have all rotated the same amount, regardless of velocity or distance traveled (see figure).

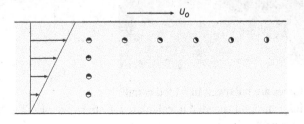

a) Sketch streaklines of particles as in the figure.
b) Sketch the field of stretching rate of a fluid element and discuss the principal axes of strain rate.

3.3 In Poiseuille flow, the velocity profile is parabolic (see figure) and rotation rates are not constant.

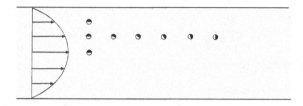

a) Sketch the streaklines of particles, as in Problem 3.2.
b) Sketch the field of stretching rate of a fluid element and discuss the principal axes of strain rate.

3.4 In the Cartesian coordinates, a strain rate is defined at a point as

$$\epsilon_{ij} = \frac{1}{2} \left(\frac{\partial v_i}{\partial x_j} + \frac{\partial v_j}{\partial x_i} \right)$$

which implicates that the strain rate and viscous stress, $\tau_{ij} = 2\mu \, \epsilon_{ij}$, depend on the given velocity field.
a) Physically interpret this statement with the Couette flow inclined at an angle θ.
b) For the horizontal Couette flow, interpret this statement with a rectangular element of fluid rotated at an angle θ.
c) Discuss the strain rates and viscous stresses with respect to the shear-plane and principal axes of the strain rate.

3.5 Mosquitos use cibarial and pharyngeal pumps to suck blood into their body through food canals (e.g. length: 1600 μm, diameter: 30 μm).

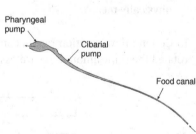

a) What forces are balanced in a food canal?
b) What information is needed to calculate the stress vector $\vec{t}\,(\vec{x}, t)$ acting on the surface of the food canal?

3.6 Sketch the viscous stress vectors acting on each fluid element in the Couette flow and show how they deform over dt.

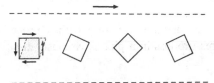

3.7 An isosceles triangle in Couette flow is depicted at time t. Note that the acute angles are 45°. Sketch the stress vectors on the three surfaces of the isosceles triangle and show the shear-deformed triangle after Δt.

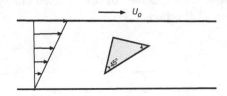

3.8 In Couette flow, the strain rate tensor ϵ_{ij} has only one component $\epsilon_{xy} = du/dy$ but the viscous stress vectors acting on a circular element are nonisotropic, as sketched below.

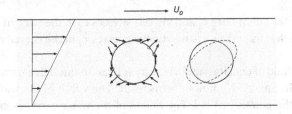

a) Construct the viscous stress vectors acting on a circular fluid element.
b) Explain why the magnitudes of the viscous stress vectors exerted on the circular element are the same.
c) The circular element deforms to an elliptical element over a finite time interval dt. Sketch the viscous stress vectors exerted on the surface of the elliptical element.
d) Comparing the results of (a) and (c), discuss the nature of the viscous stress vectors.

3.9 A stress vector acting on the surface of a unit area specified with an outward unit normal vector $\vec{n}$ is expressed as $\vec{t}\,(\vec{n}) = \vec{n}\,[\sigma]$, where the second-order stress tensor $[\sigma]$ is determined at a point $\vec{x}$ with pressure p, volumetric dilatational rate $\nabla \cdot \vec{v}$, and viscous stress tensor $[\tau]$.
a) What is the role of $\vec{n}$ to a scalar quantity (e.g. pressure or volumetric dilatational rate)?
b) What is the role of $\vec{n}$ to a tensor quantity (e.g. viscous stress tensor)?

3.10 A net surface force vector acting on a unit volume of fluid is expressed as $\nabla \cdot [\sigma]$.
a) What is the role of $\nabla \cdot$ to a tensor quantity (e.g. stress tensor)?
b) How is this operator related to the Taylor series expansion in control volume analysis?

3.11 In conservation of energy, work is done on the fluid at a rate.
a) Explain what $(\vec{v}\,[\sigma]) \cdot \vec{n}$ physically means.
b) What is the role of $\nabla \cdot$ to a vectorial quantity (e.g. $\vec{v}\,[\sigma]$)?

3.12 Using the energy conservation equation with a proper control volume, show that in Poiseuille flow, pressure loss in a pipe is viscously dissipated and increases the internal energy at a rate.

3.13 Show that in an internal flow system, the energy dissipation rate over a control volume equals the total pressure loss times the volumetric flow rate. Use the energy conservation equation with a proper control volume.

3.14 A secondary flow is often observed in open curved channels, where the fluid is shear strained at a rate in all directions.

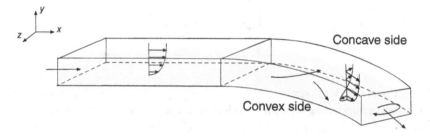

a) Specify the strain rates ϵ_{ij} and viscous stresses τ_{ij} in the straight channel.
b) Specify the strain rates ϵ_{ij} and viscous stresses τ_{ij} in the curved channel.

3.15 A viscous fluid of density ρ and viscosity μ rises up through a porous bed at a rate of q (per unit length) and slowly flows downstream. The width of the water channel is b and the length of the porous bed is l. The height of the water varies as sketched and levels at h_0 downstream.

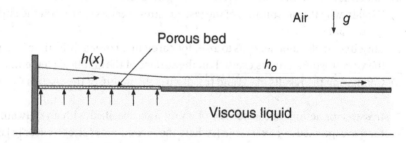

a) Express the flow rate $Q(x)$ on the porous bed.
b) Derive the equation of motion of the liquid on the porous bed and state the assumptions used.
c) Find the velocity $u(x, y)$.
d) Find the height of the liquid $h(x)$.

3.16 A viscosimeter consists of a large reservoir with a very slender tube (length: L and diameter: d) (see below). The kinematic viscosity of a liquid (density: ρ) is determined by measuring the flow rate Q. Assume that the flow is laminar and quasi-steady.

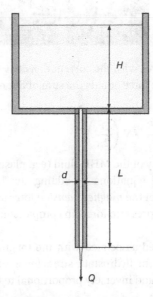

a) Derive the force balance equation in the tube.
b) Express the kinematic viscosity of the fluid in terms of the given parameters. (Hint: find $u(r)$ to obtain Q.)

3.17 Fish in water use forces in a variety of ways: blood circulation, propulsion, depth control, etc.

a) What parameters are important while balancing forces in the gills?
b) What parameters are important while balancing forces with the caudal fin?
c) What parameters are important while balancing forces with the bladder?

3.18 In magnetohydrodynamics (MHD), the Lorentz force acts as a volumetric body force to the fluid (e.g. solar winds) [5, 6].

a) Define the Lorentz force with the current density vector $\vec{J}$ and magnetic vector $\vec{B}$.

b) Show that the Lorentz force equals the sum of the magnetic tension force and magnetic pressure force:

$$\vec{J} \times \vec{B} = \frac{(\vec{B} \cdot \nabla)\vec{B}}{\mu} - \nabla \left(\frac{B^2}{2\mu} \right)$$

where μ is the viscosity of the MHD fluid (e.g. plasma).

c) Derive the governing equations by adding the Lorentz force to the Navier–Stokes equations and interpret the *magnetic tension force* and *magnetic pressure force*.

d) Discuss the magnetic pressure force, in comparison with the hydrostatic stress (or pressure) force.

e) Discuss the wave speed associated with the magnetic pressure force. (Hint: the wave speed associated with the hydrostatic stress force is the speed of sound, which is proportional to the pressure and inversely proportional to the density.)

4

Curved Motions

4.1 Centrifugal Force

Curved motion is a generic term but its implicated meaning goes beyond definition. The most fundamental physics of curved motion is the centrifugal effect, which is coupled with viscosity to make physics more complex and diverse. Curved motion is often associated with the flows of rotating machineries (e.g. fans, compressors, and turbines) and vortex flows in nature (e.g. tornados, low-pressure cyclones, and fire whirl). It is also related to the complex dynamics of wall-bounded shear flows on curved walls (e.g. Coanda effect, origin of lift).

4.1.1 Radial Force Balance

When a fluid undergoes curved motion, it is brought into a state of radial compression by two forces in action and reaction: the centrifugal force of the fluid (i.e. inertia force) versus the force exerted in the centripetal direction by reaction (Figure 4.1). The latter can be the reaction force from curved boundaries, or the fields under two different gravitational body forces (e.g. sink and surroundings). Therefore, the local pressure is determined by these two forces in action and reaction and also increases outward due to the mass accumulated in the radial direction.

For a fluid in curved motion, the equation of motion can be set in the radial direction as follows:

$$\rho \, (dn \, dA) \cdot a_n = -dp_n \cdot dA \qquad (4.1)$$

where $dm = \rho \, (dn \, dA)$, a_n is the centripetal acceleration, $dp_n = (\partial p / \partial n)dn$ is the pressure difference in the radial direction, and n denotes the local coordinates normal (outward) to the streamline.

The centripetal acceleration a_n equals the product of the tangential flow speed $v \, (= ds/dt)$ and the rate of turn of the flow direction $\dot{\theta} \, (= d\theta/dt)$

$$a_n = -v \cdot \dot{\theta} = -v \cdot \frac{ds/r_c}{dt} = -\frac{v^2}{r_c} \qquad (4.2)$$

where the negative sign indicates that the direction of action is in the centripetal direction, and r_c is the local radius of curvature of the streamline. The radial momentum equation finally reads

$$\frac{\partial p}{\partial n} = \rho \, \frac{v^2}{r_c} \qquad (4.3)$$

Equation (4.3) holds true regardless of whether the fluid in curved motion is rotated or angularly strained at a rate (Figure 4.2). It is to be noted that in fluids, curved motion does not necessarily mean rotation. It can be created by angular straining or by rotation mixed with angular straining. Similarly, linear motion does not necessarily mean irrotation; rotation can be mixed with angular

Introduction to Fluid Dynamics: Understanding Fundamental Physics, First Edition. Young J. Moon.
© 2022 John Wiley & Sons, Inc. Published 2022 by John Wiley & Sons, Inc.
Companion website: www.wiley.com/go/Moon/IntroductiontoFluidDynamics

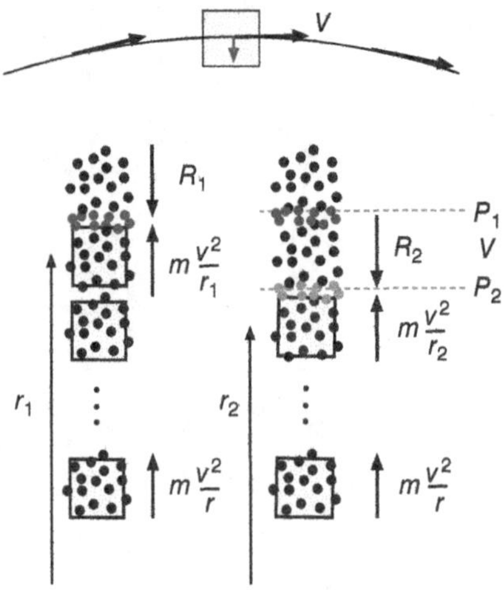

Figure 4.1 Centrifugal force and pressure on a curved streamline

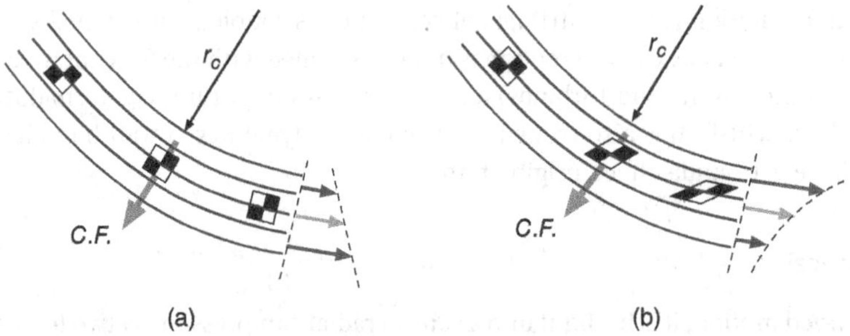

Figure 4.2 Two curved streams with rotation (a) and angular straining (b)

straining to make the motion linear (e.g. boundary layer over a flat surface). In the following sections, we describe the two most basic curved flows (i.e. forced-vortex and free-vortex), comparing the intrinsic differences between the two [1].

4.1.2 Solid Body Rotation

4.1.2.1 Shear-Driven Circular Flow (Forced-Vortex)

A fluid in solid body rotation can be obtained by rotating a cylindrical container filled with water (Figure 4.3). As soon as the container rotates at an angular speed of Ω_0 (< 0), viscous forces rotate and angularly strain the water on the concave surface of the container at very fast rates. As time proceeds, a rim of shear-straining water grows in the radial direction by molecular diffusion, while the rates of rotation and angular straining decay to Ω_0 and zero, respectively.[1]

At steady state, the entire water rotates at Ω_0 as a solid body does. Although the water has been driven by viscous forces, there is no viscous stress in the water since the rate of straining is zero. In this case, the total mechanical energy (or Bernoulli's function) is not conserved across the streamlines because the water in the container is frozen. At this stage, all the water particles in

---

1 If there is a circular surface inside the container, the angular straining rate will never become zero, no matter how small the radius of the inner circular surface is. See Section 3.7 *A concentric tube with two rotating surfaces.*

Figure 4.3 Water in solid body rotation in a cylindrical container rotating at Ω_0

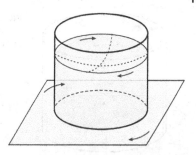

the container rotate with the same angular speed Ω_0 and the tangential velocity of this rotational field can be written as follows:

$$v(r) = -\Omega_0\, r \tag{4.4}$$

where the local radius r varies from 0 to R (radius of the container).

For the angular speed Ω_0, the centripetal acceleration of the water linearly increases with r (same as the velocity):

$$a_n = -\frac{v^2}{r} = -\Omega_0^2\, r \tag{4.5}$$

and the pressure distribution can be obtained by integrating the radial momentum equation as follows:

$$p(r) = p_0 + \frac{\rho\,\Omega_0^2}{2}\, r^2 \tag{4.6}$$

where p_0 is the pressure at the center of the bottom of the container, and it is found that the pressure of the water in solid body rotation increases as r^2.

It is also interesting to note that the pressure increase in the radial direction will change pressure in the vertical direction as well, since the water is bounded in the container and the gravitational body force acts in the vertical direction. Thus, the height of the free surface of the water can be obtained by generally expressing the pressure field as follows:

$$p(r,z) = \rho g\,(h(r) - z) \tag{4.7}$$

Now, setting $z = 0$ and equating Eq. (4.6) with (4.7) yield

$$h(r) = h_0 + \frac{\Omega_0^2}{2g}\, r^2 \tag{4.8}$$

where $h_0 = p_0/\rho g$ is the height at the center. It is shown that the free surface of the water in solid body rotation in the cylindrical container is parabolic, i.e. $h(r) \sim r^2$.

4.1.3 Swirl with a Sink

4.1.3.1 Inertia-Driven Circular Flow (Free-Vortex)

Let us suppose we have a cylindrical container with a drainage hole at the bottom, and the water is injected through a tangential slit at the outer periphery (Figure 4.4). Due to sink driven by gravitational body force,[2] the injected water draws a circular path and swirls in the container, while increasing the tangential speed as proportional to $1/r$ to preserve the angular momentum. In this

---

2 The low-pressure cyclones, fire whirl, etc., are similar swirls driven by buoyant uprising sink. More discussion continues in Chapter 5 *Vortex Dynamics*.

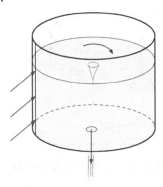

Figure 4.4 Irrotational swirl and vortex in a cylindrical container with a drainage hole at the bottom; water being injected through a tangential slit at the outer periphery

case, a radial force balance is set between the centrifugal force of the water swirling in the container and the centripetal force associated with the difference between the two gravitational body forces per unit area in the sink and container. As the water comes close to the center, it is forced to rotate at a rate by viscous torques and forms a vortex tube (or filament), which finally exits through the bottom hole.

In the container, excluding a region near the center, water particles are angularly strained at a rate. As they approach the center, the angular straining rate increases and finally equals the rate of rotation of the vortex at the center.[3] In this irrotational flow, viscous stresses are present everywhere, but rotations are not created. Thus, the total mechanical energy (or Bernoulli's function) is conserved throughout the field. Note that the viscous torques only act to produce rotation when the water particles enter the mixing layer where the irrotational swirling flow and the vortex at the center are interfaced. How the angular straining and rotation are mixed at this interfacial layer will be discussed in the next section.

The tangential speed of the irrotational swirling flow can be written as follows:

$$v(r) = -\frac{\Gamma}{2\pi r} \tag{4.9}$$

where v is inversely proportional to r due to the conservation of angular momentum. The tangential speed is also proportional to circulation Γ (< 0), which represents the strength of the circulatory flow around a vortex at the center of the container. It is interesting to note that Γ depends on various parameters of the system, e.g. viscosity and density of the fluid, radial sizes of the container and drain hole, height of the water in the container, and the volumetric flow rate of the injected water at the slit.

For a given Γ, the centripetal acceleration of the water particle decreases with r to the power of 3

$$a_n = -\frac{v^2}{r} = -\frac{\Gamma^2}{4\pi r^3} \tag{4.10}$$

and the pressure distribution can be obtained by integrating the radial momentum equation:

$$p(r) = p_0 - \frac{\rho\,\Gamma^2}{8\pi r^2} \tag{4.11}$$

where p_0 is the pressure at the bottom of the water, as $r \to \infty$.

The pressure increase in the radial direction also affects the pressure of the water in the container in the vertical direction and the height of the free surface of the water changes as follows:

$$h(r) = h_0 - \frac{\Gamma^2}{8\pi^2 g\,r^2} \tag{4.12}$$

---

3 More details can be found in Section 3.7 *A concentric tube with two rotating surfaces.*

where $h_0 = p_0/\rho g$ is the height of the water at $r \to \infty$. In this case, the height decreases as proportional to $1/r^2$ because the tangential speed decreases as proportional to $1/r$.

4.2 Spiral Vortex Flow

4.2.1 Viscous Torque

The swirling water in the container increases the tangential speed as proportional to $1/r$ to conserve the angular momentum. At some point near the center, this irrotational flow is forced to change the flow direction, reducing the tangential speed but increasing the downfall speed. This rapid change in flow direction is only possible by exerting a viscous torque on the water particles (i.e. creation of rotation) in the region close to the sink (Figure 4.5) [2]. In this case, the sink driven by gravity acts as a geometrical singular point, enforcing the tangential speed of water to be zero at the geometric center.

It is also interesting to note that the interface between the outer irrotational swirl and the inner vortex at the center is a layer in which angular straining is mixed with rotation by action of viscous forces. As the water particles enter the interface, the rotation rate starts to increase and becomes the rotation rate of the vortex at the center while the angular straining rate (negative) in the potential swirl starts to diminish with creation of new angular straining rate (positive). The angular straining rate will completely disappear as the water particles enter into the vortex. This mixing of angular straining with rotation dissipates the kinetic energy at a rate via frictional intermolecular interactions. The thickness of the mixing layer is inversely proportional to the local Reynolds number, $Re = \Gamma/\nu$, where Γ represents the circulation of the vortex, and ν $(= \mu/\rho)$ is the kinematic viscosity.

4.2.2 Vortex Strength (Circulation)

This drainage system is geometrically simple but produces a true three-dimensional spiral vortex, consisting of an outer irrotational swirl and an inner vortex at the center. If we reduce the inflow rate of the injected water, the vortex strength will decrease with a more pronounced viscous effect. The reduced tangential speed angularly strains the water at a lower rate and as a consequence, the dimple on the free surface of the water tends to weaken with reduced centrifugal effect. Conversely, the dimple will become deeper with increased inflow rate of the injected water. Furthermore, the diameters of the cylindrical container and drain hole as well as density and viscosity of the fluid will affect the strength of the vortex.

Figure 4.5 A spiral vortex flow in the drainage, consisting of an outer irrotational swirl and an inner vortex

The strength of the circulatory flow of the drainage system can be best described by a macroscopic quantity called circulation Γ. The circulation depends on the size and rotational rate of the vortex; here, the size means the total area filled with vortices (see Section 4.3 *Vorticity and circulation*). This definition of vortex strength, circulation, is self-evident for a concentrated vortex, whose tangential velocity is expressed as $v_\theta(r) = |\Omega_0|\, r$, so that the fastest tangential speed of the vortex is determined by the radius of the vortex R and the rotational rate Ω_0.

It is to be noted that the definition of circulation is equally true for linearly distributed vortex systems (e.g. boundary layer over a curved surface) in which rotation is mixed with angular straining but the velocities of the rotating fluid elements should be added up to represent a flow speed at the edge of the boundary layer. This implies that the circulation depends on the area in which the vortices are distributed. At the same time, the vortices in the boundary layer are determined by the incoming flow speed, viscosity, density, and geometrical shape of the solid body (e.g. curvature of the solid surface).

4.2.3 Rankine Vortex Model

Spiral vortex flow has a three-dimensional structure but can be broadly divided into two, in a two-dimensional sense: an outer irrotational flow (free vortex) and an inner rotational core (forced vortex). In outer irrotational flow, fluid particles are angularly strained at a rate, conserving their angular momentums. Meanwhile, near the center, they are forced to rotate at a rate as a solid body, preserving the total angular momentum.

A two-dimensional vortex model consisted of a forced vortex (inner) and a free vortex (outer) is called Rankine vortex (Figure 4.6). This vortex is a simplified model of a real vortex which cannot simply be divided into a forced vortex and a free vortex. In reality, there is a transitional interface where rotation and angular straining are mixed. The fluid in this layer is largely shear strained at a rate while dissipating the mechanical energy, and the thickness of the layer is determined by the Reynolds number defined by Γ/v, where Γ is the circulation of the vortex tube and v is the kinematic viscosity of the fluid. The radial pressure distribution of the Rankine vortex can easily be obtained by integrating the radial momentum equation with the velocity profiles given by two vortex models, Eqs. (4.4) and (4.9).

Example 4.1 Show that a fluid element is angularly strained at a rate in a free vortex while it rotates at a rate like a solid body in the forced vortex.

Free Vortex In a free vortex, the velocity field is defined as follows:

$$v = -\frac{\Gamma_0}{2\pi r} \tag{4.13}$$

where Γ_0 (< 0) is the circulation. For a fluid element with two sides ds and dn, the turning rate of dn can be obtained as follows:

$$\Omega_n = -\frac{\partial v}{\partial n} = -\frac{\Gamma_0}{2\pi r^2} = \frac{v}{r} \tag{4.14}$$

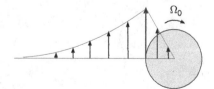

Figure 4.6 Rankine vortex; a forced vortex (inner) and a free vortex (outer)

whereas the turning rate of ds is, by definition, expressed as follows:

$$\Omega_s = -|\dot{\theta}| = -\frac{v}{r}$$ (4.15)

Thus, the fluid element does not rotate in average. It is only angularly strained at a rate.

Forced Vortex In a forced vortex, the fluid element rotates as if it is a solid body

$$v = -r\,\Omega_0$$ (4.16)

where Ω_0 (< 0) is a constant. Then, it reads

$$\Omega_n = -\frac{\partial v}{\partial n} = \Omega_0 = -\frac{v}{r}$$ (4.17)

The two sides ds and dr turn at the same rate of $-v/r$, meaning that in average, the fluid element does rotate as if it is a solid body with the angular speed of Ω_0.

4.2.4 Biot–Savart Law

A pair of spiral vortexes shed from the side-edges of the flaps are often visible when they are deployed from the wings of an aircraft in landing (Figure 4.7). Due to pressure difference between the upper and lower surfaces of the flap and the fact that the aircraft is moving forward at a certain speed, the outer irrotational swirly flow (free vortex) is continuously rolled into a spiral vortex flow (forced vortex). In fact, the strength of the vortex is determined by the speed of the aircraft, geometrical scales and shapes of the wing and its edges, kinematic viscosity of the air, etc.

As pointed out, the outer free vortex and the inner forced vortex cannot be independent from each other because a complete spiral vortex flow system is formed by both. In regard to the conjunction of these two, the Biot–Savart law links the rotational (or vortical) flow field with the irrotational (or potential) flow field, based on the physical concepts that we view the vorticity field as an aggregate of discrete point vortices and that we can superimpose each potential field induced by every element of vorticity (as the potential field is linear).

The Biot–Savart law states that if the vorticity field is known, we can find the outer potential flow field in a unique way, though the same does not go for the other way around. With Helmholtz's decomposition [3],[4] any such vector field **v** can be considered to be generated by a pair of potentials:

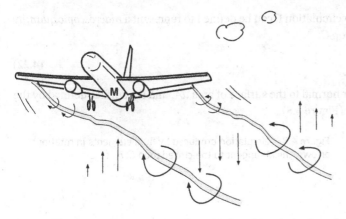

Figure 4.7 A pair of vortexes shed from the flap side-edges of an aircraft in landing

---

4 See Section 4.8.6 *Scalar and vector potentials.*

a scalar potential ϕ and a vector potential $\boldsymbol{\Psi}$

$$\mathbf{v} = \nabla\phi + \nabla \times \boldsymbol{\Psi} \tag{4.18}$$

A vector potential $\boldsymbol{\Psi}(r)$ is coupled with a vorticity field as follows:

$$\boldsymbol{\Psi}(r) = \frac{1}{4\pi} \int_V \frac{\nabla' \times \vec{v}}{|\mathbf{r} - \mathbf{r}'|}\, dV' \tag{4.19}$$

where $\boldsymbol{\Psi}(r)$ satisfies the following Poisson equation:

$$\nabla^2 \boldsymbol{\Psi}(r) = -\nabla \times \vec{v} \tag{4.20}$$

We can finally obtain the potential velocity field induced by rotational flow, $\vec{v}_r = \nabla \times \boldsymbol{\Psi}(r)$. The Biot–Savart law is a general relation that links any vorticity field to the outer potential flow. Note that the Biot–Savart law is the foundation of the vortex method that has been developed for years by fluid mechanicians and mathematicians.

4.3 Vorticity and Circulation

4.3.1 Vortex Line and Tube

A vorticity is a vector defined at a point as a curl of a velocity vector:

$$\vec{\omega} = \nabla \times \vec{v} = 2\, \vec{\Omega} \tag{4.21}$$

where the direction of the vorticity vector is parallel to the axis of rotation and follows the right hand-side rule. The magnitude of the vorticity vector is two times the rotational rate vector of the fluid element, where the rotational rate of the fluid is defined as an averaged angular velocity at a point and the vorticity is a sum of the turning rates of two mutually orthogonal sides that comprise a fluid element. In analogy to the streamline and stream tube, a vortex line is defined at an instant as a locus tangent to the vorticity vectors, and a vortex tube (or filament) is composed of a bundle of vortex lines.

4.3.2 Circulation

For a given vorticity vector field, a circulation Γ can be defined to represent a *macroscopic quantity of fluid rotation* within a closed loop C:

$$\Gamma = \int_C \vec{\omega} \cdot \vec{n}\, dA \tag{4.22}$$

where $\vec{n}$ is the unit outward vector normal to the surface of area dA, and the surface defined by the closed loop C can be of any shape (Figure 4.8).

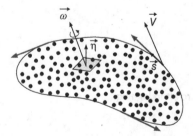

Figure 4.8 Circulation produced by fluid elements in rotation; arrows (thick) tangent to the closed loop C: $\vec{v} \cdot \vec{s}$

The circulation is defined to measure how fast a sizable fluid within the loop rotates as a solid body does for a given time interval, and it is a function of vorticity and projected cross-sectional area of the loop. With Stokes' theorem, it can also be written as a function of tangential speed and peripheral distance:

$$\Gamma = \int_C \vec{\omega} \cdot \vec{n} \, dA = \oint_C \vec{v} \cdot d\vec{s} \tag{4.23}$$

where $d\vec{s}$ is an elemental vector tangent to the closed loop.

4.3.3 Solenoidal Properties

By mathematical identity, the vorticity vector is a solenoidal vector:

$$\nabla \cdot \vec{\omega} = 0 \tag{4.24}$$

which implies that a vortex line cannot simply end in an open space but must either end on a solid surface or form a closed loop.[5] If Eq. (4.24) is integrated over a control volume V, the divergence theorem states that

$$\int_V \nabla \cdot \vec{\omega} \, dV = \oint_S \vec{\omega} \cdot \vec{n} \, dA = 0 \tag{4.25}$$

where S indicates the surrounding surfaces of the control volume V. If a vortex tube is considered as a control volume, there is no vorticity vector component normal to the side surface because the vorticity vectors are, by definition, always tangent to the side surfaces.

Equation (4.25) can be expressed as follows:

$$\int_{c_1} \vec{\omega} \cdot \vec{n} \, dA + \int_{c_2} \vec{\omega} \cdot \vec{n} \, dA = 0 \tag{4.26}$$

where c_1 and c_2 are the closed loops, denoting, respectively, the two cross-sectional boundaries of the vortex tube, and Eq. (4.26) reads

$$\Gamma_{c_1} = \Gamma_{c_2} \tag{4.27}$$

i.e. the circulation along the vortex tube remains the same. This solenoidal property of vorticity vector represents a continuous flow of fluid particles in a vortex tube, which tend to preserve the angular momentum. This is an analogy to the solenoidal velocity vector, $\nabla \cdot \vec{v} = 0$, which represents a continuous flow of mass of an incompressible fluid.[6]

Figure 4.9 shows a bend vortex tube sucked into a running jet engine of an aircraft standing on the runway, with its strength represented by the circulation. The circulation is a single value for any closed loop around the vortex tube regardless of whether it is straight or bend. Note that one side of the vortex tube ends on the runway because a negative pressure is built on the runway by the blockage effect. In fact, the blockage effect originates from the flow sucked into the jet engine, which entrains the ambient air while the runway blocks its normal component of velocity. As a result, a negative pressure is created on the runway surface by which the vortex tube tends to bend down and eventually touch the surface like a small tornado. This picture shows clear evidence that the vortex tube is solenoidal.

---

5 This is one of Helmholtz's vortex theorems. See Chapter 5 *Vortex Dynamics*.
6 In incompressible fluids, the mass conservation law preserves the solenoidal property of the velocity vector that is divergence-free, i.e. $\nabla \cdot \vec{v} = 0$.

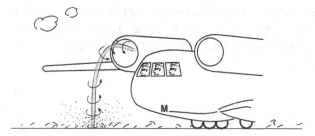

Figure 4.9 A small tornado formed by an aircraft jet engine running on the runway

[**Natural coordinates**] The divergence-free condition for the vorticity field can also be expressed in the natural coordinates (s, n) as follows:

$$\underbrace{\frac{\partial \omega_s}{\partial s}}_{A} + \underbrace{\omega_s \frac{\partial \theta}{\partial n}}_{B} = 0 \tag{4.28}$$

For a vortex tube with variable cross-sectional areas, A indicates the stretching or contraction rate of the vortex-tube and B the confluence or diffluence.

Multiplying by an infinitesimal volume $dV = ds\, dA$ to Eq. (4.28) reads

$$\underbrace{\frac{\partial \omega_s}{\partial s} \cdot ds\, dA}_{A'} \overset{\Delta \omega_s}{} + \underbrace{\omega_s \frac{\partial \theta}{\partial n} \cdot ds\, dA}_{B'} \overset{dn'/dn}{} = 0 \tag{4.29}$$

where A' represents the circulation increase (or decrease) by the streamwise vorticity difference $\Delta\, \omega_s = \omega_s(s + ds) - \omega_s(s)$. This is due to the change in cross-sectional area along the vortex tube (i.e. confluence or diffluence of the vortex tube). Meanwhile, B' represents the circulation decrease (or increase) by the confluence (or diffluence) of the vortex tube. The conservation law of angular momentum states that A' and B' cannot be independent from each other but must be equal in magnitude and opposite in sign.

Note that $d\theta \cdot ds\ (= dn')$ is the normal distance decreased by the confluence and $dn'/dn\ (= \varphi)$ is the contraction ratio, where $\varphi = 0$ conveys that there is no confluence of the vortex tube. Hence, $\varphi\ (w_s \cdot dA)$ represents how much the circulation is reduced by the confluence of the vortex tube, i.e. B'. Therefore, any influx of vorticity from the lateral direction must accompany an increase of outflux of vorticity in the streamwise direction in order to preserve the circulation (i.e. conservation of angular momentum).

4.4 Role of Viscosity on the Curved Surfaces

4.4.1 Vortices in the Boundary Layer

4.4.1.1 Convection and Diffusion

Curved motion on a solid surface cannot be explained without viscosity. The fluid in the region close to the surface is shear strained (i.e. rotated and angularly strained) at a rate by viscous forces. This layer where the fluid is retarded by viscous forces is called the boundary layer. From a vortex dynamics point of view, it can be defined as a region where vortices are created at the leading-edge of the surface and convected downstream with the fluid.

The convecting vortices are also diffused in the transverse direction via random walks of the molecules; this is a process of momentum transport by molecular diffusion.[7] Therefore, the boundary layer thickness δ grows in the streamwise direction as $\delta \sim \sqrt{x}$ and is considered as a vertical limit that the molecules carrying vortices can ever reach via molecular diffusion. It is important to note that the created vortices at the surface will never diminish unless they are diffused transversely or destructed by the deceleration of the flow.

4.4.1.2 Creation and Destruction

The vortices in the boundary layer are greatly affected by the surface curvature since it changes the inertial effect of the fluid. By local streamwise acceleration or deceleration of the fluid, the vortices in the boundary layer can be created or destroyed.[8] For example, when a fluid climbs over a convex surface, the fluid accelerates due to a pressure decrease in the streamwise direction. This pressure reduction is caused by the centrifugal effect. As a result, the boundary layer becomes thinner, newly creating vortices along the surface. The rate of creation of vortices depends on the radius of curvature of the surface, the incoming flow speed, the kinematic viscosity of the fluid, etc. It is interesting to note that the new creation of vortices is directly related to how resilient the boundary layer can be to flow separation.

If the convex surface diverges too quickly from the flow (e.g. typical diverging angle limit is 5–7°), or if the concaveness of the surface is not trivial, the fluid is forced to decelerate too quickly in a given time interval. If then, the boundary layer reaches a point where the convecting vortices are all destroyed so that it can no longer be attached to the surface. The chances of being separated from the surface depends on the rate of flow deceleration; the higher the rate, the less the boundary layer will be attached to the surface.

4.4.2 Coanda Effect

If the rate of vortex creation is sufficiently large, the boundary layer tends to be attached to the convex surface, although the surface diverges at a fast rate from the flow (e.g. typical diverging angle limit is 5–7°). One example is the leading-edge of an airfoil at an angle of attack less than 15°. The boundary layer over the upper surface of the airfoil is attached to the surface all the way close to the trailing-edge so that the pressure at the upper surface is substantially lower than the pressure at the lower surface. As a result, a lift force is created over the airfoil.

This tendency to adhere to convex surfaces becomes more evident in jets. A jet tends to remain attached to the convex surface over an extended distance, since the newly created vortices are held up over long distances. This is due to the coupled effect between the centrifugal force and the vortices in the boundary layer, and we call this jet deflection toward the convex surface a *Coanda effect*.

A simple experiment of the Coanda effect can be conducted with a spoon (Figure 4.10). When a jet tangentially impinges on a convex surface of the spoon, it is easily deflected toward the surface; the jet tends to stay over an extended distance, instead of being separated from the surface. As discussed, the Coanda effect cannot be explained without viscosity. Once the jet touches the convex surface, it greatly accelerates, creating vortices along the uphill surface. As these strong vortices are convected along the downhill surface, the boundary layer tends to be attached to the surface all the

---

7 The Rayleigh problem (also known as Stokes's first problem) is similar in the sense of vorticity creation and diffusion but without convection. The temporal growth of spanwise vorticity in the Rayleigh problem is analogous to the streamwise growth of spanwise vorticity in the boundary layer. More discussion continues in Chapter 6 *External Viscous Flows*.

8 Lighthill [4].

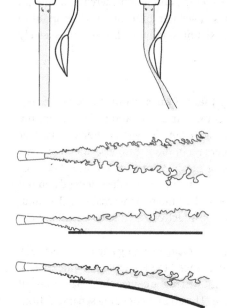

Figure 4.10 Jet deflection over a convex surface by the Coanda effect

Figure 4.11 Jet deflection in the vicinity of the surface; jet entrainment and the Coanda effect

way to the end. Another example of the Coanda effect is water dripping from a cup or bottle with rounded lips.

4.4.2.1 Jet Entrainment

Jet deflection can be observed when the jet is close to a convex or even flat surface. In this case, jet deflection can be explained by the fact that the jet entrains the ambient fluid by viscous frictions (Figure 4.11). If one side of the jet is close to a solid body of which the surface is convex or flat, a normal component of velocity of the entrained fluid makes the surface pressure lower than the ambient pressure. By the pressure difference across the jet, either the jet is attracted to the surface, or vice versa. Once the jet is attracted to surface of the body, vortices are newly created and convected downstream. This coupled effect tends to sustain a boundary layer over the surface. If the surface is convex, the jet deflection becomes more prominent due to the centrifugal effect.

4.4.3 The Origin of Lift

4.4.3.1 Kutta Condition

When an aircraft is thrust to move forward, it is lifted up by a vertical force, called lift, acting against the weight. The lift force is generated by the fluid drawn to follow a curved path over and underneath the cambered airfoil. If the angle of attack of the airfoil is not too large (e.g. less than 15°), the flow is upwashed and downwashed fore and aft the airfoil, and also attached to the surface all the way to the trailing-edge (Figure 4.12). In this case, the flow at the trailing-edge of an airfoil meets the *Kutta condition*, that is, the flow speed must be finite and tangent to the trailing-edge profile.

The origin of lift cannot be explained without the vortices in the boundary layer. The incoming flow is divided into two about the stagnation point at the leading-edge of the airfoil. The flow above is strongly accelerated, scraping the very convex surface of the leading-edge, while strong clockwise

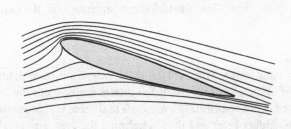

Figure 4.12 Lift generation over an airfoil at an angle of attack; upwash and downwash drawn by vortices in the boundary layers

vortices are being created in a severely retarded layer of the fluid. As these strong vortices enter into the diverging part of the surface (i.e. downhill of the upper surface), the flow starts to decelerate and produces counterclockwise vortices so that vortical strength is weakened as they are convected toward the trailing-edge. It is, however, to be noted that the boundary layer remains attached to the upper surface by two: (i) a strong convexity of the leading-edge and (ii) extensive size of the downhill surface of the airfoil.[9] As a result, a sufficient amount of vortices created at the convex surface of the leading-edge are retained to attach the boundary layer to the surface all the way to the trailing-edge.

As long as flows are attached to the upper and lower surfaces of the airfoil, there exists a difference in streamwise velocity between the two streams over and underneath the airfoil; the one above flows faster than the other below the airfoil, and this is all due to the fact that the potential flows outside the boundary layers tend to conserve their angular momentums (Figure 4.13). The inertial forces in curved potential flows (i.e. centrifugal forces) are balanced with pressure forces that act in the radial direction. As a result, pressure becomes negative and positive at the upper and lower surfaces of the airfoil, respectively, creating a lift force in the vertical direction.

Whether a boundary layer will separate from the upper surface or not depends on how strongly vortices have been created at the leading-edge of the airfoil and at what rate (or how fast) these vortices are being weakened along the downhill surface; the stronger the vortices have been created and the slower the weakening process takes place, the less likely the boundary layer will separate.

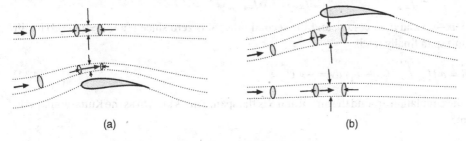

| (a) | (b) |

Figure 4.13 Centrifugal effect on the stream tubes over the curved surfaces; contraction of stream tube close to the convex surface (a); inflation of stream tube close to the concave surface (b); pressure (thin arrows), velocity (bold arrows)

---

9 If the downhill surface is too short (e.g. cylinder or sphere), the boundary layer will separate from the downhill surface because the counter-clockwise vortices are too quickly generated at the downhill surface.

If the angle of attack exceeds 15°, for example the vortices created at the leading-edge are weakened too quickly due to a strong deceleration of the flow. Thus, the airfoil encounters a stall; the entire flow is separated at the upper surface.

4.4.3.2 Kutta–Joukowsky's Lift Theorem

The pressure difference between the upper and lower surfaces of the airfoil is determined by (i) how fast the fluid in the boundary layers over and underneath the airfoil rotates per unit of time and (ii) how largely the vortices are distributed over the airfoil. The reason is that these two determine the flow speeds along the edge of the boundary layer and thus, determine the pressures on the upper and lower surfaces of the airfoil.

In this regard, we define a physical quantity called circulation Γ

$$\Gamma = \oint_c \vec{v} \cdot d\vec{s} = \int_c \vec{\omega} \cdot \vec{n} \, dA \tag{4.30}$$

where $\vec{s}$ is a unit tangent vector around a closed loop c and $\vec{n}$ is a unit outward normal vector to an incremental surface area within c. This global quantity represents an aerial size and the rate of rotation around the airfoil. The rotational rates created by viscous forces in the boundary layers over the airfoil produce upwash and downwash fore and aft the airfoil, creating lift.

A lift force (per unit span) acting on the airfoil can be found by directly integrating the surface pressure and viscous stresses acting in the vertical direction around the airfoil, or by adding all the vertical inertial forces around the airfoil:

$$\int \rho \left(u \frac{\partial v}{\partial x} + v \frac{\partial v}{\partial y} \right) \, dV \tag{4.31}$$

The momentum conservation principle states that the inertia force of the fluid within a control volume must be balanced with the external force acting on the fluid along the control surface that encloses it.

Thus, if a vertical force balance is integrated over a cambered airfoil, it produces a lift force L:

$$L = \int_{-z_0}^{+z_0} \int_{y=-\infty}^{y=+\infty} \int_{-x_0}^{+x_0} \left(\rho \, u \frac{\partial v}{\partial x} + \rho \, v \frac{\partial v}{\partial y} \right) \, dx \, dy \, dz \tag{4.32}$$

With approximations such as $u = U_\infty + u'$ and $v = U_\infty \tan \alpha$ and $u', v \ll U_\infty$, the lift per unit span is written as follows:

$$L/S = \rho \, U_\infty \int_{-\infty}^{+\infty} (v_2 - v_1) \, dy + \int_{-x_0}^{+x_0} \rho \left(\int_{y=-\infty}^{y=+\infty} d(v^2/2) \right) dx \tag{4.33}$$

where $v_1 = v(-x_0)$ and $v_2 = v(+x_0)$, and the second integral is zero since $v(y = \pm\infty) \to 0$. The lift force per unit span finally reads

$$L/S = \rho \, U_\infty \int_{-\infty}^{+\infty} (v_2 - v_1) \, dy = -\rho \, U_\infty \, \Gamma \tag{4.34}$$

where Γ is the circulation around the airfoil and S is the span. This is known as the Kutta–Joukowski lift theorem.

4.5 Decomposition of Convective Acceleration

4.5.1 Lamb–Gromeko's Form

It is interesting to note that convective acceleration written in Cartesian coordinates can be decomposed as follows. The convective acceleration in the x direction a_x, for example can be written as follows:

$$
\begin{aligned}
a_x &= u\,\frac{\partial u}{\partial x} + v\,\frac{\partial u}{\partial y} + w\,\frac{\partial u}{\partial z} \\
&= \left(u\,\frac{\partial u}{\partial x} + v\,\frac{\partial v}{\partial x} + w\,\frac{\partial w}{\partial x} \right) + v\left(\frac{\partial u}{\partial y} - \frac{\partial v}{\partial x} \right) + w\left(\frac{\partial u}{\partial z} - \frac{\partial w}{\partial x} \right) \\
&= \frac{1}{2}\left(\frac{\partial u^2}{\partial x} + \frac{\partial v^2}{\partial x} + \frac{\partial w^2}{\partial x} \right) - v\,\omega_z + w\,\omega_y \\
&= \frac{1}{2}\frac{\partial\, v^2}{\partial x} - v\,\omega_z + w\,\omega_y
\end{aligned}
\tag{4.35}
$$

If we decompose the convective accelerations in the y and z directions, a_y and a_z, in a similar way, the decomposed expressions can be case into a vector form as follows:

$$
\vec{a}_c = (\vec{v}\cdot\nabla)\,\vec{v} = \frac{1}{2}\nabla\, v^2 + \vec{\omega}\times\vec{v}
\tag{4.36}
$$

where $v^2 = u^2 + v^2 + w^2$ and $\vec{\omega}$ is a vorticity vector.

Equation (4.36) shows that convective acceleration can be decomposed into two parts: (i) a gradient vector of the specific kinetic energy and (ii) Lamb acceleration vector. The first term represents the acceleration associated with the fluid being strained and/or rotated at a rate. It is always present regardless of whether the flow is irrotational or rotational. However, the second term, called the Lamb acceleration vector, disappears if the flow is irrotational, or if $\vec{\omega}$ is parallel to $\vec{v}$ in rotational flows (e.g. Beltrami flows) [5].[10] The Lamb vector represents the acceleration in a direction normal to the streamline (always toward the lower speed side) when the fluid convects and rotates at the same time.

4.5.2 Lamb Acceleration Vector

When a fluid element in rotation is moving at a certain speed, there is an acceleration created in the direction normal to the streamline (toward the lower-speed side) (Figure 4.14). We call this the Lamb acceleration vector. It is associated with the superposition of rotation with translation of a vortex filament. Thus, the magnitude of the Lamb acceleration vector is proportional to two

Figure 4.14 Lamb acceleration of a rotating fluid element in convection; v < −R Ω (a), v = −R Ω (b), v > −R Ω (c); v (convection speed), R (radius of a fluid element), Ω (< 0) (rotation rate)

(a)   (b)   (c)

---

10 Beltrami flows are highly helical flows in well-developed tornadic flows.

times the angular speed of the vortex and its convection speed v. Note that the normal acceleration associated with the Lamb vector is attributed not only by the turning rate of ds but also to that of dn, where ds and dn are the horizontal and vertical sides of a squared element of a fluid.

The Lamb vector can be better described by casting the Euler equation into the Crocco equation (see Section 4.8.3). In viscous flows, the role of the Lamb vector can be more clarified by taking two gradient operators, divergence and curl, on the Lamb vector. Divergence shows how it is associated with the rate of viscous shear work, while curl is related to the rate of change of vorticity when it is convected, stretched, tilted, and isotropically expanded.

4.5.2.1 Divergence of Lamb Vector
A divergence of the Lamb vector can be expressed as follows:

$$\nabla \cdot \vec{L} = \nabla \cdot (\vec{\omega} \times \vec{v}) = \underbrace{\vec{v} \cdot (\nabla \times \vec{\omega})}_{\dot{K}} + \underbrace{(-\vec{\omega} \cdot \vec{\omega})}_{I} \tag{4.37}$$

where $\dot{K} = \vec{v} \cdot (\nabla \times \vec{\omega}) = \vec{v} \cdot \vec{f}_v / \mu$, and $\vec{f}_v = \mu (\nabla \times \vec{\omega})$ is the viscous force vector per unit volume.

The divergence of the Lamb vector $\nabla \cdot \vec{L}$ is a scalar quantity that represents the sum of the squares of the two rates $[1/s^2]$: (i) $\dot{K}$ is associated with the rate of shear work done on a fluid per unit volume and per unit viscosity while vorticity is being diffused and convected with the fluid, and (ii) I is related to the rate of dissipation of energy of a fluid per unit volume and per unit viscosity (called *enstrophy*), while angularly straining the rotational fluids against frictions. In Couette flow, for example the divergence of the Lamb vector is not zero because $|I| \neq 0$, although $|\dot{K}| = 0$. Other examples showing the role of $\nabla \cdot \vec{L}$ include shear flows, boundary layers, etc. [6].[11]

4.5.2.2 Curl of Lamb Vector
The curl of the Lamb acceleration vector can be expressed as follows:

$$\nabla \times (\vec{\omega} \times \vec{v}) = \underbrace{(\vec{v} \cdot \nabla) \vec{\omega}}_{(0)} - \underbrace{(\vec{\omega} \cdot \nabla) \vec{v}}_{(1)} + \underbrace{\vec{\omega}(\nabla \cdot \vec{v})}_{(2)} - \vec{v}(\nabla \cdot \vec{\omega})^0 \tag{4.38}$$

which is a vectorial quantity that represents the rate of change of the vorticity vector while a vortex is convected (term 0), stretched, and titled by local velocity components (term 1), and isotropically expanded by compressibility effect (term 2).[12]

4.5.3 Streamline Coordinates

To understand the nature of the curved motions of irrotational and rotational flows, it is important to compare the convective acceleration decomposed in Crocco–Gromeko's form to the convective acceleration written in streamline coordinates, particularly the centripetal acceleration.

A gradient operator ∇ can be written in streamline coordinates (in two-dimensions) as follows:

$$\nabla = \vec{i}_s \frac{\partial}{\partial s} + \vec{i}_n \frac{\partial}{\partial n} \tag{4.39}$$

where $\vec{i}_s$ is a local unit vector tangent to the streamline and $\vec{i}_n$ is the local unit vector normal (outward) to $\vec{i}_s$.[13]

---

11 More discussion continues in Section 4.7 *Divergence of Lamb vector*.
12 More discussion continues in Chapter 5 *Vortex Dynamics*.
13 In three-dimensions, $\nabla = \vec{i}_s \frac{\partial}{\partial s} + \vec{i}_n \frac{\partial}{\partial n} + \vec{i}_b \frac{\partial}{\partial b}$ where $\vec{i}_b$ is a binormal unit vector.

With this gradient operator, we can express the convective acceleration of the total velocity vector $\vec{v} = v\,\vec{i}_s$ as follows:

$$\vec{a}_c = (\vec{v} \cdot \nabla)\,\vec{v} = v\,\frac{\partial}{\partial s}(v\,\vec{i}_s) = \vec{i}_s\,\frac{\partial}{\partial s}(v^2/2) + v^2\,\frac{\partial \vec{i}_s}{\partial s} \tag{4.40}$$

where the rate of change of the magnitude of a unit vector tangent to the streamline is inversely proportional to the local radius of curvature r_c with the direction pointing opposite to $\vec{i}_n$:

$$\frac{\partial \vec{i}_s}{\partial s} = -\frac{\vec{i}_n}{r_c} \tag{4.41}$$

Equation (4.40) finally reads

$$\vec{a}_c = \vec{i}_s\,\frac{\partial}{\partial s}(v^2/2) + \vec{i}_n\left(-\frac{v^2}{r_c}\right) \tag{4.42}$$

4.6 Physical Interpretations of a_s and a_n

To show how two expressions of convective acceleration are mutually related, we apply the gradient operator Eq. (4.39) to Eq. (4.36):

$$\vec{a}_c = \vec{i}_s\,\frac{\partial}{\partial s}(v^2/2) + \vec{i}_n\,\frac{\partial}{\partial n}(v^2/2) + \vec{i}_n\,((\vec{\omega} \times \vec{v}) \cdot \vec{i}_n) \tag{4.43}$$

By equating Eq. (4.43) with (4.42), the streamwise and normal components of the convective acceleration are written as follows:

$$a_s = \frac{\partial}{\partial s}(v^2/2) \tag{4.44}$$

$$a_n = -\frac{v^2}{r_c} = \frac{\partial}{\partial n}(v^2/2) + (\vec{\omega} \times \vec{v}) \cdot \vec{i}_n \tag{4.45}$$

where the normal component of the convective acceleration, i.e. the centripetal acceleration consists of (i) the rate of change of specific kinetic energy of the fluid in the normal direction, and (ii) the Lamb acceleration vector.

4.6.1 Irrotational Flow

In irrotational (or potential) flows, the fluid momentum is changed at a rate by forces that act on the center of mass of the fluid particles, e.g. pressure forces, gravitational body forces. In this case, the rate of change of specific kinetic energy of the fluid along a streamline simply represents the convective acceleration of the fluid in the streamwise direction:

$$a_s = \frac{\partial}{\partial s}(v^2/2) = v\,\frac{\partial v}{\partial s} \tag{4.46}$$

which is associated with the rate of change of cross-sectional area in a stream tube.

The rate of change of specific kinetic energy of the fluid in the normal direction can be rearranged as a product of the local velocity and its normal gradient:

$$a_n = \frac{\partial}{\partial n}(v^2/2) = v\,\frac{\partial v}{\partial n} \tag{4.47}$$

where the normal gradient of the total velocity represents the turning rate of the vertical side of the fluid element dn. In irrotational flow, it is the same as that of the horizontal side ds but in the opposite direction, i.e. $\dot{\theta} = -v/r_c$ (i.e. irrotationality condition). Therefore, the product of the two turns out to be the centripetal acceleration of the fluid:

$$v \frac{\partial v}{\partial n} = v \left(-\frac{v}{r_c} \right) = -\frac{v^2}{r_c} \tag{4.48}$$

From this argument, we can understand that the rate of change of specific kinetic energy of the fluid in the normal direction is associated with the fluid element in angular straining while the fluid element moves along a curved path, preserving the angular momentum. For a fluid element ($ds \times dn$), for instance, the turning rate of ds is

$$\dot{\theta}_1 = -\frac{v}{r_c} \tag{4.49}$$

while the turning rate of dn is

$$\dot{\theta}_2 = -\frac{\tan \theta_2}{dt} = -\left[\frac{\{v + (\partial v/\partial n)dn - v\} \, dt}{dn} \right] / dt = -\frac{\partial v}{\partial n} \tag{4.50}$$

If $\dot{\theta}_1 + \dot{\theta}_2 = 0$, the fluid is simply angularly strained at a rate. Note also that the inertial forces associated with convective accelerations a_s and a_n are balanced with pressure forces and gravitational body force in the steamwise and normal directions.

4.6.2 Rotational Flow

4.6.2.1 Shear-Straining Fluids
In rotational (or viscous shear) flows, the streamwise component of convective acceleration is the rate of change of specific kinetic energy of the fluid along the streamline:

$$a_s = \frac{\partial}{\partial s}(v^2/2) = v \frac{\partial v}{\partial s} \tag{4.51}$$

and it is again associated with the rate of change of cross-sectional area in a stream tube. In this case, a viscous force is included in the forces that react to the momentum change in the streamwise direction (e.g. a boundary layer over a flat or curved surface, or a boundary layer in the entrance region of a pipe or channel).

Meanwhile, the rate of change of specific kinetic energy in the normal direction expresses the centripetal acceleration with the Lamb acceleration vector as follows:

$$a_n = -\frac{v^2}{r_c} = \frac{\partial}{\partial n}(v^2/2) + (\vec{\omega} \times \vec{v}) \cdot \vec{i}_n \tag{4.52}$$

where r_c is the radius of curvature of the streamline. In shear flows, fluids are simultaneously rotated and angularly strained but not necessarily at the same rate, except in simple shear flow. By defining the angular strain rate ϵ and rotation rate Ω in streamline coordinates as follows:

$$\epsilon = \frac{1}{2} \left(\frac{\partial v}{\partial n} - \frac{v}{r_c} \right), \qquad \Omega = -\frac{1}{2} \left(\frac{\partial v}{\partial n} + \frac{v}{r_c} \right) \tag{4.53}$$

the right-hand side of Eq. (4.52) can be expressed as follows:

$$v \frac{\partial v}{\partial n} + (\vec{\omega} \times \vec{v}) \cdot \vec{i}_n = v \, (\underbrace{\epsilon - \Omega}_{①} + \underbrace{2\Omega}_{②}) = v \, (\epsilon + \Omega) \tag{4.54}$$

Then, Eq. (4.52) finally shows

$$\epsilon + \Omega = -\frac{v}{r_c} \tag{4.55}$$

It is interesting to note that in curved rotational flows, centripetal acceleration is associated with a fluid element, angular strained at a rate of ϵ and rotated at a rate of Ω. Furthermore, the rates of angular straining and rotation always differ by $|v/r_c|$; their difference disappears only when the two rates become the same.

4.6.2.2 Physical Interpretations

It is to be noted that $\partial v/\partial n$ in Eq. (4.54), the term ①, corresponds to the turning rate (e.g. clockwise if it is positive) of dn of a fluid element, angularly strained at a rate of ϵ and rotated at a rate of $-\Omega$. This term also represents the turning rate of ds of the fluid element because ϵ and Ω are the averaged values of rates of the angular straining and rotation. Thus, $v\,\partial v/\partial n$ corresponds to the acceleration in the normal direction of a fluid element in a curved rotational flow. Meanwhile, 2Ω in Eq. (4.54) is the sum of the turning rates of ds and dn of a fluid element, which rotates at a rate of Ω. Therefore, $v\,(2\,\Omega)$ also represents the acceleration in the normal direction of a fluid element, which simultaneously rotates and convects in a curved rotational flow. Note that the actual contribution to the centripetal acceleration is the sum of these two.

In a simple shear flow, for example $r_c \to \infty$, and Eq. (4.55) shows

$$\epsilon + \Omega = 0 \tag{4.56}$$

i.e. the rate of rotation and the rate of angular straining of the fluid are exactly the same so that there is no net normal acceleration, i.e. $a_n = 0$. It is to be noted again that, even in parallel or almost parallel shear flows (e.g. Couette flow or boundary layer over a flat surface), the viscous forces act not only in the streamwise direction but also in the normal direction so that the two terms in Eq. (4.55) are to be the same finite value with opposite signs. In this case, the centripetal acceleration is zero or almost zero with $r_c \to 0$.

Over a convex surface, r_c is finite and positive, and Eq. (4.55) shows

$$\epsilon + \Omega = -\frac{v}{r_c} \; (<0) \tag{4.57}$$

where the rate of rotation exceeds the rate of angular straining by $|v/r_c|$. Examples of convex surfaces are the Coanda effect, or generation of lift over a cambered airfoil

On the other hand, over a concave surface, r_c is finite and negative, and Eq. (4.55) shows

$$\epsilon + \Omega = -\frac{v}{r_c} \; (>0) \tag{4.58}$$

where the rate of angular straining exceeds the rate of rotation by $|v/r_c|$. This case was already demonstrated in the rotating concentric tube where only the inner tube rotates at a rate. Another example is the thickening of the boundary layer, destroying the rotation rate and at some point, separating the boundary layer from the concave surface when the rotation rate is diminished completely. It is to be noted again that the Lamb acceleration vector is the inertia force of a fluid of a unit mass, rotating and convecting in rotational flows. It acts in the direction normal to the streamline and is part of the centripetal acceleration.

Example 4.2 *Forced vortex and free vortex*

There are two specific cases, i.e. a fluid in solid body rotation (or a frozen fluid) with a zero angular strain rate (see Section 4.1.2), and a fluid in irrotational swirl with a zero rotation rate (see Section 4.1.3). In the former, there is no viscous resistance to the fluid in rotation, but the Lamb vector is still responsible for the centripetal acceleration of the fluid, in which case the rate of change of specific kinetic energy represents a counter rotation of the rotating fluid. In the latter, the fluid is angularly strained against viscous resistances, but viscous forces cannot produce torques due to the fluid tending to preserve the angular momentum; therefore, viscous dissipation of energy does occur. The centripetal acceleration is associated with the angularly straining fluid.

4.7 Divergence of Lamb vector[†]

4.7.1 Curl of Vorticity Vector

A curl of a vector $\vec{\varphi} = (\varphi_x, \varphi_y, \varphi_z)$ is a vector that represents a rotational measure at a point with the axis of rotation. It is mathematically defined as follows:

$$(\nabla \times \vec{\varphi}) \cdot \vec{n} = \lim_{A \to 0} \frac{1}{A} \oint_c \vec{\varphi} \cdot d\vec{s} \tag{4.59}$$

where c represents a closed loop on the plane of area A specified with an outward unit normal vector $\vec{n}$.

If $\vec{\varphi}$ is a velocity vector, its curl represents a vorticity vector, $\vec{\omega} = \nabla \times \vec{v}$, of which the direction is normal to the plane on which the velocity vectors are nonuniformly distributed, and the magnitude is a measure of *rotational rate* Ω:

$$(\nabla \times \vec{v}) \cdot \vec{n} = \lim_{A \to 0} \frac{1}{A} \oint_c \vec{v} \cdot d\vec{s} = 2\,\Omega \tag{4.60}$$

If $\vec{\varphi}$ is a vorticity vector, its curl represents a vector, $\nabla \times \vec{\omega}$, of which the direction is normal to the plane on which the vorticity vectors are nonuniformly distributed, and the magnitude is an intensity measure of *diffusional mixing of vorticity* ζ

$$(\nabla \times \vec{\omega}) \cdot \vec{n} = \lim_{A \to 0} \frac{1}{A} \oint_c \vec{\omega} \cdot d\vec{s} = \zeta \tag{4.61}$$

In a field with uniform vorticity (e.g. Couette flow, fluid in solid body rotation), $\zeta = 0$ means no diffusional mixing of vorticity. In a viscous boundary layer or free shear layer, $\zeta \neq 0$ means diffusional mixing of vorticity and thereby growth of boundary layer or free shear layer. A nonzero ζ in the field indicates that the diffusional mixing of vorticity is produced by a net viscous force exerted on the fluid.

4.7.2 Rate of Change of Kinetic and Internal Energies

A divergence of the Lamb vector is written as follows:

$$\nabla \cdot \vec{L} = \nabla \cdot (\vec{\omega} \times \vec{v}) = \underbrace{\vec{v} \cdot (\nabla \times \vec{\omega})}_{\dot{K}} + \underbrace{(-\vec{\omega} \cdot \vec{\omega})}_{\dot{I}} \tag{4.62}$$

where $\dot{K}$ represents the rate of shear work done on a fluid per unit volume and per unit viscosity, while vorticity is being diffused and convected with the fluid, and $\dot{I}$ represents the rate of dissipation of energy of a fluid per unit volume and per unit viscosity (called *enstrophy*), while angularly straining the rotating fluids against frictions.

It is to be noted that a viscous force per unit volume can be defined by the product of the viscosity μ and the diffusional mixing of vorticity vector $\nabla \times \vec{\omega}$:

$$\vec{f}_v = \mu\,(\nabla \times \vec{\omega}) \tag{4.63}$$

Therefore, an inner product of the curl of vorticity vector with a local velocity vector:

$$\dot{K} = \vec{v} \cdot (\nabla \times \vec{\omega}) = \vec{v} \cdot \vec{f}_v / \mu \tag{4.64}$$

represents the rate of energy transfer by viscous force (or shear) work done on the fluid per unit volume and per unit viscosity (or the rate of change of kinetic energy per unit volume and per unit viscosity).

---

† For advanced studies.

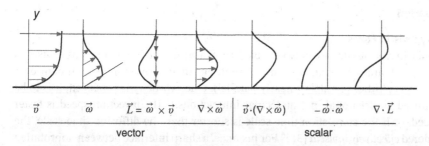

$$\vec{v} \qquad \vec{\omega} \qquad \vec{L} = \vec{\omega} \times \vec{v} \qquad \nabla \times \vec{\omega} \qquad \vec{v} \cdot (\nabla \times \vec{\omega}) \qquad -\vec{\omega} \cdot \vec{\omega} \qquad \nabla \cdot \vec{L}$$

vector   scalar

Figure 4.15 Lamb vector-related quantities in a laminar boundary layer; vectors: (i) velocity, (ii) vorticity, (iii) Lamb acceleration, and (iv) curl of vorticity; scalars: (v) rate of change of kinetic energy $\dot{K}$, (vi) rate of change of internal energy $\dot{I}$ (called enstrophy), and (vii) divergence of Lamb vector (from left to right); $\dot{K}$ and $\dot{I}$ (per unit volume and per unit viscosity)

If $\vec{\omega}$ is uniform in space, then $\dot{K} = 0$. In this case, there is no diffusion of vorticity in the flow field. The vorticity is simply preserved by $\dot{I}$, as in Couette flow or in a fluid of solid body rotation. If $\vec{\omega}$ is not uniform in space, $\nabla \times \vec{\omega}$ is nonzero such that $\dot{K} > 0$ when $\nabla \times \vec{\omega}$ and $\vec{v}$ are in the same direction, or that $\dot{K} < 0$ when $\nabla \times \vec{\omega}$ and $\vec{v}$ are in the opposite direction. In a viscous boundary layer (or in a concentrated vortex), for example $\dot{K} > 0$ from the edge of the vorticity layer to the wall (or to a geometrical symmetry point), with a local maxima in the middle. Meanwhile, $\dot{I}$ is negative all the way from 0 at the edge to a finite value at the wall or at the symmetry point (Figure 4.15). In this case, $\dot{K}$ represents the rate of shear work done on the fluid per unit volume per unit viscosity that advects and diffuses the vorticity of the fluid from the wall (or symmetry point) against frictions.

The second term $\dot{I}$ $(= -\vec{\omega} \cdot \vec{\omega})$ is called *enstrophy*. It is always negative (thus, a sink) and represents a rate of dissipation of energy per unit volume and per unit viscosity, while angularly straining the rotational fluids against frictions, e.g. unidirectional shear flow, boundary layer. In a solid body rotation of a fluid, it represents a rotational rate of the fluid attained after dissipating the energy per unit volume per unit viscosity. In fact, one can evaluate how much energy has been used to set the fluid in solid body rotation by integrating the enstrophy over the entire area of rotation with multiplication of the viscosity and the time taken to reach a complete frozen rotating fluid.

Figure 4.15 shows the variations of the Lamb vector-related quantities in a laminar boundary layer: (i) velocity, (ii) vorticity, (iii) Lamb acceleration, (iv) curl of vorticity, (v) rate of change of kinetic energy $\dot{K}$, (vi) rate of change of internal energy $\dot{I}$ (called enstrophy), and (vii) divergence of Lamb vector ($\dot{K}$ and $\dot{I}$ are per unit volume and per unit viscosity). It is interesting to note that the divergence of Lamb vector is positive toward the edge of the boundary layer ($\dot{I} < \dot{K}$) while negative toward the wall ($\dot{I} > \dot{K}$), where $\dot{K} > 0$ across the boundary layer.

4.8 Momentum Equations in Selective Forms

4.8.1 Navier–Stokes Equations

The Navier–Stokes equations can be written in Lamb–Gromeko's form as follows:

$$\frac{\partial \vec{v}}{\partial t} + \frac{1}{2} \nabla v^2 + \vec{\omega} \times \vec{v} = \frac{1}{\rho} \left(-\nabla p + \vec{f}_v + \vec{f}_b \right) \tag{4.65}$$

where $\vec{f}_v$ is the viscous force vector per unit volume and $\vec{f}_b$ is the body force vector per unit volume, and the convective acceleration vector is expressed in Lamb–Gromeko's form with two terms: a gradient vector of specific kinetic energy, $\nabla(v^2/2)$, and the Lamb vector, $\vec{\omega} \times \vec{v}$. Equation (4.65) is the most general form of momentum equation that includes all fluid motions with all forces, e.g. pressure force, volumetric body force, and viscous force.

4.8.2 Euler Equation

4.8.2.1 Effectively Inviscid Flows

Sometimes, we deal with viscous flows where viscous stresses are present, but there is no net viscous force acting on the fluid, and where vortices are present, but their convection is dominant over diffusion (thus, flows are either parallel or nearly parallel). Once vorticity is produced, it is to be convected and diffused via frictional intermolecular interactions.[15] If convection speed is faster than diffusion speed, or if the convection time scale is shorter than the diffusion time scale, the flow can be considered *effectively inviscid* [3].[16] For instance, a sharp interface between nonrotating and rotating fluids can be sustained if viscous diffusion is negligible within a time interval of our interest.[17] Slip streams (also called contact surfaces) are also a similar example.

In this case, we can drop off the viscous force vector but not the Lamb acceleration vector in the Navier–Stokes equation:

$$\frac{\partial \vec{v}}{\partial t} + \frac{1}{2}\nabla \, v^2 + \vec{\omega} \times \vec{v} = \frac{1}{\rho}\left(-\nabla p + \vec{f}_b\right) \tag{4.66}$$

This Euler equation can handle such effectively inviscid flows but matters with the followings. The right amount of vorticity must be introduced into the field as an initial condition, since vorticity generation is entirely based on the physics of viscous flows. Furthermore, resolution is an issue for convection and diffusion of vorticity.

The use of the Euler equation is not exactly a precise way of modeling the physics of viscous flows, but it has been used for a wide range of highly localized vortical flows at high Reynolds numbers because the computational cost of the Euler equation is much lower than that of the Navier–Stokes equation. In many high-speed flows (or high Reynolds number flows, in general), the dynamics of vortex can be solved by the Euler equation within the time interval of our interest. In marine applications, vortices created in water often last for a fairly long time since the viscosity of water is one hundred times higher than the viscosity of air whilst its density is also one thousand times higher.

4.8.3 Crocco's Theorem

Let us suppose we have an incompressible inviscid fluid under conservative body forces. Then, the Euler equation is written as follows:

$$\frac{\partial \vec{v}}{\partial t} + \frac{1}{2}\nabla \, v^2 + \vec{\omega} \times \vec{v} = -\frac{1}{\rho}\,\nabla p - g\,\nabla h \tag{4.67}$$

where the conservative body force is expressed as a gradient vector of a gravitational potential, $\vec{f}_b = -\rho\,g\,\nabla h$.

It is interesting to note that the terms in the Euler equation, except the Lamb acceleration vector, can be expressed as a gradient vector of a scalar quantity B(t) (called the Bernoulli function):

$$\nabla B(t) = -\left(\vec{\omega} \times \vec{v}\right) \tag{4.68}$$

---

15 Viscous diffusion of vorticity is only possible with a net viscous force acting on the fluid. More discussion continues in Section 5.4.3. *Viscous diffusion*, Chapter 5 *Vortex Dynamics*.

16 It is to be noted that the effectively inviscid flow does not necessarily mean that the flow is irrotational; the other way around is, of course, always valid.

17 To be more precise, this interface is not a discontinuity; this is a buffer layer of finite thickness where two motions (rotation and angular straining) are superimposed. In this layer, a net viscous force is acting so that vorticity diffuses through random walks of the molecules.

where the Bernoulli function B(t) is defined as follows:

$$B(t) = \underbrace{\int \frac{\partial \vec{v}}{\partial t} \cdot d\vec{r}}_{\text{Ⓐ}} + \underbrace{\frac{1}{2} v^2}_{\text{Ⓑ}} + \frac{p}{\rho} + gh \tag{4.69}$$

and $d\vec{r}$ denotes an arbitrary incremental distance vector in space. The local time acceleration can also be expressed as follows:

$$\frac{\partial \vec{v}}{\partial t} = \frac{\partial}{\partial t} \left(\int \nabla \vec{v} \cdot d\vec{r} \right) = \nabla \left(\int \frac{\partial \vec{v}}{\partial t} \cdot d\vec{r} \right) \tag{4.70}$$

Note that the Bernoulli function B(t), which has a dimension of energy per unit mass (i.e. specific energy), is comprised of two groups. One group of specific energies is pressure and gravitational potential, of which gradients are the forces that accelerate the fluid of a unit mass in time and space. The other group of specific energies is an indefinite integral of local acceleration over a distance Ⓐ and a specific kinetic energy at a point Ⓑ, of which gradients represent the inertial forces of the fluid of a unit mass accelerating in time and space.

By taking an inner-product with $d\vec{r}$, Eq. (4.68) can be expressed as the difference of Bernoulli's function B(t) over an incremental distance vector $d\vec{r}$:

$$\nabla B(t) \cdot d\vec{r} = -(\vec{\omega} \times \vec{v}) \cdot d\vec{r} \tag{4.71}$$

and if $d\vec{r} = d\vec{s}$, it becomes

$$\nabla B(t) \cdot d\vec{s} = 0 \tag{4.72}$$

since the Lamb vector is always orthogonal to $d\vec{s}$.

Equation (4.72) shows that at any instant, an incremental difference of the scalar quantity B(t) along a streamline $\Delta B(t)_s$ is zero:

$$\Delta B(t)_s = \frac{\partial \vec{v}}{\partial t} \cdot d\vec{s} + \nabla \left(\frac{1}{2} v^2 + \frac{p}{\rho} + gh \right) \cdot d\vec{s} = 0 \tag{4.73}$$

or that B(t) is constant, i.e. $B(t) = B_0$ along the streamline, even with the vorticity in the field; the presence of vorticity does not contribute any on the force balance along the streamline because it always acts normal to the streamline. A surface with B(t) = constant is also called *Lamb's surface*. It is interesting to note that the first term is the product of the average local acceleration and the incremental streamwise distance, whereas the other terms are expressed as spatial differences.

Bernoulli's function B(t), however, differs from streamline to streamline. In other words, the total mechanical energy is not conserved across the streamlines, i.e. $\partial B(t)/\partial n \neq 0$. As indicated in Eq. (4.72), the gradient vector of Bernoulli's function always points in the direction normal to the streamline and also opposite to the Lamb acceleration vector. The Lamb vector can be interpreted as an indicator for the break of total mechanical energy conservation, which might have been preserved if there were no rotation in the flow field.

[**Notes**] If we take the divergence on both sides of Eq. (4.68), it reads

$$\nabla^2 B = -\nabla \cdot (\vec{\omega} \times \vec{v}) = -\nabla \cdot \vec{L} \tag{4.74}$$

If $\nabla \cdot \vec{L} > 0$, B has a local maxima, while if $\nabla \cdot \vec{L} < 0$, B has a local minima. If $\nabla \cdot \vec{L} = 0$, i.e. $|\vec{L}|$ is constant, B becomes a planar function.

4.8.4 Crocco's Theorem Extended

Bernoulli's function B(t) can be discussed in a more general context by rearranging the Navier–Stokes equations as follows:

$$\nabla B(t) - \frac{1}{\rho} \vec{f}_v = -\vec{\omega} \times \vec{v} \qquad (4.75)$$

If we take the inner-product of Eq. (4.75) with $d\vec{r} = (d\vec{s},\ d\vec{n})$, the resulting energy equation per unit mass has two components.

The one in the streamwise direction shows that an incremental difference of Bernoulli's function is expressed as follows:

$$\nabla B(t) \cdot d\vec{s} = \Delta B(t)_s = \frac{1}{\rho} \vec{f}_v \cdot d\vec{s} \qquad (4.76)$$

and if it is integrated along the streamline from point 1 to point 2, the difference of Bernoulli's function between 1 and 2 is the integrated viscous shear work done by the fluid of a unit mass between points 1 to 2 against frictional resistances:

$$B(t)_{s2} - B(t)_{s1} = \underbrace{\int_1^2 \frac{1}{\rho} \vec{f}_v \cdot d\vec{s}}_{w_s^v} \qquad (4.77)$$

where w_s^v equals the loss of total mechanical energy per unit mass.

Similarly, the other equation in the normal direction can be written as

$$\nabla B(t) \cdot d\vec{n} = \Delta B(t)_n = \frac{1}{\rho} \vec{f}_v \cdot d\vec{n} - (\vec{\omega} \times \vec{v}) \cdot d\vec{n} \qquad (4.78)$$

and if it is integrated from points 1 to 2 across the streamlines, the difference of Bernoulli's function between 1 and 2 comprises of two integrals:

$$\int_1^2 dB_n^v(t) + \int_1^2 dB_n^L(t) = \underbrace{\int_1^2 \frac{1}{\rho} \vec{f}_v \cdot d\vec{n}}_{w_n^v} - \underbrace{\int_1^2 (\vec{\omega} \times \vec{v}) \cdot d\vec{n}}_{w_n^L} \qquad (4.79)$$

where the first term w_n^v is the integrated viscous shear works done by the fluid per unit mass from 1 to 2 across the streamlines, and the second term w_n^L is the work done by the Lamb acceleration vector (or Lamb work, for short) on the fluid per unit mass from 1 to 2 across the streamlines. These two works are associated with the rates of rotation and angular straining of the fluid, which always destroy the uniformity of the total mechanical energy across the streamlines.

If we consider a transient Couette flow, the energy supplied by the plate has been used in part to accelerate the fluid in time and has been dissipated into heat in part to viscously strain the fluid at a rate. A net viscous force acting in the streamwise direction will do work on the fluid per unit mass as follows:

$$w_s^v = B(t)_{s2} - B(t)_{s1} = \int_1^2 \frac{\partial \vec{v}}{\partial t} \cdot d\vec{s} \qquad (4.80)$$

where the integrated viscous shear work from points 1 to 2 along the streamline is used to accelerate the fluid in time between 1 and 2.

Meanwhile, $w_n^v = 0$ since flows are all parallel to the plates, i.e. $\vec{v} \cdot \vec{n} = 0$. In this case, there is a net viscous force acting in the normal directions, but it is balanced with the normal force externally applied to the plates. However, the energy loss (per unit mass) between the top and bottom plates, $\int_1^2 dB_n^L(t)$, has been produced in the normal direction by Lamb work w_n^L:

$$w_n^L = \int_1^2 dB_n^L(t) = B(t)_{n2} - B(t)_{n1} = (v_2^2 - v_1^2)/2 \tag{4.81}$$

Once steady state is reached, w_s^v becomes zero because a net viscous force acting on the fluid in the streamwise direction disappears, whereas the Lamb work w_s^L remains the same for all times, though $\vec{\omega}$ and $\vec{v}$ are a function of time. The Lamb work can be considered a resulting work done through viscous dissipation of energy.

4.8.5 Potential Flows

4.8.5.1 Bernoulli's Equation

In potential (or irrotational) flow, the right-hand side Eq. (4.68) is zero for all $d\vec{r}$ taken

$$\nabla B(t) \cdot d\vec{r} = 0 \tag{4.82}$$

which means that at any instant, $B(t)$ is constant throughout the fluid:

$$B(t) = \int \frac{\partial \vec{v}}{\partial t} \cdot d\vec{r} + \frac{1}{2} v^2 + \frac{p}{\rho} + gh = B_0 \tag{4.83}$$

This well-known form of Bernoulli's equation states that the total mechanical energy of the fluid B_0 is conserved entirely since there is no rotation in the flow field.

4.8.5.2 Velocity Potential

In irrotational flow, the velocity vector satisfies at all points

$$\nabla \times \vec{v} = 0 \tag{4.84}$$

to which Stokes' theorem shows that

$$\Gamma = \int_C (\nabla \times \vec{v}) \cdot \vec{n} \, dA = \oint_C \vec{v} \cdot d\vec{x} = 0 \tag{4.85}$$

for a simply connected region defined by a closed curve C, i.e. the circulation is free.

If A and B are two points on the curve and if C_1 and C_2 are two curves joining A to B, Eq. (4.85) proves that

$$\int_{C_1} \vec{v} \cdot d\vec{x} = \int_{C_2} \vec{v} \cdot d\vec{x} \tag{4.86}$$

Equation (4.86) shows that the line integral from A to B is path-independent, meaning that the velocity vector field is *conservative*.[18] If then, it is possible to define a potential function $\phi(\vec{x})$ such that

$$\phi(\vec{x}) = \phi(\vec{x}_0) + \int_A^B \vec{v} \cdot d\vec{x} \tag{4.87}$$

where the integral is taken over one of the paths. Thus, Eq. (4.87) gives

$$\nabla \phi = \vec{v} \tag{4.88}$$

where ϕ is termed the *velocity potential* for the velocity vector field.

---

18 The external force vectors per unit mass such as $\nabla(p/\rho_0)$ and $-\nabla(gh)$ are conservative, since they can be expressed as gradient vectors of the scalar specific energies, i.e. p/ρ_0 and $-gh$.

In summary, the velocity potential necessarily exists when the fluid motion is irrotational. Conversely, when the velocity potential exists, the motion is necessarily irrotational. Therefore,

$$\nabla \times \vec{v} = \nabla \times \nabla \phi = 0 \tag{4.89}$$

where $\nabla \times \nabla \phi = 0$ is one of the mathematical identities.

4.8.5.3 Potential Equation

The importance of velocity potential lies in the fact that a pressure field and a velocity field can be decoupled from each other, as long as the flow is irrotational. For example, the imposition of the continuity equation for the incompressible flow (i.e. $\nabla \cdot \vec{v} = 0$, a kinematic condition) on Eq. (4.88) yields a potential equation for ϕ

$$\nabla^2 \phi = 0 \tag{4.90}$$

Once the potential equation for ϕ is solved with proper boundary conditions, a velocity field can be obtained, and a pressure field can directly be obtained by Bernoulli's equation expressed in terms of the velocity potential:

$$B(t) = \frac{\partial \phi}{\partial t} + \frac{1}{2}(\nabla \phi)^2 + \frac{p}{\rho} + g\,h = B_0 \tag{4.91}$$

where the integral of local acceleration over a distance is expressed as a local time derivative of the velocity potential, and the specific kinetic energy becomes its spatial gradient.

The question is how a unique solution of velocity field is sufficient enough to yield the corresponding pressure field (knowing in advance the field of gravitational potential), or how a unique velocity field can be defined by the potential equation with boundary conditions such as the slip condition, periodic condition, and free stream condition (but not the nonslip condition). The answer to these questions are found in the irrotationality condition that the flow can have.

As long as the flow is irrotational, this is a very efficient way of obtaining the flow solution, i.e. velocity and pressure fields. This is, in fact, a complete solution procedure for irrotational flows. Note that an irrotational flow field can be constructed for an arbitrary configuration by combining the potential flow models such as source, sink, doublet, free-vortex. A compressible version of the potential equation (called a full potential equation) also exists with density correction in the continuity equation, isentropic equation of state, and the modified Bernoulli equation. These allow compressibility effects in the flow so that the full potential equation solutions are valid up to mid-subsonic range. However, when a shock wave appears (e.g. in transonic cases), shock location and strength are not correctly represented due to the lack of the energy equation and equation of state.

4.8.5.4 Stream Function

A similar approach can be taken for a stream function ψ defined as follows:

$$(u, v) = \left(\frac{\partial \psi}{\partial y} , -\frac{\partial \psi}{\partial x} \right) \tag{4.92}$$

that ψ automatically satisfies the continuity equation for the incompressible flow, $\nabla \cdot \vec{v} = 0$. Now, imposition of the vorticity-free condition (i.e. $\nabla \times \vec{v} = 0$) on Eq. (4.92) yields a Laplace equation for ψ

$$\nabla^2 \psi = 0 \tag{4.93}$$

Once the stream function is solved with proper boundary conditions, the pressure field can be obtained again by solving the Bernoulli equation, Eq. (4.83).

4.8.6 Scalar and Vector Potentials

4.8.6.1 Helmholtz's Decomposition

Helmholtz's theorem, also known as the fundamental theorem of vector calculus, states that any sufficiently smooth, rapidly decaying vector field in three dimensions can be resolved into the sum of an irrotational (curl-free) vector field (longitudinal field) and a solenoidal (divergence-free) vector field (transverse field).

This implies that any such vector field **v** can be considered to be generated by a pair of potentials: a *scalar potential* ϕ and a *vector potential* $\mathbf{\Psi}$.

$$\mathbf{v} = \underbrace{\nabla \phi}_{\mathbf{v}_l} + \underbrace{\nabla \times \mathbf{\Psi}}_{\mathbf{v}_t} \tag{4.94}$$

where, by mathematical identities,

$$\nabla \times (\nabla \phi) = 0 \tag{4.95}$$

$$\nabla \cdot (\nabla \times \mathbf{\Psi}) = 0 \tag{4.96}$$

We can obtain the solution of the Poisson equation for ϕ using Green's function

$$\phi = -\frac{1}{4\pi} \int_V \frac{\nabla' \cdot \mathbf{v}}{|\mathbf{r} - \mathbf{r}'|} dV' \tag{4.97}$$

where ϕ satisfies the Poisson equation:

$$\nabla \cdot (\nabla \phi) = \nabla^2 \phi = \nabla \cdot \mathbf{v} \tag{4.98}$$

4.8.6.2 Biot–Savart Law

The Poisson equation for the vector potential $\mathbf{\Psi}$ reads

$$\nabla \times (\nabla \times \mathbf{\Psi}) = \nabla(\nabla \cdot \mathbf{\Psi}) - \nabla^2 \mathbf{\Psi} = \nabla \times \mathbf{v} \tag{4.99}$$

and if $\nabla \cdot \mathbf{\Psi} = 0$,

$$\nabla^2 \mathbf{\Psi} = -\nabla \times \mathbf{v} \tag{4.100}$$

We obtain the solution of the Poisson equation Eq. (4.100):

$$\mathbf{\Psi} = \frac{1}{4\pi} \int_V \frac{\nabla' \times \mathbf{v}}{|\mathbf{r} - \mathbf{r}'|} dV' \tag{4.101}$$

which is known as the *Biot–Savart* law.

The Biot–Savart law simply links the vorticity field with the potential flow field. The vorticity field is based on the concept of solid body rotation. Meanwhile, the potential field is based on the fact that each fluid particle conserves the angular momentum, without frictional interactions between particles. Since potential field is linear, we superimpose the potential fields induced by every element of vorticity (point vortex) in the rotational field.

References

1 Kundu, P.K., *Fluid Mechanics*, San Diego, CA: Academic Press, 1990.
2 Shapiro, A.H., "Vorticity," *Illustrated Experiments in Fluid Mechanics*, National Committee for Fluid Mechanics Films, 1980.

3 Batchelor, G.K., *An Introduction to Fluid Dynamics*, Cambridge: Cambridge University Press, 1964.

4 Lighthill, J., *An Informal Introduction to Theoretical Fluid Mechanics*, Oxford: Oxford University Press, 1988.

5 Wu, J.Z., H.Y. Ma, and M.D. Zhou, *Vorticity and Vortex Dynamics*, 2006th ed. Berlin: Springer-Verlag, 2006.

6 Hamman, C.W., J.C. Klewicki, and R.M. Kirby, "On the Lamb Vector Divergence in Navier–Stokes Flows," *J. Fluid Mech.* 610, 261–284, 2008.

7 Talyor, E.S., "Secondary Flow," *Illustrated Experiments in Fluid Mechanics*, National Committee for Fluid Mechanics Films, 1980.

8 Howe, M.S., *Theory of Vortex Sound*, 1st ed. Cambridge: Cambridge University Press, 2003.

9 Cowern, D. PBS digital studios channel "Physics Girl", *Crazy pool vortex*, YouTube.

10 Vogel, S., C.P. Ellington, Jr., and D.L. Kilgore, Jr., "Wind-induced Ventilation of the Burrow of the Prairie-dog, *Cynomys ludovicianus*," *J. Comp. Physiol.*, 85, 1–14, 1973.

Problems

4.1 When a cup of tea is stirred with a spoon and then stopped, the tea leaves gather at the center of the bottom of the cup [7]. Explain this famous observation with centrifugal force acting on water. (Hint: the water circulating in the cup is viscously resisted on the walls.)

4.2 We have two vortex-generating systems. Figure (a) shows the vortex in a cylindrical container with a drainage hole at the bottom (water is injected at the side slit). Figure (b) shows a bended vortex sucked into the running jet engine of an aircraft standing on the runway.

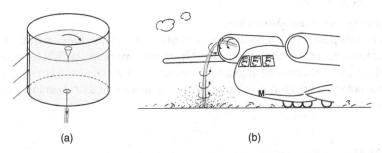

(a)   (b)

a) Explain how the vortex in the cylindrical container attains energy at a rate.

b) Explain how the vortex in the running engine attains energy at a rate.

c) Discuss the similarities in the vortex-generating mechanism between the two systems.

4.3 Show that the circulation Γ defined by

$$\Gamma = \int_C \vec{\omega} \cdot \vec{n} \, dA = \oint_C \vec{v} \cdot d\vec{s}$$

is independent of the circuit enclosing the vortex filament.

4.4 The Biot–Savart law states that if a vorticity field is given, the induced velocity field $\vec{v} = \nabla \times \mathbf{\Psi}$ is obtained by

$$\mathbf{\Psi} = \frac{1}{4\pi} \int_V \frac{\nabla' \times \mathbf{v}}{|\mathbf{r} - \mathbf{r}'|} \, dV'$$

where $\mathbf{\Psi}$ is the vector potential function. For a vortex filament, show the followings.
a) The induced velocity field can be expressed as

$$\vec{v} = \nabla \times \mathbf{\Psi} = \frac{\Gamma}{4\pi} \int_{-\infty}^{+\infty} \frac{d\vec{l} \times \vec{v}}{|\vec{v}|^3}$$

where $d\vec{l}$ is the arc length element vector along the vortex filament and Γ is the circulation.
b) If the vortex filament is a straight line, the induced velocity field in 2D is

$$v_\theta(r) = \frac{\Gamma}{2\pi r}$$

where r is the radial distance.

4.5 (Problem 4.4 is continued). If the vortex line extends from 0 to $+\infty$ (i.e. the vortex line is semi-infinite), show that the induced velocity field is half that induced by an infinite vortex line:

$$v_\theta(r) = \frac{\Gamma}{4\pi r}$$

4.6 A B-2 stealth bomber produces thrust with two jet engines hidden inside the main body (see figure below). Under what physical principle is the exhaust duct designed for efficient production of thrust and for concealing the engine nozzle from direct rear view?

4.7 A cambered airfoil at a small angle of attack is tested in a wind tunnel.
a) Explain the generation of lift by the streamlines around the wing.
b) Explain how the vortices in the boundary layers over and underneath the airfoil surfaces determine the streamline patterns around the wing.

c) If the airfoil were tested at a high-altitude condition, how would lift be changed?
d) If the wind tunnel were operated to produce a higher volumetric flow rate, how would lift be changed?

4.8 In unsteady vortical flows, the Lamb acceleration vector, $\vec{L} = \vec{\omega} \times \vec{v}$, can represent a sound source [8]. Physically explain the sound generation mechanism in the followings.
   a) Explain how sound is generated when turbulent eddies impinge upon the leading-edge of the airfoil.
   b) Explain how sound is generated when turbulent eddies scatter at the trailing-edge of the airfoil.

4.9 Discuss the physical significances of (i) Lamb vector divergence $\nabla \cdot \vec{L}$ and (ii) Lamb vector curl $\nabla \times \vec{L}$, where the Lamb acceleration vector is defined as $\vec{L} = \vec{\omega} \times \vec{v}$.

4.10 Sketch the following quantities in the free shear layer (see below): (i) velocity vector $\vec{v}$, (ii) vorticity vector $\vec{\omega}$, (iii) the Lamb vector $\vec{L} = \vec{\omega} \times \vec{v}$, (iv) curl of vorticity vector $\nabla \times \vec{\omega}$, (v) rate of change of kinetic energy $\vec{v} \cdot (\nabla \times \vec{\omega})$, (vi) rate of change of internal energy (or enstrophy) $-\vec{\omega} \cdot \vec{\omega}$, and (vii) divergence of the Lamb vector $\nabla \cdot \vec{L}$.

4.11 In curved motion, the normal component of the convective acceleration in streamline coordinates represents the centripetal acceleration. It can also be expressed as the sum of $v\, \partial v/\partial n$ and $(\vec{\omega} \times \vec{v}) \cdot \vec{i}_n$.
   a) For a forced vortex, find the centripetal acceleration with $v\, \partial v/\partial n$ and $(\vec{\omega} \times \vec{v}) \cdot \vec{i}_n$.
   b) For a free vortex, find the centripetal acceleration with $v\, \partial v/\partial n$ and $(\vec{\omega} \times \vec{v}) \cdot \vec{i}_n$.

4.12 Let us suppose a half-ring vortex is created by suddenly pushing a dish in a swimming pool, as sketched below. Surprisingly, the half-ring vortex travels fairly long distances without significantly losing its strength [9].

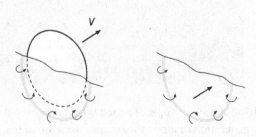

a) Explain how the half-ring vortex propels by itself.
b) What equation of motion is most suitable to model the movement of the vortex in water? Physically explain the reason.
c) What equation of motion should be used if the modeling includes vortex created at the dish side-edges? Physically explain the reason.

4.13 In a potential flow, the Bernoulli function B(t) is defined as

$$B(t) = \underbrace{\int \frac{\partial \vec{v}}{\partial t} \cdot d\vec{r}}_{Ⓐ} + \underbrace{\frac{1}{2} v^2 + \frac{p}{\rho} + gh}_{Ⓑ}$$

satisfies

$$\nabla B(t) \cdot d\vec{s} = 0$$

This means that B(t) is constant on the streamline. Explain how the term Ⓐ can physically be interpreted, comparing with the term Ⓑ.

4.14 In an acoustic field, the instantaneous density and pressure can be decomposed as

$$\rho = \rho_0 + \rho', \qquad p = p_0 + p'$$

where ρ_0 and p_0 are the density and pressure of the ambient air.
a) Show that the momentum equation can be linearized as

$$\rho_0 \frac{\partial \vec{v}}{\partial t} = -\nabla p'$$

b) Using the velocity potential ϕ, where $\vec{v} = \nabla \phi$, show that the acoustic pressure p' can be represented by the local time derivative of the velocity potential:

$$p' = -\rho_0 \frac{\partial \phi}{\partial t}$$

4.15 The gap between two circular disks of radius R is filled with liquid of density ρ. If the disk above moves at a constant speed of V toward the disk below, which is stationary, the liquid is squeezed out between the two disks. If we assume that the gap width h is very small relative to R and the flow is inviscid, the velocity of the liquid $u(r)$ can be considered uniform and parallel to the disks. Note that the pressure at $r = R$ is the ambient pressure.

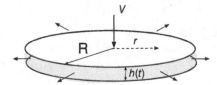

a) Derive an expression for velocity $u(r)$.
b) Derive an expression for static pressure $p(r)$ in the gap.
c) Calculate the force applied to the disk on the top.
d) Sketch the pressure distribution $p(r)$ versus r.
e) With respect to pressure distribution, discuss the difference between this problem and the example presented in Section 1.5.1 *Transient inertia force*.

4.16 Prairie dogs build burrows with some knowledge on wind aerodynamics [10].

a) Discuss their air-ventilation system, sketching their burrows. Explain how it works regardless of wind direction.
b) Exercise the Bernoulli equation to estimate the volumetric flow rate of ventilated air in a tunnel, which is simplified by a U-tube.
c) What parameters are important for prairie dogs to build the burrows for efficient ventilation?

4.17 In potential flow, the velocity potential ϕ is associated with two specific energies: $\partial\phi/\partial t$ and $(\nabla\phi)^2/2$.
a) Physically describe these two terms.
b) Show that in compressible fluids,

$$p - p_\infty = -\rho_\infty \frac{\partial\phi}{\partial t}$$

represents the generation of sound, where p_∞ and ρ_∞ are the pressure and density in undisturbed parts of the fluid.

5

Vortex Dynamics

5.1 What is Vortex?

Vortex refers, in a broad sense, to a fluid in rotation. It can either be concentrated in space to form a tube (or filament), or be distributed in space as a sheet (Figure 5.1). The strength of the vortex in a closed circuit C can be defined by the circulation Γ_c:

$$\Gamma_c = \int_C \vec{\omega} \cdot \vec{n} \, dA = \oint_C \vec{v} \cdot d\vec{s} \tag{5.1}$$

where $\vec{\omega} = \nabla \times \vec{v}$ is the vorticity vector, $\vec{n}$ is the outward unit vector normal to an elemental area dA, and $d\vec{s}$ is the unit tangent vector along the closed circuit C. The circulation Γ_c measures how fast a sizable fluid within the closed circuit rotates at a rate like a solid body [1].

5.1.1 Helmholtz's Vortex Theorems

In a barotropic fluid under the conservative body forces and in the absence of viscous forces, Helmholtz's vortex theorems state that a vortex tube moves with the fluid and that its strength (or circulation) in a closed circuit C is the same in all cross-sections and remains constant [2]:

$$\frac{D\Gamma_c}{Dt} = 0 \tag{5.2}$$

If there were no circulation in the beginning, it would be kept that way at all times.[1] Otherwise, the magnitude of the vorticity vector that preserves the solenoidal properties is proportional to the density ρ times the total length l^* of the vortex line of our interest, i.e.

$$|\vec{\omega}| \sim \rho \, l^*, \quad \text{or} \quad |\vec{\omega}|/\rho \, l^* = \text{constant} \tag{5.3}$$

while the vortex is isotropically expanded by compressibility or stretched (or contracted) along the axis of rotation. This is due to the fact that the mass of the vortex tube is invariant and the total angular momentum is to be conserved.[2]

---

1 This condition is often called *persistency of circulation* [3]. More discussion on this continues in Section 5.4.2 *Impulsive motion of a body at rest*.
2 One of Helmholtz's vortex theorems, Eq. (5.3), will be proven in Section 5.3.4 *Proof of one of Helmholtz's vortex theorems*.

Introduction to Fluid Dynamics: Understanding Fundamental Physics, First Edition. Young J. Moon.
© 2022 John Wiley & Sons, Inc. Published 2022 by John Wiley & Sons, Inc.
Companion website: www.wiley.com/go/Moon/IntroductiontoFluidDynamics

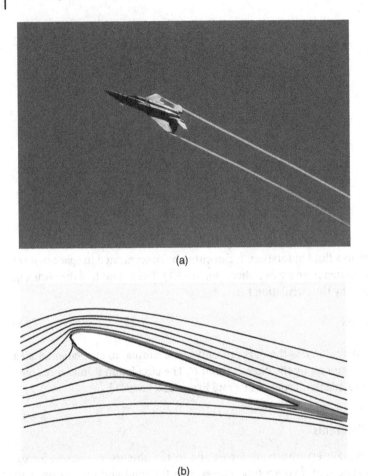

(a)

(b)

Figure 5.1 A concentrated-vortex pair shed from the wingtips of an aircraft (a); vortices distributed in the boundary layers over and underneath the airfoil (b). Source: Timothy Newman/Unsplash

5.2 Kelvin's Circulation Theorem

Kelvin's circulation theorem states that the circulation Γ_c in a closed circuit C can be changed over time by a net external torque applied to the fluid. The total rate of change of the circulation in the closed circuit C can be expressed as follows:

$$\frac{D\Gamma_c}{Dt} = \frac{D}{Dt}\oint_C \vec{v} \cdot d\vec{s} = \oint_C \frac{D\vec{v}}{Dt} \cdot d\vec{s} + \oint_C \vec{v} \cdot \frac{D\,d\vec{s}}{Dt} \tag{5.4}$$

where the second term on the right-hand side is zero [3], since

$$\oint_C \vec{v} \cdot \frac{D\,d\vec{s}}{Dt} = \oint_C \vec{v} \cdot d\vec{v} = \oint_C d\left(\frac{1}{2}\vec{v}\cdot\vec{v}\right) = 0 \tag{5.5}$$

By employing the Navier–Stokes equations that relate an acceleration vector to the net force vector acting on the fluid per unit mass

$$\frac{D\vec{v}}{Dt} = \vec{f}_{net} = \frac{1}{\rho}(-\nabla p + \vec{f}_{cbf}) + \frac{1}{\rho}(\vec{f}_v + \vec{f}_{ncbf}) \tag{5.6}$$

we can show that the line integral of the acceleration vector tangent to the closed circuit (i.e. rotational kinetic energy of the fluid per unit mass) equals the mechanical work done on the fluid per unit mass by the net torque:

$$\oint_C \frac{D\vec{v}}{Dt} \cdot d\vec{s} = \oint_C \frac{D\vec{v}}{Dt} \cdot (d\vec{\theta} \times \vec{r}) = \oint_C \vec{f}_{net} \cdot (d\vec{\theta} \times \vec{r}) \tag{5.7}$$

where $\vec{r}$ is a vector of moment arm.

With Eqs. (5.6) and (5.7), the total rate of change of the circulation in the closed circuit C can finally be written as a line integral of the torque vectors acting on the fluid per unit mass:

$$\frac{D\Gamma_c}{Dt} = \oint_C \frac{1}{\rho}\left(-\nabla p + \vec{f}_v + \vec{f}_{ncbf}\right) \cdot d\vec{s} \tag{5.8}$$

where the conservative body force term disappears since it can be expressed as a gradient vector of a scalar potential, i.e. $\vec{f}_{cbf} = -\nabla\phi$. Note that the unit of Eq. (5.8) $[m^2/s^2]$ is the specific energy of the fluid in the closed circuit C.

Equation (5.8), known as Kelvin's circulation theorem, states that at any instant, the rate of change of the circulation within the closed circuit C is proportional to the net torque exerted on the fluid by the three forces on the right-hand side of Eq. (5.8). For instance, viscous force exerts a torque on the fluid via frictional intermolecular interferences, while pressure force exerts a *baroclinic torque* when the fluid is not of a uniform density and its density gradient is not parallel to the pressure gradient. A nonconservative body force (e.g. Coriolis force or Lorentz force) also produces a torque when the fluid moves in the rotating reference frame, or in magnetohydrodynamic flows; the reason is that the line of action of such forces does not go through the center of mass of the particles.

If all the forces on the right-hand side of Eq. (5.8) are absent along the closed circuit C (e.g. a barotropic fluid under the conservative body forces and in the absence of viscous forces), the circulation in the closed loop C does not change over time:

$$\frac{D\Gamma_c}{Dt} = 0 \tag{5.9}$$

which corresponds to one of Helmholtz's vortex theorems. One example is the circulation within a Couette flow. Although the flow is rotational and viscous stresses are present everywhere, the circulation does not change over time because there is no net viscous force acting on the fluid.

5.3 Vorticity Transport Equation

In contrast to the circulation theorem of Kelvin, the vorticity transport equation shows how the total rate of change of a vorticity vector at a point is related to the torques exerted on the fluid of a unit mass and also to the local gradients of the velocity components, namely, the rates of rotation and strain of the fluid (e.g. a curl of the Lamb vector). The vorticity transport equation can be derived by taking a curl on the Navier-Stokes equations. Before deriving the vorticity transport equation, the physical meanings of the curl vectors taken on the velocity, acceleration, and force vectors are explained in the following section.

5.3.1 Curl of Velocity, Acceleration, and Force (Per Unit Mass) Vectors

5.3.1.1 Curl of Velocity Vector

A curl of a velocity vector $\nabla \times \vec{v}$ represents the sum of the turning rates of the two orthogonal sides, dx and dy, of an infinitesimally small fluid element:

$$\frac{\tan d\theta_1 - \tan d\theta_2}{dt} = \left\{ \left(\underbrace{\frac{\partial v}{\partial x} dx \cdot dt}_{v'} \right)/dx - \left(\underbrace{\frac{\partial u}{\partial y} dy \cdot dt}_{u'} \right)/dy \right\}/dt$$

$$= \frac{\partial v}{\partial x} - \frac{\partial u}{\partial y} = \omega_z \tag{5.10}$$

where $v'\,dt$ and $u'\,dt$ are the distances of travel in the y and x directions at $(x + dx, y)$ and $(x, y + dy)$ over dt, respectively. Note that Eq. (5.10) corresponds to the vorticity vector in the z direction, ω_z. Similarly, ω_x and ω_y can be defined in three-dimensional space.

5.3.1.2 Curl of Acceleration Vector

It is to be noted that a curl of an acceleration vector, $\nabla \times (D\vec{v}/Dt)$, does not equal the total rate of change of a vorticity vector. In fact, it consists of (i) the local rate of change of the vorticity vector and (ii) the curl of the Lamb acceleration vector:

$$\nabla \times \frac{D\vec{v}}{Dt} = \nabla \times \frac{\partial \vec{v}}{\partial t} + \nabla \times (\vec{\omega} \times \vec{v}) \tag{5.11}$$

where the curl of the gradient vector of the specific kinetic energy, $\nabla \times \nabla(v^2/2)$, is be zero since it is one of the mathematical identities:

$$\nabla \times \nabla \phi = 0, \qquad \nabla \cdot (\nabla \times \vec{\varphi}) = 0 \tag{5.12}$$

The curl of the Lamb acceleration vector, $\nabla \times (\vec{\omega} \times \vec{v})$,[3] can be expressed as follows:

$$\nabla \times (\vec{\omega} \times \vec{v}) = \underbrace{(\vec{v} \cdot \nabla)\vec{\omega}}_{(0)} - \underbrace{(\vec{\omega} \cdot \nabla)\vec{v}}_{(1)} + \underbrace{\vec{\omega}(\nabla \cdot \vec{v})}_{(2)} - \cancel{\vec{v}(\nabla \cdot \vec{\omega})}^{0} \tag{5.13}$$

which shows that the curl of the Lamb vector represents the local rate of change of the vorticity vector while the vortex is convected (term 0), stretched, and tilted by local velocity components (term 1),[4] and isotropically expanded by compressibility effect (term 2).

The curl of the acceleration vector is finally written as follows:

$$\nabla \times \frac{D\vec{v}}{Dt} = \frac{D\vec{\omega}}{Dt} - (\vec{\omega} \cdot \nabla)\vec{v} + \vec{\omega}(\nabla \cdot \vec{v}) \tag{5.14}$$

which represents (i) the total rate of change of the vorticity vector (via its local and convective rates of change), (ii) the rates of stretching and tilting (via straining of the fluid), and (iii) the rate of isotropic expansion (via compressibility effects) [4].

5.3.1.3 Curl of Force Vector (Per Unit Mass)

A curl of a force vector per unit mass, $\nabla \times (\vec{f}/\rho)$, represents the sum of the works done by two components of torque exerted on the fluid per unit mass, by which the two sides of a fluid element,

---

3 $\nabla \times (\vec{a} \times \vec{b}) = (\vec{b} \cdot \nabla)\vec{a} + \vec{a}(\nabla \cdot \vec{b}) - (\vec{a} \cdot \nabla)\vec{b} - \vec{b}(\nabla \cdot \vec{a})$.

4 More discussion continues in Sections 5.5 and 5.6.

dx and dy, have been turned $d\theta_1$ and $d\theta_2$ degrees. In this case, each work is normalized by the area expressed by the product of moment arm and arc length. It mathematically reads

$$M'_{z1} \, d\theta_1/(dx \, dy') - M'_{z2} \, d\theta_2/(dx' \, dy)$$

$$= \left(\underbrace{\frac{\partial(f_y/\rho)}{\partial x} dx \, dx}_{(f_y/\rho)'} \right) \left(\frac{v' \, dt}{dx} \right) \Big/ (dx \, dy') - \left(\underbrace{\frac{\partial(f_x/\rho)}{\partial y} dy \, dy}_{(f_x/\rho)'} \right) \left(\frac{u' \, dt}{dy} \right) \Big/ (dx' \, dy)$$

$$= \frac{\partial(f_y/\rho)}{\partial x} - \frac{\partial(f_x/\rho)}{\partial y} = M_z \tag{5.15}$$

where $M'_{z1} = (f_y/\rho)' \, dx$ and $M'_{z2} = (f_x/\rho)' \, dy$ are the torques per unit mass, produced by the forces per unit mass exerted at $(x + dx, y)$ and $(x, y + dy)$ with moment arms dx and dy. Note that $u'dt = dx' = dy \, d\theta_2$ and $v'dt = dy' = dx \, d\theta_1$ are the arc lengths drawn in the x and y directions, and that the working angles are thus $d\theta_1 = v'dt/dx$ and $d\theta_2 = u'dt/dy$, respectively. Therefore, $M'_{z1} \, d\theta_1$ and $M'_{z2} \, d\theta_2$ represent the works done by torques M'_{z1} and M'_{z2} per unit mass exerted over the angles $d\theta_1$ and $d\theta_2$, respectively. The curl of the force vector per unit mass then becomes a net torque vector (in the z-direction) per unit mass denoted by M_z.

5.3.2 Curl of the Navier–Stokes Equations

We have shown that the curl of the acceleration vector and the curl of the force vector per unit mass represent the rate of change of the vorticity vector and the torque vector per unit mass, respectively. Therefore, to describe how or at what rate a vorticity vector can change at a point, a curl is directly taken on the Navier–Stokes equations:

$$\nabla \times \frac{D\vec{v}}{Dt} = \nabla \times \left(\frac{1}{\rho}(-\nabla p + \vec{f}_{cbf}) + \frac{1}{\rho}(\vec{f}_v + \vec{f}_{ncbf}) \right) \tag{5.16}$$

where the acceleration vector is expressed in Crocco's form:

$$\frac{D\vec{v}}{Dt} = \frac{\partial \vec{v}}{\partial t} + \frac{1}{2}\nabla v^2 + \vec{\omega} \times \vec{v} \tag{5.17}$$

Using the curls of velocity, acceleration, and force vectors, Eqs. (5.16) and (5.17) can be written as follows:

$$\frac{D\vec{\omega}}{Dt} = \underbrace{(\vec{\omega} \cdot \nabla)\vec{v}}_{(1)} - \underbrace{\vec{\omega}(\nabla \cdot \vec{v})}_{(2)} - \underbrace{\nabla \times (\frac{1}{\rho}\nabla p)}_{(3)} + \underbrace{\nabla \times (\frac{1}{\rho}\vec{f}_v)}_{(4)} + \underbrace{\nabla \times (\frac{1}{\rho}\vec{f}_{ncbf})}_{(5)} \tag{5.18}$$

where the source terms on the right-hand side represent the followings:

(1) vortex stretching and tilting
(2) isotropic expansion
(3) pressure (or baroclinic) torque
(4) viscous torque
(5) nonconservative body-force torque

This is called the *vorticity transport equation*. The first two terms, (1) and (2), show how the total rate of change of the vorticity vector can change with the local velocity gradients and the compressibility of the fluid, respectively. The last three terms (3), (4), and (5) represent three different torques that can produce vorticity at a rate.

In a two-dimensional, inviscid, incompressible fluid under the conservative body forces, the vorticity transport equation simply reads

$$\frac{D\vec{\omega}}{Dt} = 0 \tag{5.19}$$

Equation (5.19) implies that if there were vorticity in the field, it remains unchanged, and if there were no vorticity at the beginning, the field has to be constantly vorticity-free.

5.3.3 In Connection with Kevin's Circulation Theorem

In Kelvin's circulation theorem, the total rate of change of the circulation in a closed circuit C can be decomposed as follows:

$$\frac{D\Gamma_c}{Dt} = \int_C \frac{D\vec{\omega}}{Dt} \cdot \vec{n}\, dA + \int_C \vec{\omega} \cdot \left(\frac{D\vec{n}}{Dt}\, dA + \vec{n}\, \frac{D\, dA}{Dt} \right) \tag{5.20}$$

where the first integral on the right represents the rate of change of the circulation created by the total rate of change of the vortices in the closed circuit C.

The second integral is the sum of the rate of change of the circulation by vortex tilting and the same done by vortex stretching and isotropic expansion (Figure 5.2). It is to be noted that the second integral can also be expressed as follows:

$$\int_C \vec{\omega} \cdot \left(\frac{D\vec{n}}{Dt}\, dA + \vec{n}\, \frac{D\, dA}{Dt} \right) = \int_C \left(-(\vec{\omega}\cdot\nabla)\vec{v} + \vec{\omega}(\nabla\cdot\vec{v}) \right) \cdot \vec{n}\, dA \tag{5.21}$$

If we combine Eqs. (5.20) and (5.21), we have

$$\frac{D\Gamma_c}{Dt} = \int_C \left(\frac{D\vec{\omega}}{Dt} - (\vec{\omega}\cdot\nabla)\vec{v} + \vec{\omega}(\nabla\cdot\vec{v}) \right) \cdot \vec{n}\, dA \tag{5.22}$$

If Stokes' theorem is applied to the right-hand side of Kelvin's circulation theorem, it is written as follows:

$$\oint_C \frac{1}{\rho}\left(-\nabla p + \vec{f}_v + \vec{f}_{ncbf} \right) \cdot d\vec{s} = \int_C \nabla \times \left\{ \frac{1}{\rho}\left(-\nabla p + \vec{f}_v + \vec{f}_{ncbf} \right) \right\} \cdot \vec{n}\, dA \tag{5.23}$$

By combining Eqs. (5.22) and (5.23) and dropping off the integral, we can obtain the vorticity transport equation:

$$\frac{D\vec{\omega}}{Dt} - (\vec{\omega}\cdot\nabla)\vec{v} + \vec{\omega}(\nabla\cdot\vec{v}) = \nabla \times \left\{ \frac{1}{\rho}\left(-\nabla p + \vec{f}_v + \vec{f}_{ncbf} \right) \right\} \tag{5.24}$$

In summary, the total rate of change of the circulation in the closed loop C is equivalent to three area-integral terms that represent (i) the total (local and convective) rate of change of the vorticity

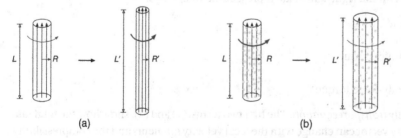

(a)   (b)

Figure 5.2 Moment of inertia changes via vortex stretching in incompressible flow (a) and isotopic volume expansion in compressible flow (b)

vector, (ii) the rate of change of the vorticity vector by vortex tilting and stretching, and (iii) the rate of change of the vorticity vector by volumetric dilatation of the fluid.

5.3.4 Proof of One of Helmholtz's Vortex Theorems

In three-dimensional space, one of Helmholtz's vortex theorems states that while preserving the solenoidal properties, the magnitude of the vorticity vector is proportional to the density ρ times the length l^* of the vortex:

$$|\vec{\omega}| \sim \rho\, l^*, \quad \text{or} \quad |\vec{\omega}|/\rho\, l^* = \text{constant} \tag{5.25}$$

as the vortex is stretched (or contracted) at a rate along the axis of rotation with the fluid or isotropically expanded by compressibility.

To prove this theorem, we consider a barotropic fluid under the conservative body forces and in the absence of viscous forces. In this case, Kelvin's circulation theorem states that the circulation in the closed circuit C does not change over time:

$$\frac{D\Gamma_c}{Dt} = 0 \tag{5.26}$$

or that the vorticity transport equation is written as follows:

$$\frac{D\vec{\omega}}{Dt} = (\vec{\omega} \cdot \nabla)\,\vec{v} - \vec{\omega}\,(\nabla \cdot \vec{v}) \tag{5.27}$$

With the continuity equation

$$\nabla \cdot \vec{v} = -\frac{1}{\rho}\frac{D\rho}{Dt} \tag{5.28}$$

and by dividing both sides of the equation by ρ, Eq. (5.27) reads

$$\frac{D(\vec{\omega}/\rho)}{Dt} = \left(\frac{\vec{\omega}}{\rho} \cdot \nabla\right)\vec{v} \tag{5.29}$$

To consider only stretching and isotropic expansion of the vortex, the source term in Eq. (5.29) is modified as follows:

$$\frac{D(\vec{\omega}/\rho)}{Dt} = \left(\vec{\omega}/\rho\right)^T [A] \tag{5.30}$$

where the matrix $[A]$ reads

$$[A] = \left[\nabla \vec{v}\right][I] \tag{5.31}$$

If we use the definition of linear strain rate, the matrix $[A]$ can be expressed as follows:

$$[A] = \left[\nabla \vec{v}\right][I] = \left[\frac{1}{l_j}\frac{Dl_i}{Dt}\right][I] \tag{5.32}$$

where $Dl_i = l_i(t + dt) - l_i(t)$ is an incremental change of the length of the vortex in the i direction over a time interval dt.

If we combine Eq. (5.30) with Eq. (5.32), Eq. (5.30) reads

$$\frac{D(\vec{\omega}/\rho)}{Dt} = \left(\vec{\omega}/\rho\right)^T \left[\frac{1}{l_j}\frac{Dl_i}{Dt}\right][I] \tag{5.33}$$

and can be simplified as follows:

$$\frac{D(\vec{\omega}/\rho)}{\vec{\omega}/\rho} = \frac{D\vec{l}}{\vec{l}} \tag{5.34}$$

If we integrate Eq. (5.34), one of Helmholtz's vortex theorems is proven, i.e.

$$\vec{\omega}/\rho \sim \vec{l} \quad \text{or} \quad |\vec{\omega}|/\rho \sim l^* \tag{5.35}$$

where $l^* = |\vec{l}|$ is the length of the vortex and $|\vec{\omega}|$ is its magnitude.

In two-dimensional flows, Eq. (5.30) shows that vorticity per unit density $\vec{\omega}/\rho$ is an invariant quantity, i.e.

$$\frac{D(\vec{\omega}/\rho)}{Dt} = 0 \tag{5.36}$$

which is analogous to an incompressible counterpart, to which $\vec{\omega}$ is an invariant quantity in two-dimensional flows.

5.4 Viscous Torque

In a barotropic viscous fluid under the conservative body forces, the vorticity transport Eq. (5.18) reads

$$\frac{D(\vec{\omega}/\rho)}{Dt} = \left(\frac{\vec{\omega}}{\rho} \cdot \nabla\right) \vec{v} + \nabla \times \left(\frac{1}{\rho}\vec{f}_v\right) \tag{5.37}$$

where the first term on the right representing vortex stretching and tilting is associated with the changes in strength and direction of a vortex filament.[5]

The second term is the net viscous torque vector acting on the fluid per unit mass, which creates and diffuses vorticity at a rate. Note also that the importance of the convection of vorticity vs. its diffusion is determined by the local Reynolds number. In fluid dynamics, viscous torques are involved in many interesting physics such as vortex roll-up, stretching and tilting, diffusion, formation of secondary flow, turbulence production, lift generation. In this section, we are going to describe flow physics related to viscous torque.

5.4.1 Vortex Roll-up

5.4.1.1 Physics at the Sharp Edge
When a fluid at rest is suddenly forced to move, the fluid close to the solid surface is simultaneously rotated and angularly strained at a rate by viscous forces acting on the solid surface. However, the fluid at the edge of the surface encounters a geometrical singularity; it is instantly rotated and forced to move out to the ambient fluid so that a shear layer emanated from the edge rolls into a concentrated vortex. In a more descriptive expression, a concentrated viscous force creates a concentrated vortex.[6]

Figure 5.3 shows an interesting sketch of vortex roll-up created by an impulsive jet from the mouth of a dolphin. The dolphin's mouth and lips act like a tube with a rounded edge. The strength of the concentrated vortex depends on various parameters such as fluid acceleration, geometric

---

5 More discussion continues in Sections 5.5 *Vortex stretching* and 5.6 *Vortex tilting*.
6 Shapiro [5].

Figure 5.3 Vortex ring of air produced by an impulsive jet from a dolphin's mouth

scales of the system, viscosity and density of the fluid. Note that the created vortex ring travels quite a distance in the water because the time scale of molecular diffusion is much larger than that of convection.

5.4.1.2 Vortex Rings

In nature, vortex roll-ups are often used to propel a body or change the direction of fluid inertia. Fish are known to generate propulsion with the wave motion of their body. In particular, the swinging motion of the tail fin creates a series of vortex rings, through which a linear momentum can be directed in the streamwise direction (Figure 5.4). What is interesting is that the formation of a vortex ring at the tail fin is a clear evidence that a vortex is solenoidal; a vortex tube ends at the surface of the body or forms a closed loop. The same principle of thrust applies to birds as they fly forward by flapping wings to create a series of vortex rings to produce propulsive force. In fact, the flapping motion of the birds' two wings is similar to the swinging motion of the fish's tail fin.

5.4.1.3 Wingtip Vortex

A wingtip vortex is created at the tip of the wings of an aircraft through a vortex roll-up at the sharp edge (Figure 5.5). The pressure difference between the upper and lower surfaces of the aircraft wing causes the air to be spilled from the lower surface of the wingtip to the upper surface when the aircraft wing moves forward. As a result, a shear layer shed from the wingtip rolls into a three-dimensional spiral vortex, as shown in Figure 5.5. It is to be noted that the wingtip vortices are undesirable for the system due to mechanical energy loss. The wingtip vortex accompanies a strong downwash behind the wings, which induces drag by changing the effective angle of attack of the wing. This is the so-called *induced-drag*.

5.4.2 Impulsive Motion of a Body at Rest

The vortex roll-up at the sharp edge accompanies a force, which may or may not be desirable for the system. A classic example in aerodynamics is a starting vortex (counterclockwise) generated

Figure 5.4 Fish propulsion by swinging motions of the tail fin; linear momentum directed in the streamwise direction through a series of vortex rings

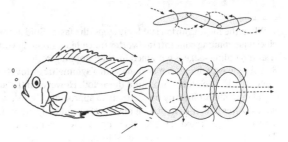

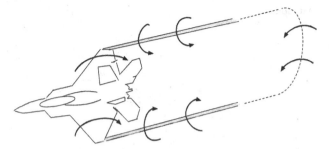

Figure 5.5 Wingtip vortices of an aircraft; vortex roll-up in three-dimensional space

by an airfoil that suddenly moves forward with a small angle of attack (Figure 5.6).[7] If we consider this impulsive motion of the airfoil as an action of force, the fluid at rest reacts to it with viscosity. To be more specific, the fluid creates a circulatory flow around the airfoil and the strength of the circulatory flow (i.e. circulation) equals that of the starting vortex (direction is opposite).

This reaction of the fluid strictly obeys Kelvin's circulation theorem; under the conservative body forces, the circulation of an inviscid, uniform-density fluid in the closed circuit C is preserved in time:

$$\frac{D\Gamma_c}{Dt} = 0 \tag{5.38}$$

This is often referred to as *persistence of circulation*. If there were to be no circulation in the circuit at the beginning, it must remain unchanged for all times, and if there were a circulation inside the circuit, the circulation should not be changed in time.

The circulatory flow around the airfoil produces a force on the airfoil in the transverse direction, namely, the lift force. In this case, the flow is forced to be divided over and underneath the airfoil, which scrapes the surfaces and creates vortices in the boundary layers, while imposing the Kutta condition at the trailing-edge. The strength and distribution of the vortices in the boundary layers are directly coupled with the strength of the starting vortex shed from the trailing-edge.[8] When

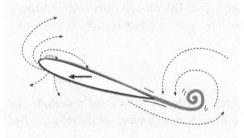

Figure 5.6 A starting vortex (counterclockwise) formed at the trailing-edge by an impulsive motion of an airfoil from rest (persistency of circulation)

---

7 When the moving airfoil suddenly stops, the faster fluid at the upper surface spills again and rolls into another clockwise-rotating concentrated vortex (or called a stopping vortex), while the fluid that stops at the lower surface near the trailing-edge is entrained into the stopping vortex.

8 A simple experiment can be done with a symmetric airfoil. If it is suddenly moved forward at zero angle of attack, there will be no circulation around the airfoil. However, if the angle of attack is increased from 0 to 90°, the circulation almost linearly increases up to approximately 15°, with linear increase of the strength of the starting vortex. After that, the circulation of the airfoil suddenly drops because of sudden appearance of a counter-rotating starting vortex created at the leading-edge of the upper surface of the airfoil with separation of boundary layer. Therefore, the net circulation of the starting vortices is decreased, and we call this stall.

circulation is created around the airfoil (also known as the bound vortex inside the airfoil), the two flows above and below the airfoil must take on different speeds (flow above is faster than flow below) and consequently maintain different surface pressures. Thus, a lift force is generated over the airfoil.

5.4.3 Viscous Diffusion

5.4.3.1 Convection-Diffusion Equation of Vorticity

A curl of a vorticity vector $\nabla \times \vec{\omega}$ is a transverse spatial gradient vector of the vorticity vector and is related to the net viscous force vector per unit mass as follows:

$$\vec{f}_v = \mu \left(\nabla \times \vec{\omega} \right) \tag{5.39}$$

where μ is the viscosity of the fluid.

Similarly, a curl of a viscous force vector per unit mass is related to the total rate of change of the vorticity vector, as presented in the vorticity transport equation:

$$\frac{D(\vec{\omega}/\rho)}{Dt} = \left(\frac{\vec{\omega}}{\rho} \cdot \nabla \right) \vec{v} + \nu \nabla \times \left(\nabla \times \vec{\omega} \right) \tag{5.40}$$

With a vector relation and the solenoidal property of vorticity, the right-hand side of Eq. (5.40) is written as follows:

$$\nabla \times \left(\nabla \times \vec{\omega} \right) = \nabla (\underbrace{\nabla \cdot \vec{\omega}}_{0}) - \nabla^2 \vec{\omega} \tag{5.41}$$

and the vorticity transport equation can be cast into a convection–diffusion equation of vorticity:

$$\frac{D(\vec{\omega}/\rho)}{Dt} = \left(\frac{\vec{\omega}}{\rho} \cdot \nabla \right) \vec{v} - \nu \nabla^2 \vec{\omega} \tag{5.42}$$

where the kinematic viscosity ν can generally be referred to as the hydrodynamic diffusivity.

5.4.3.2 Reynolds Number

In unbounded flows, the vorticity field will never be uniform since it is not possible to reach a static force equilibrium. In a boundary layer, for example, the vorticity produced at the wall is being diffused away from the wall while being convected with the flow. Thus, the boundary layer continuously grows with its thickness δ determined by the Reynolds number, that is, the ratio between the convection speed u_c and the diffusion speed u_d:

$$Re = \frac{u_c}{u_d} = \frac{U_\infty}{\mu/(\rho l)} = \frac{\rho U_\infty l}{\mu} \tag{5.43}$$

It is interesting to note that the edge of the boundary layer is the boundary of viscous diffusion that ever reaches the incoming flow. Thereby, the Reynolds number can be used to characterize the boundary layer thickness. If the viscous effect is more significant, the diffusion process is faster than the inertia effect and thus, the boundary layer becomes thicker. In the case of an inertia-dominated flow, the opposite holds true.

5.5 Vortex Stretching

A rate of change of vorticity via vortex stretching can be expressed as follows:

$$\frac{D(\vec{\omega}/\rho)}{Dt} = \left(\frac{\vec{\omega}}{\rho} \right)^T [A] \tag{5.44}$$

where the matrix [A] representing the linear strain rate of the fluid element is written as follows:

$$[A] = [\nabla \vec{v}] \, [I] \tag{5.45}$$

Integration of Eq. (5.44) yields one of Helmholtz's vortex theorems:

$$\vec{\omega}/\rho \sim \vec{l} \qquad \text{or} \qquad |\vec{\omega}|/\rho \sim l^* \tag{5.46}$$

where $l^* = |\vec{l}|$ is the length of the vortex and $|\vec{\omega}|$ is its magnitude.

The vortex stretching is concerned with the conservation of the total angular momentum of a vortex filament or tube. If no external torque is applied, the mass $\int \rho \, dV$ and the total angular momentum $\int \rho \, (\vec{r} \times \vec{v}) \, dV$ of the vortex filament must be conserved; thus, a stretching of the vortex filament increases the angular speed $|\vec{\omega}|/2$, as the radius decreases.[9] The principle of the conservation of the total angular momentum can be observed in various sunken or uprising swirls, e.g. the bathtub vortex, fire whirl (or flame vortex), low-pressure cyclones. A rationale common to these cases is that there is a sink flow at the center where the vortex strength is intensified by vortex stretching.[10]

5.5.1 Vortex in an Upside-Down Bottle

If a bottle of water is turned upside down with the lid open, the water will not drain out easily for several reasons: (i) surface tension of the water, (ii) blockage by the bottle neck, (iii) negative pressure of the air trapped in the bottle due to the weight of the water, etc. However, if we draw a circle several times with the bottle before turning it upside down, the water will drain out more easily, forming a spiral vortex (Figure 5.7).

Drawing a circle with the bottle supplies angular momentum to the water in the outer area, while the water at the center tends to fall through the opening by gravity. As a result, a viscous torque is produced at the interface between these two fluids, which then rotates the draining water at a rate. At this point, the vortex formed at the center is stretched at a rate by the axial velocity gradient of the water, while the intensified vortex filament enhances the centrifugal effect, lowering

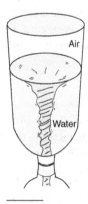

Figure 5.7 Vortex in an upside-down bottle; vortex stretching in gravity-driven sink flow

---

9 The most common example of the conservation of the total angular momentum is a figure skater who spins faster when the arms are pulled toward the body or spins slower when they are stretched out.

10 Vortex stretching can also occur in viscous shear flows, where the normal viscous stresses stretch and contract the fluid at a rate along the principal axes. In turbulent shear flows, vortex stretching is the primary mechanism causing the so-called *turbulent energy cascade*. When the axis of the vorticity vector is aligned with the positive principal axis, the turbulent eddy is stretched at a rate in that direction, intensifying the energy while reducing the size. The smaller eddy follows the same process, which continues until it is viscously dissipated into heat. More discussion on this continues in Section 6.5.3 *Energy cascade process*.

the pressure of the water at the center. Eventually, an air-tunnel is created, making the air pressure inside the bottle equal the ambient pressure.

5.5.2 Fire Whirl

A fire whirl is often observed in wildfire (Figure 5.8). The fire creates an updraft due to the buoyant effect, and the neighboring air is then radially sucked into the fire by continuity. In certain wind conditions, the axi-symmetry of the radially converging upward air is broken by imposition of the tangential velocity of the wind, which then makes the air in the fire rotate at a rate via viscous torque. Once the fire rotates, its rotation is intensified through vortex stretching which will strengthen the fire whirl, and so on and so forth. The basic physics creating a fire whirl is based on the fact that the pressure in the sink is lower by the density difference between the two fluids in the surroundings and the flame, and on the fact that the centrifugal effect decreases the pressure in the center of the flame.

This fire whirl can be reproduced in a student laboratory by rotating the flame of an alcohol lamp with a mesh-type basket. As the basket rotates, the meshes entrain and rotate the ambient air by exerting the viscous torques. Even with a small angular speed of the basket, the vortex in the flame can be intensified via vortex stretching. Note that this fire whirl can more easily be reproduced compared to that in the wildfire because the outer swirl is more uniformly generated by the circular basket.

5.5.3 Propeller Blade-Tip and Hub Vortices

A multiple series of vortex roll-ups are produced at the blade tips of a ship propeller by the pressure difference across the suction and pressure surfaces of the blade (Figure 5.9). The blade tip vortex

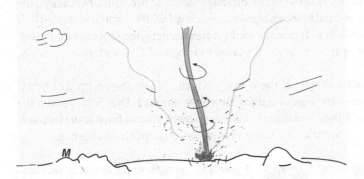

Figure 5.8 Fire whirl in wildfire; vortex stretching in buoyancy-driven sink flow

Figure 5.9 Cavitating propeller blade-tip vortices

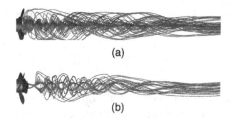

(a)

(b)

Figure 5.10 Propeller hub vortex control; 5% energy saving and 2% speed increase by propeller boss cap fins (PBCF); w/o PBCF (a), w/ PBCF (b)

is responsible for degrading the propeller performance and causing cavitation. The cavitation is sometimes fatal to the blades, as the rupture of the air bubbles erodes the surface and creates cracks.

The ship propeller also generates a three-dimensional spiral vortex flow at the hub (Figure 5.10a). Viscous torques are exerted between the outer swirly jet and the separated flow at the hub (i.e. a recirculating flow at the backward-step). To be more specific, the circumferential component of the swirly jet velocity applies torques, whereas the axial-component velocity difference makes the separated flow a *virtual sink*. The vortex strength is then intensified through stretching. We can confirm this virtual sink with a coordinate transformation. If we take off the axial velocity of the swirly jet from the entire flow field, the separated flow at the hub will become an upstream-moving sink surrounded by the outer swirl, as visualized in Figure 5.10a, which has the exact same structure as the spiral vortex flow in the drainage vortex or fire whirl. The outer swirl jet is radially entrained into the sink, and the vortex in the axial direction is produced by viscous torques that act along the viscously sheared interface between the swirly jet and the separated flow at the hub.

5.5.4 Vortex Flow Control

The spiral vortex at the rotating hub cap of the marine ship propeller or at the wingtips of the aircraft is a source of energy loss for transportation vehicles. The hub cap vortex and wingtip vortex are caused by the same flow physics as those in the drainage vortex or fire whirl. Nowadays, the ship propeller uses a boss cap to minimize the energy loss associated with the spiral vortex formed by the rotor hub cap (Figure 5.10b). By reducing the spiral vortex strength, the propeller torque can be reduced while thrust is increased; for instance, 5% energy saving and 2% speed increase can be achieved.

The aircraft wing also employs a winglet at the wingtip, which reduces the energy loss by 5% (Figure 5.11). Note that there are two ways to reduce the vortex strength. One is to weaken the stretching effect to modify the sink flow, and the other is to reduce the viscous force at the interface by modifying the pressure difference across the sharp edge so that it gradually diminishes.

Figure 5.11 Wingtip vortex control by winglets

5.6 Vortex Tilting

A rate of change of vorticity via vortex tilting can be written as follows:

$$\frac{D(\vec{\omega}/\rho)}{Dt} = \left(\frac{\vec{\omega}}{\rho}\right)^T [B] \tag{5.47}$$

where the matrix $[B]$ representing the straining rate of a fluid element reads

$$[B] = \left[\nabla \vec{v}\right] - \left[\nabla \vec{v}\right] [I] \tag{5.48}$$

The underlying physics of vortex tilting is centrifugal force coupled with viscous effects in three-dimensional space [6]. There are various examples of vortex tilting associated with secondary flow in curved vessels (e.g. curved pipes and channels, cylindrical containers).

5.6.1 Nonuniformity in the Centrifugal Forces

In curved vessels, the fluid is centripetally accelerated while exerting a centrifugal force to the curved boundary. In response to this inertial force, pressure increases as the radius increases, satisfying the local force balance in the radial direction. If we ignore the viscous effect, the flow is essentially two-dimensional, maintaining the uniformity of centrifugal forces in the direction normal to the curvature plane. Here, the curvature plane is the plane on which the streamlines are curved.

If the fluid is sheared by viscous forces, the uniformity of the centrifugal force in the direction normal to the curvature plane will be broken. As shown in Figure 5.12, the higher-momentum fluid will over-exercise the inertial effect because the lower-momentum fluid cannot compensate for the same centrifugal effect that the higher-momentum fluid produces. As a result, the higher-momentum fluid tends to migrate from the convex side to the concave side, whereas the lower-momentum fluid migrates from the concave side to the convex side.

The shear flow in the curved vessel then rotates (clockwise) at a rate about an axis of rotation parallel to the flow direction. This rotational flow perpendicular to the primary flow in the vessel is called *secondary flow*. In this case, the spanwise vortex $\vec{\omega}$ is tilted to $\vec{\omega}'$ ($= \vec{\omega} + \vec{\omega}_s$) with the creation of a streamwise vortex $\vec{\omega}_s$.

In rivers, for example, the secondary flow associated with vortex tilting takes the sand and pebbles at the concave side of the bank bottom and piles them up at the convex side. As time proceeds,

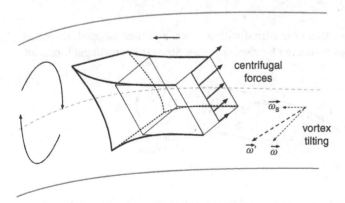

Figure 5.12 Secondary flow and vortex tilting by nonuniform centrifugal forces in the curved vessel; shear deformation of a fluid element w/o curvature effect (thin line) and w/ curvature effect (thick line); vortex tilting: $\vec{\omega}' = \vec{\omega} + \vec{\omega}_s$ ($\vec{\omega}_s$: streamwise vortex)

Figure 5.13 River meandering; secondary flow and vortex tilting in an open curved channel. Source: Imp5pa/Pixabay

the convex bank becomes more convex, whereas the concave bank becomes more concave, which makes the river more curvy. This phenomenon is called river meandering (Figure 5.13).

5.6.2 Counter-Rotating Cells in a Curved Pipe Flow

A similar type of secondary flow can be observed in curved pipe flows, where the pipe curvature creates two counter-rotating cells in the cross section (Figure 5.14). Since the centrifugal forces are nonuniformly acting on the sheared flow, the higher-momentum fluid in the center will over-exercise the centrifugal effect, whereas the lower-momentum fluids along the peripheral side will under-exercise the inertial effect. Thus, the fluid in the center tends to move outwards to the concave surface, whereas the fluid in the outer periphery moves inward to the convex surface of the curved pipe. Two counter-rotating cells are then formed at the cross section of the pipe. As depicted in Figure 5.14, a vortex loop in the boundary layer is tilted, while two counter-rotating streamwise components of vortex are being created. The two counter-rotating spiral flows result in an increase of frictional loss since the higher-momentum fluid in the center is forced to move along the periphery of the curved surface of the pipe.[11]

5.6.3 Tea Leaves in a Cup

When the water in the cylindrical container is stirred with a spoon and then stopped, the water close to the bottom and side surfaces is sheared by viscous forces. Since the centrifugal forces are

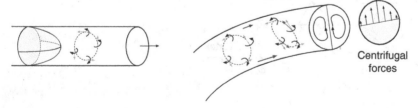

Centrifugal forces

Figure 5.14 Formation of secondary flow (two counter-rotating cells) by vortex tilting in a curved pipe; nonuniform centrifugal forces

---

11 More discussion on this continues in Section 7.5.4 *Curved pipes*.

Figure 5.15 Secondary flow with upwelling in a cylindrical container (a) and tea leaves gathering in the center of the bottom of the cup (b)

not uniformly distributed in the vertical direction, a secondary flow starts to form close to the bottom surface, creating a streamwise vortex. As time proceeds, the vortex lines tilted over the bottom surface gradually merge to form a concentrated vortex at the center of rotation, called upwelling (Figure 5.15a). Once the spiral vortex at the center touches the free surface of the fluid, the secondary flow will fully occupy the entire cross section. This merging of the vortex lines describes the process by which the tea leaves gather at the center of the bottom of the cup, after the tea has stopped being stirred (Figure 5.15b).

5.7 Pressure Torque

5.7.1 Baroclinicity

In a baroclinic fluid under the conservative body forces and in the absence of viscous forces, the vorticity transport equation reads

$$\frac{D\vec{\omega}}{Dt} = (\vec{\omega} \cdot \nabla)\vec{v} - \vec{\omega}(\nabla \cdot \vec{v}) + \frac{1}{\rho^2}(\nabla\rho \times \nabla p) \tag{5.49}$$

where the last term on the right-hand side represents the pressure (or baroclinic) torque. If the pressure gradient is not parallel to the density gradient, the line of action of the pressure force does not go through the center of mass of a fluid element and produces a torque (Figure 5.16).

If the fluid is incompressible but not of a uniform density and the density gradient vector is not parallel to the pressure gradient vector, Eq. (5.49) reads

$$\frac{D\vec{\omega}}{Dt} = (\vec{\omega} \cdot \nabla)\vec{v} + \frac{1}{\rho^2}(\nabla\rho \times \nabla p) \tag{5.50}$$

where the second term on the right produces a baroclinic torque per unit mass.

Figure 5.16 Pressure (or baroclinic) torque, $(\nabla\rho \times \nabla p)/\rho^2$, acting on a fluid element at the interface between two fluids of different densities; $\nabla p \times \nabla \rho = 0$ (a), $\nabla p \times \nabla\rho \neq 0$ (b); arrows: hydrostatic pressure; lines: iso-density lines

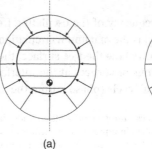
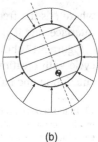

(a) (b)

If the fluid is compressible and baroclinic, Eq. (5.49) can be written with the continuity equation, $\nabla \cdot \vec{v} = -(1/\rho)D\rho/Dt$, as follows:

$$\frac{D(\vec{\omega}/\rho)}{Dt} = \left(\frac{\vec{\omega}}{\rho} \cdot \nabla\right) \vec{v} + \frac{1}{\rho^3} (\nabla\rho \times \nabla p) \tag{5.51}$$

With the Gibbs equation, $T\,\nabla s = C_p\nabla T - \nabla p/\rho$,[12] Eq. (5.51) can be cast into Vazsonyi's vorticity equation:

$$\frac{D(\vec{\omega}/\rho)}{Dt} = \left(\frac{\vec{\omega}}{\rho} \cdot \nabla\right) \vec{v} + \frac{1}{\rho} (\nabla T \times \nabla s) \tag{5.52}$$

which shows that a pressure torque can be produced if the temperature gradient is not parallel to the entropy gradient. Note also that the pressure torque can be produced when the pressure gradient is not parallel to the entropy gradient. For example since $T = T(p, s)$,

$$\nabla T \times \nabla s = \left.\frac{\partial T}{\partial p}\right|_s (\nabla p \times \nabla s) + \left.\frac{\partial T}{\partial s}\right|_p \underbrace{(\nabla s \times \nabla s)}_{0} \tag{5.53}$$

where a pressure gradient can be created in many different ways aside from entropy production such as bomb explosions, preheating of piston engine with acoustic waves, and shock-bubble interaction. The production of vorticity with a pressure torque can be engaged in multiple thermodynamic processes.

5.7.2 Incompressible Interfacial Flows

5.7.2.1 Vortex Sheet at the Free Surface

Let us suppose a cup of water is filled with two immiscible incompressible fluids of different densities, for example water and air. At the interface (or free surface), there is a sharp variation of density between the two fluids. This interface is mathematically defined as a discontinuity but, from a microscopic point of view, it is a smooth transitional layer of finite thickness.

If we look at a tiny fluid particle at the interface, it is comprised of high-density water molecules and low-density air molecules. In this case, the center of mass of the particle is on the vertical line and below the middle horizontal line. When the cup of water is suddenly tilted counter-clockwise, for example the center of mass of the fluid element is instantly shifted to the right and therefore, the line of action of the buoyant force (or hydrostatic pressure force) does not go through the center of mass of the particle. In consequence, a clockwise pressure (or baroclinic) torque starts to act on the fluid particle at the interface.

If we look at the circulation around the free surface, it harmonically varies with the vortices produced at the interface by the pressure torque; the maximums occur at the most tilted positions, and the signs are switched at the horizontal position. It can be expressed as follows:

$$\Gamma(t) = \oint_C \vec{\omega}(t) \cdot \vec{n}\, dA = -|\Gamma_0| \cos(2\pi ft) \tag{5.54}$$

where f is the frequency of the oscillating free surface, C is the closed circuit that encloses the interface, and $|\Gamma_0|$ is the maximum circulation at the most tilted positions.

Figure 5.17 shows how the free surface, the water underneath, and the air above move in one cycle with creations of vortices at the interface. The top seven figures show the changes of circulation in one half cycle (phase I: from the top to the fourth and phase II: from the fourth to the

---

12 In thermodynamics, entropy represents an unavailability of a system's thermal energy for conversion into mechanical work – often interpreted as the degree of disorder or randomness in the system. According to the Gibbs equation, increase of temperature will increase entropy, while increase of pressure will decrease it.

Figure 5.17 Time-dependent vorticity production by a baroclinic torque at the free surface; a transient slug motion (vertical direction) and a sheared motion (tangential direction) are superimposed; the vertical speed is at the maximum, when the free surface is at the horizontal position, while the sheared speed is at the maximum, when it is at the most tilted position; one full cycle of oscillation (figures from top to bottom).

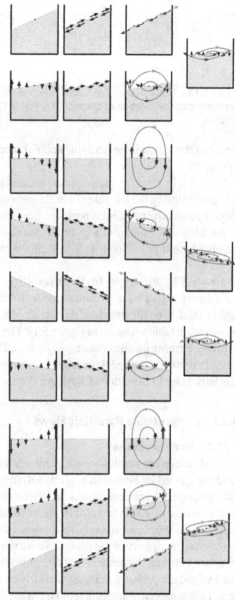

seven). Note that the velocity vectors are discontinuous across the interface except at the horizontal position, due to time-dependent vortices linearly distributed at the free surface. The water and air attain maximum kinetic energy in the vertical direction when the interface is at the horizontal position, while shear straining at the free surface stops completely. Meanwhile, they completely lose the kinetic energy, while shear straining of the fluid at the interface is the maximum at the most tilted positions. This oscillatory motion continues until the kinetic energy of the fluid is completely dissipated by viscous frictions at the side walls.

If we look at the free surface of the water, the linearly distributed shear straining fluid particles is another form of free shear layer of very small but finite thickness. These fluid particles move not only in the vertical direction but also in the horizontal direction with three motions superimposed: (i) slug motion in the vertical direction, (ii) solid body rotation with respect to the

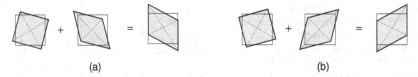

(a) (b)

Figure 5.18 Shear straining of fluid at the free surface; clockwise rotation at phase I $(-\Gamma_0 < \Gamma(t) < 0)$ (a); counter-clockwise rotation at phase II $(0 < \Gamma(t) < \Gamma_0)$ (b); Γ_0 is the maximum circulation at the most tilted position

center of the free surface, and (iii) angular straining as the fluid particles move tangentially to the side walls.

As shown in Figure 5.18, the fluid particles are vertically shear strained, pulling down the right side and uplifting the left side of the element during phase I $(-|\Gamma_0| < \Gamma(t) < 0)$. As soon as the free surface passes the horizontal position, the center of mass of the fluid element is now being shifted to the left and a counter-clockwise torque starts to act to decrease the inertial motion of the fluid during phase II $(0 < \Gamma(t) < |\Gamma_0|)$, pulling down the left side and uplifting the right side.

Example 5.1 *Rayleigh–Taylor instability*

Let us suppose we have two immiscible fluids of different densities. If a heavier fluid is under a lighter fluid, then the interface is in stable condition. If the heavier fluid is at the top, the interface becomes unstable as shown in Figure 5.19. The baroclinic torque causes the vortices at the interface to rotate, forming mushroom-shaped fingers. This phenomenon is called the Rayleigh–Taylor instability. In density stratified flows, *Atwood* number $A = (\rho_h - \rho_l)/(\rho_h + \rho_l)$ is an important dimensionless parameter in the study of Rayleigh–Taylor instability [7].

5.7.3 Compressible Baroclinic Flows

5.7.3.1 Bomb Explosion

A bomb explosion creates strong gradients in temperature, pressure, and density. The pressure gradient caused by explosion will immediately disappear as strong pressure waves propagate, but the temperature gradient cannot be attenuated quickly. It leaves the density gradients not only in the vertical direction but also in the horizontal direction.

As soon as the plume rises up, the buoyant force (or hydrostatic pressure force) produces a torque at the outer periphery of the plume, where the center of the mass of the air particles at the vertical interface between the hot and cold air columns is shifted to the cold side. As expressed in Eq. (5.53), the hydrostatic pressure gradient is nearly orthogonal to the entropy gradient at the plume interface. In this case, the uprising jet-like plume rolls into a huge vortex ring, as shown in Figure 5.20. This vortex roll-up occurs with the same physics that occurs at the sharp edge of a solid body, where a concentrated viscous torque creates a concentrated vortex.

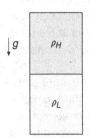 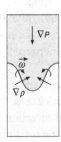 

Figure 5.19 Rayleigh–Taylor instability at the interface between two immiscible fluids of different densities $(\rho_H > \rho_L)$ by pressure (or baroclinic) torque

Figure 5.20 A huge mushroom-like vortex ring formed by pressure (or baroclinic) torque in bomb explosion. Source: U.S. Army/Wikimedia Commons/Public Domain

5.7.3.2 Mixing with Acoustic Waves

Let us suppose a cylindrical tube is preheated to build up a gradient of temperature (or entropy) and then, the piston is forced to oscillate at high frequency but with small amplitude to generate acoustic waves, as shown in Figure 5.21. By heating, the fluid becomes lighter toward the tube side surfaces and a baroclinic (or pressure) torque will be produced with the acoustic waves in the cylindrical tube. For example, compression waves produced by the piston moving forward create the vortices in the clockwise direction, whereas the vortices in the counter-clockwise direction are produced by the rarefaction waves with the piston moving backward. The sign changes of the pressure gradient vector ∇p in acoustic waves will change the rotational direction of the vorticity and make the mixing more efficient inside the tube.

5.7.3.3 Supersonic Flow Over a Blunt Body

If an incompressible flow approaches a blunt body, velocity and pressure are continuously changed by a force balance between the pressure force and the inertia force, all while conserving the mass. Since the density is uniform, a pressure gradient vector is always in line with the center of mass of the fluid particles, and thus, vorticity cannot be produced by a change of velocity with pressure.

If the flow approaching the blunt body is subsonic, the velocity, pressure, density, and temperature are continuously changed by a force balance between the pressure force and the inertia force, while conserving the mass and total energy and obeying the equation of state (e.g. ideal gas law). Density can be changed by pressure and temperature, and the Gibbs equation shows that the temperature gradient vector is parallel to the pressure gradient vector in an isentropic process, since there is no spatial change of entropy. If then, $p = c_0 \, \rho^\gamma$ and thus $\nabla p = c^2 \, \nabla \rho$, meaning that the pressure gradient is also parallel to the density gradient. In this case, vorticity cannot be produced by the same reason as in incompressible flows.

Figure 5.21 Mixing by pressure (or baroclinic) torque in a preheated cylinder with acoustic waves

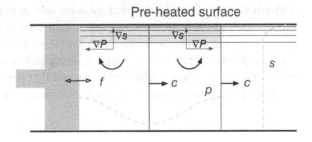

Pre-heated surface

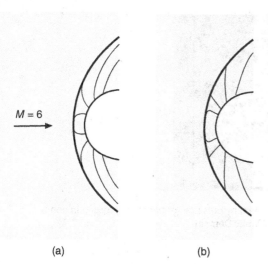

Figure 5.22 Vorticity production by pressure (or baroclinic) torque in supersonic flow past a circular cylinder at $M = 6$; iso-density lines (a), isobaric lines (b)

$M = 6$

(a)   (b)

If the flow approaching the blunt body is supersonic, the variables cannot continuously be changed because the flow speed is greater than the speed of sound. As a result, a bow shock stands off the blunt body and the flow adjusts the variables, crossing the shock wave (Figure 5.22). Note that the bow shock formation is based on the superposition of the sound waves emitted from the blunt body surface but with different magnitudes. Thus, the shock strength varies along the bow shock, across which the pressure, density, and temperature abruptly increase with different jump conditions along the bow shock wave.[13]

What is interesting is that the density increase with the pressure jump far exceeds the density decrease with the temperature jump in magnitude, and from this reason, there exists a substantial density gradient created along the bow shock. The density gradient vector pointing in the direction toward the centerline and the pressure gradient vector acting downstream produce vortices within the bow shock; the fluid particles passing through the bow shock are rotated at a rate by the baroclinic torque.

This baroclinic vorticity production is important for aerodynamic problems in supersonic flows such as eddy-shocklets and turbulences, and reduction of wave drag in supersonic or hypersonic blunt body flows – which is associated with energy deposition, nonequilibrium instabilities, etc. [8].

5.8 Nonconservative Body-Force Torque

5.8.1 Time Derivatives in the Inertial and Rotating Reference Frames

A time derivative of an arbitrary vector $\vec{A}$ is considered in two reference frames. The first reference frame is the inertial reference frame and is expressed in terms of the Cartesian coordinates $\vec{r}' = (x', y', z')$. The second reference frame is the rotating reference frame and is expressed in terms of the Cartesian coordinates $\vec{r} = (x, y, z)$. Let us suppose that the rotating reference frame shares the same origin as the inertial reference frame and that the angular velocity vector $\vec{\Omega}$ of the rotating reference frame (with respect to the fixed reference frame) has three components $(\Omega_x, \Omega_y, \Omega_z)$.

---

13 More discussion continues in Chapter 8 *Compressible Flows*.

If we decompose the vector $\vec{A}$ in terms of the components A_i in the rotating reference frame (with unit vectors e^i), it is expressed as $\vec{A} = A_i\, e^i$ and the time derivative of $\vec{A}$ as observed in the inertial reference frame reads

$$\frac{d\vec{A}}{dt} = \frac{dA_i}{dt}\, e^i + A_i\, \frac{de^i}{dt} \tag{5.55}$$

The interpretation of the first term is that of the time derivative of $\vec{A}$ as observed in the rotating reference frame (where the unit vectors e^i are constant), while the second term involves the *time-dependence of the relation between the inertial and rotating reference frames*. If we express de^i/dt as a vector in the rotating reference frame, it follows that

$$de^i/dt = \epsilon^{ijk}\, \Omega_k\, e_j \tag{5.56}$$

and the second term in Eq. (5.55) reads

$$A_i\, \frac{de^i}{dt} = A_i\, \epsilon^{ijk}\, \Omega_k\, e_j = \vec{\Omega} \times \vec{A} \tag{5.57}$$

The time derivative of an arbitrary rotating reference frame vector $\vec{A}$ in an inertial reference frame is finally expressed as follows:

$$\left(\frac{d\vec{A}}{dt}\right)_i = \left(\frac{d\vec{A}}{dt}\right)_r + \vec{\Omega} \times \vec{A} \tag{5.58}$$

where $(d/dt)_i$ denotes the time derivative as observed in the inertial (i) reference frame, while $(d/dt)_r$ denotes the time derivative as observed in the rotating (r) reference frame.

An important application of this formula relates to the time derivative of the angular velocity $\vec{\Omega}$ itself. One can easily see that

$$\left(\frac{d\vec{\Omega}}{dt}\right)_i = \dot{\vec{\Omega}} = \left(\frac{d\vec{\Omega}}{dt}\right)_r \tag{5.59}$$

since the second term in Eq. (5.58) vanishes for $\vec{A} = \vec{\Omega}$; the time derivative of $\vec{\Omega}$ is, therefore, the same in both frames of reference and is denoted as $\dot{\vec{\Omega}}$ in what follows.

5.8.2 Accelerations in the Rotating Reference Frame

5.8.2.1 Centrifugal and Coriolis Accelerations

Let us suppose that the position vector of a point P with respect to the inertial reference frame is denoted by $\vec{r}\,'$ and the position vector of the same point with respect to the rotating reference frame is denoted by $\vec{r}$. Then, the two position vectors $\vec{r}\,'$ and $\vec{r}$ are related as follows:

$$\vec{r}\,' = \vec{R} + \vec{r} \tag{5.60}$$

where $\vec{R}$ denotes the position vector of the origin of the rotating reference frame with respect to the inertial reference frame.

The velocities of point P as observed in the inertial and rotating reference frames are expressed, respectively, as follows:

$$\vec{v}_i = \left(\frac{d\vec{r}\,'}{dt}\right)_i \quad \text{and} \quad \vec{v}_r = \left(\frac{d\vec{r}}{dt}\right)_r \tag{5.61}$$

and the relation between $\vec{v}_i$ and $\vec{v}_r$ is written as follows:

$$\vec{v}_i = \left.\frac{d(\vec{R}+\vec{r})}{dt}\right|_i = \vec{V} + \vec{v}_r + \vec{\Omega} \times \vec{r} \tag{5.62}$$

where $\vec{V} = (d\vec{R}/dt)_i$ denotes the translation velocity of the origin of the rotating reference frame, as observed in the inertial reference frame.

It is important to note that we need two velocity vectors to express the time derivative of the position vector $\vec{r}$ with respect to the inertial reference frame, $(d\vec{r}/dt)_i$. First, we measure the velocity in the rotating frame by subtracting the position vector $\vec{r}(t + dt)$ from $\vec{r}(t)$, while tracking the point P in time and dividing it by the time interval dt. In the limit, this becomes the time derivative of the position vector $\vec{r}$ with respect to the rotating frame and is denoted by $\vec{v}_r$.

Besides, we have to take into account the rotational effect of the reference frame by measuring how much the position vector $\vec{r}$ at t is *angularly displaced*, as the reference frame rotates over dt with $\vec{\Omega}$ with respect to the inertial reference frame. We then subtract the two position vectors $\vec{r}(\theta + d\theta)$ and $\vec{r}(\theta)$ and divide by $dt \,(= d\theta/|\vec{\Omega}|)$; in the limit, it becomes $\vec{\Omega} \times \vec{r}$.

We now evaluate expressions for the acceleration of point P as observed in the inertial and rotating reference frames:

$$\vec{a}_i = \left(\frac{d\vec{v}_i}{dt}\right)_i \quad \text{and} \quad \vec{a}_r = \left(\frac{d\vec{v}_r}{dt}\right)_r \tag{5.63}$$

Taking the time derivative of Eq. (5.62) reads

$$\vec{a}_i = \left(\frac{d\vec{V}}{dt}\right)_i + \left(\frac{d\vec{v}_r}{dt}\right)_i + \left(\frac{d\vec{\Omega}}{dt}\right)_i \times \vec{r} + \vec{\Omega} \times \left(\frac{d\vec{r}}{dt}\right)_i$$

$$= \vec{A} + \underbrace{(\vec{a}_r + \vec{\Omega} \times \vec{v}_r)}_{\text{(A)}} + \dot{\vec{\Omega}} \times \vec{r} + \underbrace{\vec{\Omega} \times (\vec{v}_r + \vec{\Omega} \times \vec{r})}_{\text{(B)}}$$

$$= \vec{A} + \vec{a}_r + \underbrace{\dot{\vec{\Omega}} \times \vec{r}}_{\text{(i)}} + \underbrace{2\,\vec{\Omega} \times \vec{v}_r}_{\text{(ii)}} + \underbrace{\vec{\Omega} \times (\vec{\Omega} \times \vec{r})}_{\text{(iii)}} \tag{5.64}$$

where $\vec{A} = (d\vec{V}/dt)_i$ represents the translational acceleration of the origin of the rotating reference frame, as observed in the inertial reference frame. The last three terms in Eq. (5.64) represent the following:

(i) $-\dot{\vec{\Omega}} \times \vec{r}$ (angular acceleration),

(ii) $-2\vec{\Omega} \times \vec{v}_r$ (Coriolis acceleration),

(iii) $-\vec{\Omega} \times (\vec{\Omega} \times \vec{r})$ (centrifugal acceleration).

5.8.3 Physical Interpretations

The centrifugal acceleration acts in the outward-radial direction, increasing its magnitude as r increases, since the reference frame rotates at an angular speed with respect to the inertial reference frame. Therefore, any object away from the center of rotation is subject to the centrifugal force, though it is stationary in the rotating reference frame.

In contrast, the Coriolis acceleration only acts when the object is moving in the rotating reference frame. As shown in Eq. (5.64), the Coriolis acceleration arises from two sources. One is associated with Ⓐ changes in direction of $\vec{v}_r$ since the reference frame rotates with respect to the inertial reference frame. Therefore, the magnitude of this acceleration vector is proportional not only to $\vec{v}_r$ itself but also to the rotational speed $\vec{\Omega}$, and the direction is biorthogonal to these two vectors.

The other is associated with the fact that as the object moves across the radius of rotation, it encounters Ⓑ *different rotational effects of the reference frame* (or radial difference in the circumferential speed of the rotating reference frame), which rotates with respect to the inertial frame because the rotating frame is in solid body rotation. Hence, the magnitude of this acceleration will be proportional to $\vec{v}_r$ and $\vec{\Omega}$ as well, and the direction will be biorthogonal to these two vectors.

The Coriolis acceleration can more intuitively be explained with the position vector $\vec{r}$ of a point P at time t in the rotating frame. If this point P moves to another position over dt, the difference of these two position vectors over dt represents a velocity vector at time t in the rotating reference frame. As the reference frame rotates with respect to the inertial frame, these two position vectors will be angularly displaced (counted by Ⓐ) so that the differences over dt represent two velocity vectors that have taken into account the rotational effect of the reference frame. It is important to note that the rotational effect also depends on the radius of rotation (counted by Ⓑ). If we take a difference of these two difference vectors (i.e. velocity vectors) and divide by dt, it represents the Coriolis acceleration vector [9].

A simple experiment in the rotating carousel can easily explain the Coriolis acceleration; if a person moves to change seats from one horse to another in a different radius of rotation, he or she is forced to move to the right (assuming the carousel rotates in the counter-clockwise direction). Another example is the *inertia circles* drawn by a small steel ball, constantly changing the radius of rotation on an anti-clockwise-rotating parabolic surface [10, 11]. In Figure 5.23, the one on the left shows oscillating motion of the ball on the rotating parabolic surface when observed in an inertial frame, while the one on the right shows the same motion observed in a rotating frame of reference.

Example 5.2 *Motion of fluid under the Coriolis force*

Let us suppose we pour cold milk into a cup of coffee (YouTube, *The Full Monty: Laboratory Demonstrations of Planetary-Style Fluid Dynamics*). As soon as the milk hits the bottom of the cup, it quickly spreads (Figure 5.24a). If the cup, however, rotates at a fast rate in the counter-clockwise direction, the column of poured milk does not collapse. It rotates at a rate with the coffee, while holding its shape. The column of poured milk eventually scatters, but forms a two-dimensional structure like a curtain (Figure 5.24b).

This simple experiment demonstrates the Coriolis effect very well in rotating systems. As soon as the poured milk occupies a vertical space in the coffee which rotates at a fast rate, the milk at the bottom tends to spread out. However, it will wrap the poured milk column in the clockwise direction and hold its shape, since the Coriolis force resists displacement by trying to return the motion to the origin. Due to the Coriolis inertial force, the milk at the top is radially pulled in and tends

Figure 5.23 Inertia circles; oscillating motion of a small ball on the rotating the parabolic surface in the inertial frame (a) and in the rotating frame of reference (b); the parabola is rotating in an anticlockwise (cyclonic) sense

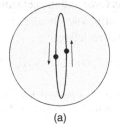
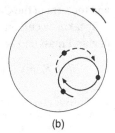

(a)   (b)

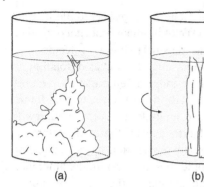

Figure 5.24 Coriolis inertial force resists changes in a body's motion; cold milk poured into a cup of coffee; stationary (a), rotating (b)

to rotate at a rate in the counter-clockwise direction, which will then intensify the rotation [12]. The poured milk column then does not collapse, holding up its two-dimensional structure.

5.8.4 Coriolis-force Torque

The Navier–Stokes equations in the rotating reference frame are written as follows:

$$\frac{D\vec{v}_r}{Dt} = \underbrace{-2\,\vec{\Omega}\times\vec{v}_r}_{\text{Ⓐ}}\ \underbrace{-\vec{\Omega}\times(\vec{\Omega}\times\vec{r})}_{\text{Ⓑ}} + \frac{1}{\rho}(-\nabla p + \vec{f}_{cbf}) + \frac{1}{\rho}\vec{f}_v \tag{5.65}$$

where the Coriolis and centrifugal accelerations, Ⓐ and Ⓑ, appear to act as a force exerted on the fluid per unit mass.

A curl of the centrifugal force per unit mass (or centrifugal acceleration) becomes zero:

$$\nabla\times\text{Ⓑ} = \nabla\times\left(-\vec{\Omega}\times(\vec{\Omega}\times\vec{r})\right) = 0 \tag{5.66}$$

since it is a conservative body force like a gravitational body force; i.e. the line of action of the net centrifugal force goes through the center of mass of the fluid particle so that torques cannot be produced.[14]

The Coriolis force is, however, a nonconservative body force; the line of action of its net force does not go through the center of mass of the particle and thus, a torque will be produced. As shown in Figure 5.25a, a counter-clockwise rotating low-pressure cyclone (e.g. typhoons and hurricanes) is generated in the tropics of the northern hemisphere by the Coriolis-force torque. Due to updraft driven by buoyancy, the surrounding air moving inward to the sink and the Coriolis forces exerted to the right along the circumference of the sink produce a counter-clockwise torque (Figure 5.25b). It is also clear that the line of action of the Coriolis force does not go through the center of mass of the fluid particles because there is a gradient of the inward radial velocity in the sink and so is the Coriolis force. In this case, a radial force balance is set between the pressure gradient force (inward) and the sum of the Coriolis force and the centrifugal force (outward).[15]

By taking a curl on the Coriolis acceleration vector, the Coriolis-force torque exerted on the fluid of a unit mass can be written as follows:

$$\nabla\times\text{Ⓐ} = \nabla\times(-2\vec{\Omega}\times\vec{v}_r) = (2\vec{\Omega}\cdot\nabla)\,\vec{v}_r - 2\vec{\Omega}\,(\nabla\cdot\vec{v}_r) \tag{5.67}$$

---

14 The issue of conservativeness of a vector field depends on whether its line integral is path-independent or not.
15 A relative importance between the inertial force of the fluid and the Coriolis force is discussed in Section 5.9.1 *Rossby number*.

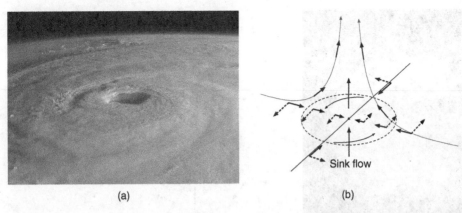

Figure 5.25 A counter-clockwise rotating low-pressure cyclone observed in the northern hemisphere (a); velocity vectors (solid arrows), Coriolis acceleration vectors (dashed arrows) (b). Source: NASA/Wikimedia Commons/Public Domain

where the Coriolis acceleration acts as a force per unit mass and its curl represents a torque exerted on the fluid per unit mass. This Coriolis-force torque arises from two sources: (i) the stretching and tilting of a stream tube in the rotating reference frame, and (ii) the isotropic expansion of the stream tube via compressibility effect.

As illustrated in Figure 5.25b, the Coriolis-force torque is produced by the radial gradient of the Coriolis force. This torque is caused by the inertial effect of fluid particles that cross a radius in a rotating reference frame. Hence, a circulation is produced by this virtual torque, according to the Kelvin circulation theorem. Meanwhile, this circulation should be identical to the circulation produced by the vortex filaments stretching at a rate in the direction of rotation with updraft driven by buoyancy.

5.8.4.1 Anti-Cyclones
The high-pressure anti-cyclones such as the one observed in the great red-spot in the southern hemisphere of Jupiter (Figure 5.26a) are produced with sink flow driven by gravity. The high-pressure flow at low altitude is forced to move outward from the center of the source, producing Coriolis-force torque in the counter-clockwise direction by the net Coriolis force acting on the left in the southern hemisphere of the planet (Figure 5.26b). It is interesting to note that anti-cyclone tornados are occasionally created by the high-pressure anti-cyclones.

5.8.5 Vorticity Transport Equation in the Rotating Reference Frame

If a curl is taken on the Navier–Stokes equations in the rotating reference frame Eq. (5.65), we obtain the vorticity transport equation in the rotating reference frame

$$\frac{D(\vec{\omega}_r + 2\vec{\Omega})}{Dt} = \left((\vec{\omega}_r + 2\vec{\Omega}) \cdot \nabla\right)\vec{v}_r - (\vec{\omega}_r + 2\vec{\Omega})(\nabla \cdot \vec{v}_r) + \frac{1}{\rho^2}(\nabla\rho \times \nabla p) + \nabla \times \left(\frac{1}{\rho}\vec{f}_v\right)$$

(5.68)

In a barotropic, compressible fluid under the Coriolis force and in the absence of viscous forces, Eq. (5.68) can be expressed with the continuity equation, $\nabla \cdot \vec{v}_r = \nabla \cdot \vec{v}_i = -(1/\rho)D\rho/Dt$, as follows:

$$\frac{D(\vec{\omega}_r + 2\vec{\Omega})/\rho}{Dt} = \left((\vec{\omega}_r + 2\vec{\Omega})/\rho \cdot \nabla\right)\vec{v}_r$$

(5.69)

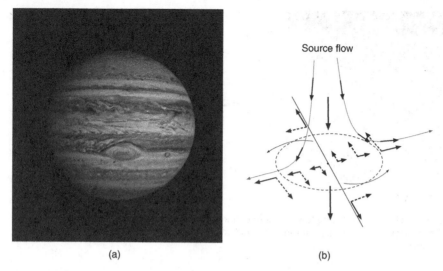

(a)   (b)

Figure 5.26 A counter-clockwise rotating anti-cyclone, the great red-spot, observed in the southern hemisphere of Jupiter (a); velocity vectors (solid arrows), Coriolis acceleration vectors (dashed arrows) (b). Source: Marcel/Adobe Stock

If the fluid is incompressible, it can further be reduced to

$$\frac{D(\vec{\omega}_r + 2\,\vec{\Omega})}{Dt} = \left((\vec{\omega}_r + 2\,\vec{\Omega}) \cdot \nabla\right)\,\vec{v}_r \tag{5.70}$$

It is interesting to note that the vorticity in the rotating reference frame can be changed at a rate (i) by tilting and stretching the vortex tube and (ii) by tilting and stretching the stream tube. Note also that the latter depends on the latitude of the rotating reference frame. In a two-dimensional planar flow, Eqs. (5.69) and (5.70) show

$$\frac{D(\vec{\omega}_r/\rho)}{Dt} = 0\,, \quad \text{or} \quad \frac{D\vec{\omega}_r}{Dt} = 0 \tag{5.71}$$

indicating that $\vec{\omega}_r/\rho$ or $\vec{\omega}_r$ is an invariant quantity in compressible or incompressible flow, respectively.

5.8.5.1 Planetary Vorticity

It is interesting to note that a relation between the vorticity vectors in the inertial and rotating reference frames can be found by taking a curl on Eq. (5.62):

$$\vec{\omega}_i = \vec{\omega}_r + 2\,\vec{\Omega} \tag{5.72}$$

where $2\,\vec{\Omega}$, called *planetary vorticity*, represents the vorticity due to rotation of the frame. The planetary vorticity is a constant vector, but its magnitude depends on the planet's latitude, $\vec{\Omega} = \vec{\Omega}_0 \sin\phi$, where ϕ is the elevation angle and $\vec{\Omega}_0$ of the earth is $360.9856°$ per day or 7.292115×10^{-5} radians per second.

Using the vorticity in the inertial reference frame $\vec{\omega}_i$ and velocity in the rotating reference frame $\vec{v}_r$, the vorticity transport equation in the rotating reference frame Eq. (5.70) can be written as follows:

$$\frac{D\vec{\omega}_i}{Dt} = \left(\vec{\omega}_i \cdot \nabla\right)\,\vec{v}_r - \vec{\omega}_i\,(\nabla \cdot \vec{v}_r) + \frac{1}{\rho^2}\,(\nabla\rho \times \nabla p) + \nabla \times \left(\frac{1}{\rho}\vec{f}_v\right) \tag{5.73}$$

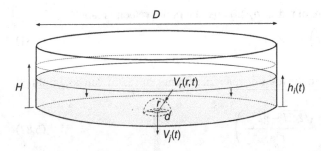

Figure 5.27 A drainage vortex produced by Coriolis-force torque (latitude: 40°)

Example 5.3 *A circular tank experiment: the Coriolis-force torque*

Let us suppose that a large cylindrical tank is filled with water (Figure 5.27). A science-minded student attempts to set up an experimental facility to demonstrate a vortex created by the Coriolis-force torque (YouTube, Fluid Dynamics Vorticity 1961 Ascher Shapiro, MIT PSSC Vortex Physics) [5].

From the mass conservation,

$$-\frac{d}{dt}\left(\frac{\pi}{4}D^2 h_i\right) = \frac{\pi}{4}d^2 v_j \tag{5.74}$$

where dh_i/dt is the velocity at the free surface of the water.

If $\beta = d/D$ is very small, we can assume

$$v_j \approx \sqrt{2gh_i} \tag{5.75}$$

neglecting the unsteady inertial force in the Bernoulli equation, $\int_i^e d\vec{v}/dt \cdot d\vec{s}$, where i indicates the free surface of the water and e the exit.

Thus,

$$-\frac{dh_i}{dt} = \beta^2\sqrt{2gh_i} \tag{5.76}$$

If we integrate this differential equation with the initial condition, i.e. at $t = 0$, $h_i = H$, we obtain the height of the free surface as a function of time:

$$h_i(t) = \left(\sqrt{H} - \sqrt{g/2}\,\beta^2\, t\right)^2 \tag{5.77}$$

Note that a time T to empty the tank reads

$$T = \frac{\sqrt{2H/g}}{\beta^2} \tag{5.78}$$

where T is proportional to the inverse of β^2, i.e. $T \sim (D/d)^2$.

From $v_j(t) = \sqrt{2gh_i(t)}$, we can find the exit velocity of the water as a function of time:

$$v_j(t) = \sqrt{2gH} - g\beta^2\, t \tag{5.79}$$

In order to estimate the rotational rate of the fluid at a radial distance r from the bottom hole, we need to find a radial inflow velocity v_r at the hemisphere of radius r. For the control volume of the hemisphere of radius r, the mass conservation states

$$\frac{4\pi r^2}{2} v_r(r, t) = \frac{4\pi(d/2)^2}{2} v_j(t) \tag{5.80}$$

where the exit surface is the hemisphere of radius $d/2$. Thus, the radial velocity reads

$$v_r(r,t) = \frac{1}{4}(d/r)^2 \left(\sqrt{2gH} - g\beta^2\, t \right) \tag{5.81}$$

where $v_r(r,t)$ is a function of (r,t).

Now, the vorticity transport equation in the rotating reference frame, Eq. (5.70), can be simplified as follows:

$$\frac{D\omega_z}{Dt} = -\frac{\partial}{\partial r}\left(2\,\Omega_z^*\, v_r\right) = \Omega_z^*\, d^2 \left(\frac{\sqrt{2gH} - g\beta^2\, t}{r^3} \right) \tag{5.82}$$

where $\Omega_z^* = \Omega_E \sin\phi$, $\Omega_E = 7.292115 \times 10^{-5}\ rad/s$ is the rotation rate of the earth and ϕ is the latitude. By integrating this equation over time, we can finally estimate the rotational rate of the fluid at a radial distance r and time t:

$$\Omega_z(r,t) = \omega_z(r,t)/2 = \frac{\Omega_z^*\, d^2}{2} \left(\frac{\sqrt{2gH} - (g\beta^2/2)\, t}{r^3} \right) t \tag{5.83}$$

A numerical experiment is conducted for a cylindrical tank of diameter $D = 1.55\ m$ located at the latitude of $40°$ in the northern hemisphere. The tank discharges water of height $H = 0.07\ m$ through a drain hole of diameter $d = 1\ cm$ at the bottom. Figure 5.28 shows $h_i(t)$ and $\Omega_z(r = 0.015\ m, t)$ against time (min) for $\beta = d/D = 1/155$.[16] For example, the rotation rates at $t = 8, 16$, and $24\ (min)$ are predicted as $20, 37$, and $50\ (°/s)$. The rotation rate at $t = T$ (when the tank is emptied) is the maximum at $\Omega_z = 66.5\ (°/s)$. This value is found to be quite close to $67.33\ (°/s)$, calculated by the conservation law of angular momentum.

The conservation law of angular momentum states

$$v_0\, d/2 = v_1\, D/2 = \text{const.} \tag{5.84}$$

where $v_0 = (d/2)\,\Omega_0$ and $v_1 = (D/2)\,\Omega_E \sin\phi$ are the tangential speeds at $d/2$ and $D/2$, respectively. From this, the rotation rate at the drain hole is

$$\Omega_0 = (D/d)^2\, \Omega_{\text{Earth}} \sin\phi = 1/\beta^2\, \Omega_E \sin\phi \tag{5.85}$$

For $\beta = 1/155$, Eq. (5.85) yields $\Omega_0 = 67.33°/s$.

It is interesting to test Eq. (5.83) for other geometrical conditions. If a cylindrical cup (e.g. $D = 10\ cm$, $d = 1\ cm$, and $H = 10\ cm$) were used in the experiment, the Coriolis force torque would

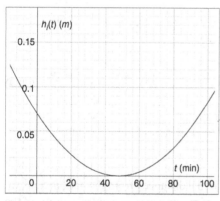
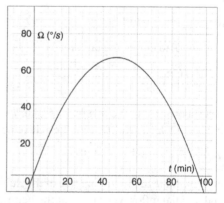

Figure 5.28 $h_i(t)$ and $\Omega_z(r = 0.015\ m, t)$ vs. time (min); $d = 0.01\ m$, $H = h_i(t = 0) = 0.07\ m$, $\beta = d/D = 1/155$, $\Omega_E = 7.292115 \times 10^{-5}\ rad/s$, and latitude $= 40°$; Ω_z in degrees per second

---

16 The numerical results are found to be quite close to the experiment conducted in Boston, USA [5].

Figure 5.29 $\Omega_z(r = 0.015\ m, t)$ vs. time (min); $d = 0.01\ m$, $H = h_i(t = 0) = 0.1\ m$, $\beta = d/D = 1/10$, $\Omega_z^* = 7.292115 \times 10^{-5}\ rad/s$, and latitude = $40°$; Ω_z in degrees per second

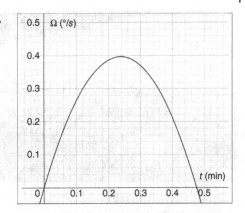

hardly produce rotation, as illustrated in Figure 5.29. The maximum rotation rate predicted by Eq. (5.83) at $t = 0.24$ (min) is about $0.4\ (°/s)$, while the conservation law of angular momentum predicts the rotation rate of $0.28\ (°/s)$.

5.9 Rossby Waves

5.9.1 Rossby Number

In geophysical fluid dynamics, large-scale circulatory flows such as jet streams, low-pressure cyclones (e.g. typhoons and hurricanes), tornados, are some of our great importance. In this case, the Coriolis and centrifugal forces (outward-directed) are in balance with the pressure gradient force (inward-directed), but these three forces are not always of equal importance [13].

To determine the significance of these three forces, the vorticity transport equation in the rotating reference frame for barotropic fluids are nondimensionalized by scaling the length and velocity as $\vec{x}^* = \vec{x}/l$ and $\vec{v}_r^* = \vec{v}_r/U_0$ and by defining the planetary vorticity as $\vec{\Omega} = \Omega \vec{k}$. It then follows that

$$R_0 \frac{D\vec{\omega}_r^*}{Dt^*} = R_0 \left(\vec{\omega}_r^* \cdot \nabla^*\right) \vec{v}_r^* + \left(2\vec{k} \cdot \nabla^*\right) \vec{v}_r^*$$
$$- R_0\, \vec{\omega}_r^* \left(\nabla^* \cdot \vec{v}_r^*\right) - 2\vec{k}\left(\nabla^* \cdot \vec{v}_r^*\right) + Ek\, \nabla^{*2}\vec{\omega}_r^* \tag{5.86}$$

where the Rossby number is defined as the ratio of the inertia force to the Coriolis force

$$R_0 = U_0/l\,\Omega \tag{5.87}$$

or the ratio between the two angular frequencies of the system and fluid element, f_{sys} and f_{elmt}:

$$R_0 = \frac{f_{elmt}}{f_{sys}} \tag{5.88}$$

The significance of the Rossby number is demonstrated in Figure 5.30 with three different conditions:

- If $R_0 \ll 1$ ($f_{elemt} \ll f_{sys}$), the Coriolis force is more significant than the centrifugal force (*geostrophic* balance). In this case, the pressure gradient force is balanced only with the Coriolis force, e.g. jet stream ($T_{sys} \approx T_{elmt}/3$) where T is the period.
- If $R_0 \sim O(1)$ ($f_{elemt} \sim f_{sys}$), the centrifugal force and Coriolis force are equally important (*gradient wind* balance). In this case, the pressure gradient force is balanced with the sum of the two, e.g. low-pressure cyclones such as typhoons and hurricanes.

(a)

Figure 5.30 Rossby number R_O determines significance among the Coriolis force, centrifugal force, and pressure gradient force; jet stream ($R_O \ll 1$), tropical typhoon ($R_O \sim O(1)$), and tornado ($R_O \gg 1$) (from a to b). Source: (a) NASA/Unsplash. (b) WikiImages/Pixabay. (c) Yasioo/Pixabay

(b)

(c)

- If $R_O \gg 1$ ($f_{elemt} \gg f_{sys}$), the centrifugal force is more significant than the Coriolis force (*cyclostrophic* balance). In this case, the pressure gradient force is balanced with the centrifugal force, e.g. tornados.

5.9.1.1 Ekmann Number

It is also to be noted that in Eq. (5.86), an Ekman number is defined as the ratio of the viscous force to the Coriolis force:

$$Ek = v/l^2 \, \Omega = R_O/Re \tag{5.89}$$

where in vortex dynamics, this number is usually very small. However, it must be pointed out that Ekman spirals and costal Ekman transports such as upwelling and downwelling are of great importance in oceanic fluid dynamics.

5.9.2 Conservation of Potential Vorticity

5.9.2.1 Potential Vorticity

If only the vortex stretching in the z direction is concerned, Eq. (5.70) can be expressed as follows:

$$\frac{d\,\omega_r}{dt} = (\omega_r + 2\,\Omega)\,\frac{d\,v_{zr}}{dz} = (\omega_r + 2\,\Omega)\,\frac{d}{dz}\frac{dz}{dt} \tag{5.90}$$

where z is the vertical coordinate.

If we integrate both sides from 0 to h, assuming that ω_z is uniform in the z direction (i.e. *shallow water theory*), Eq. (5.90) becomes

$$\int_0^h d\omega_r\, dz = \int_0^h (\omega_r + 2\,\Omega)\, d^2z \tag{5.91}$$

and it yields

$$h\, d\omega_r = (\omega_r + 2\,\Omega)\, dh \tag{5.92}$$

Integrating both sides gives

$$\int \frac{d\,(\omega_r + 2\,\Omega)}{\omega_r + 2\,\Omega} = \int \frac{dh}{h} \tag{5.93}$$

and it follows that

$$\text{PV} = \frac{\omega_r + 2\,\Omega}{h} = \text{constant} \tag{5.94}$$

which is the conservation of *potential vorticity* (PV) of a barotropic fluid.

5.9.2.2 Interpretation of Rossby Waves with PV

It is interesting to note that two factors can change the relative vorticity ω_r: (i) planetary vorticity $2\,\Omega$ which is a function of the latitude, i.e. $\Omega = \Omega_0 \sin\phi$, and (ii) depth h. The Rossby waves can be explained by the conservation of PV. If a large-scale stream tube moves northward, for example, the relative vorticity, which is initially zero, will increase its strength in the clockwise direction to conserve the PV. If it moves southward, then it will increase the strength in the counter-clockwise direction for the same reason. Similarly, ω_r will change the strength with its orientation when the stream tube gets stretched or contracted with depth h of the terrain. This conservation of PV corresponds to one of Helmholtz's vortex theorems expressed in the rotating reference frame. It is useful for understanding the generation of vorticity in *cyclogenesis*, e.g. polar fronts and oceanic circulations.

5.9.2.3 Isentropic Potential Vorticity

In meteorology, the conservation of PV is written as follows:

$$\text{PV} = -g\,(\omega_r + 2\,\Omega)\,\frac{d\theta}{dp} \approx (\omega_r + 2\,\Omega)\,\frac{d\theta}{dz} = \text{constant} \tag{5.95}$$

where the potential temperature θ is defined as follows:

$$\theta = T\left(\frac{p_0}{p}\right)^{(\gamma-1)/\gamma} \tag{5.96}$$

where γ is the specific heat ratio of an ideal gas. The potential temperature of a parcel of fluid at pressure p is the temperature that the parcel would attain if adiabatically brought a standard reference pressure p_0 (1000 *mbar*) [14].

Since potential temperature is conserved under adiabatic or isentropic air motions, adiabatic flow lines or surfaces of constant potential temperature act as streamlines or flow surfaces, respectively; thus, it is useful in visualizing a large-scale vertical motion of fluid. The PV is the *absolute circulation* of an air parcel that is enclosed between two *isentropic surfaces*. Note that PV is simply the product of absolute vorticity on an isentropic surface and static stability (or stratification).

5.9.2.4 Static Stability

In an incompressible fluid, static stability is solely determined by the density since it is a constant. If the density of a fluid parcel is lower than the ambient density, it rises up due to buoyancy, and we call this state an unstable condition. If the density is higher, it sinks down, and this state is called stable. The fluid is in neutral condition if the density is uniform throughout the fluid.

In a compressible fluid, however, static stability is not solely determined by the density because density changes with temperature and pressure. In a neutrally stable condition, entropy is uniform throughout the fluid. In a perfect gas, a parcel of fluid rises or sinks following the isentropic relation:

$$\frac{T}{T_0} = \left(\frac{p}{p_0}\right)^{(\gamma-1)/\gamma} \tag{5.97}$$

If we take a logarithm and differentiate with respect to the vertical height z, then

$$\frac{dT_a}{dz} = -\frac{g}{C_p} \tag{5.98}$$

where the subscript a denotes an adiabatic atmosphere and $g/C_p = 9.8\ (m/s^2)/1.005\ (kJ/kg\,^\circ K) = 10\,^\circ C/km$ is called the dry adiabatic lapse rate. When condensation occurs in the adiabatic process of the parcel, g/C_p changes to $5\,^\circ C/km$ and is called the wet (or moist) adiabatic lapse rate. The decrease of lapse rate is due to the fact that condensation releases energy in the adiabatic process.

The discussion of static stability can further be extended, using the potential temperature θ. Using the hydrostatic pressure $p = -\rho\, gz$ and the perfect gas equation of state $p = \rho RT$, the potential temperature gradient $d\theta/dz$ can be expressed as follows:

$$\frac{T}{\theta}\frac{d\theta}{dz} = \frac{dT}{dz} + \frac{g}{C_p} = \frac{d}{dz}(T - T_a) \tag{5.99}$$

Note that the static stability of the atmosphere is determined by the following conditions:

$d\theta/dz > 0,$ stable

$d\theta/dz = 0,$ neutrally stable

$d\theta/dz < 0,$ unstable

In stable condition, potential temperature increases upwards in the atmosphere, unlike actual temperature, which decreases. Potential temperature is conserved for all dry adiabatic processes. In unstable condition, potential temperature decreases upwards in the atmosphere (Figure 5.31).

5.9.2.5 Interpretation of Rossby Waves with IPV

The Rossby waves are often observed in the weather maps all over the country where mountains such as the Rocky Mountains block the westerly winds [15]. As illustrated in Figure 5.32, when a column of air between two potential temperatures (or isentropic surfaces) approaches the mountains, it is vertically stretched at a rate, increasing its relative vorticity in the counter-clockwise direction. Hence, it turns its direction to the north. As it ascends, it is substantially contracted at a

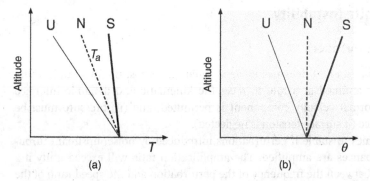

Figure 5.31 Static stability; temperature (a) and potential temperature (b) vs. altitude (T_a: adiabatic atmosphere); U: unstable, N: neutral, S: stable

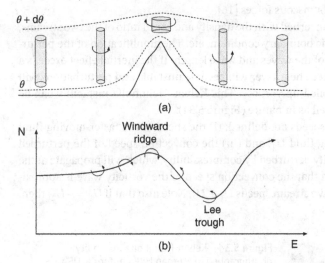

Figure 5.32 Changes of the relative vorticity through stretching and contraction (a); Rossby waves with IPV (b)

rate by the mountains, which block the air column. Along this pass, the gradient of the potential temperature gets stiffer and produces clockwise rotation to conserve the PV. It also turns direction to the south. As the air column descends the mountains, it gets stretched at a rate and overly produces the counter-clockwise rotation. It is also important to note that the Rossby waves also affect air pollution on the ground (Figure 5.33).

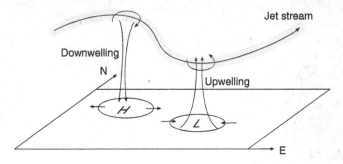

Figure 5.33 Rossby waves in jet streams affect air pollution on the ground via downwelling and upwelling of air

5.10 Kelvin–Helmholtz Instability

5.10.1 Amplification of Perturbations

When two interfacial fluids of different densities move at different speeds, the two fluids at the interface are viscously shear strained at a rate to meet the kinematic and dynamic interface conditions: (i) no relative normal velocity component is permitted, and (ii) pressure must be continuous across the interface (if surface tension is neglected).

This free shear layer becomes unstable if perturbations introduced by noise, upstream turbulences, or intentional disturbances are amplified. The amplification ratio will exceed unity if a certain condition is satisfied between the frequency of the perturbation and the speed ratio of the two streams. It is well known that the waves of not all frequencies, nor all speed ratios of the two streams will survive; the waves within a certain range of these two parameters will survive, otherwise damping out by surface tensions or viscous forces [16].

The stability condition is generally concerned with the density and speed ratio of the two streams, viscosities, Reynolds number, geometric boundary condition, etc. The amplification of the perturbation will depend on the phase speed of the waves and wave length. If the inertial effect exceeds a critical condition (i.e. $Re > Re_{cr}$), the free shear layer will become unstable and perturbations will be amplified, developing a series of rolled-up vortices. This Kelvin–Helmholtz instability can be observed in many technical flows as well as in nature (Figure 5.34).

In free shear layer instability, three speeds are defined: (i) the speed of the faster-moving fluid U_1, (ii) the speed of the slower-moving fluid U_2, and (iii) the convection speed of the perturbed vorticity wave U_c, excluding any quantity perturbed by compressibility, which will propagate at the speed of sound c. It is generally known that the convection speed of the vorticity wave is approximately 0.6 to 0.7 times the sum of the two stream speeds $U_1 + U_2$. Note also that if $U_1 = -U_2$, then, $U_c = 0$.

Figure 5.34 Kelvin-Helmholtz instability; photographed in Morgan Hill, California, USA (10 December 2021)

5.10.2 Physics on the Formation of Rolled-Up Vortices

In free shear layer, the fluid is rotated and angularly strained at the same rate. Thus, the two flows above and below the shear layer are parallel. Once the free shear layer becomes unstable, the vortex sheet is sinusoidally perturbed. Through the potential interactions, the velocity and pressure sinusoidally vary in the streamwise direction, while satisfying the condition of pressure continuity at the interface. From this reason, the maximum and minimum values of the velocity and pressure must exist at the inflection points (e.g. A and C) in Figure 5.35a.

The fastest speed exists at the inflection point C between the crest B and the trough D (where the pressure is the lowest), while the lowest speed exists at the inflection point A between the trough and the crest (where the pressure is the highest). The fluid accelerates from the inflection point A to the inflection point C and decelerates from C to A. It is interesting to note that while the fluid is being accelerated from A to C, the vortices migrating away from the inflection point A are accumulated at the inflection point C. The vortex sheet is then displaced further upwards and downwards in the crests and troughs by the vortices redistributed along the interface, according to the Biot–Savart law (Figure 5.35). Note that the migration and accumulation of vortices at inflection points A and C are related to the strain and rotation rates, $\varepsilon + \Omega$, on the concave and convex surfaces of the free shear layer (see Section 4.6.2.1 *Shear-straining fluids*).

5.10.2.1 Vortex Coalescence

Once the vortices in the shear layer are rolled at the inflection points of the shear layer, the rolled-up vortices cannot maintain their formation. They are paired and begin to revolve in the

(a)

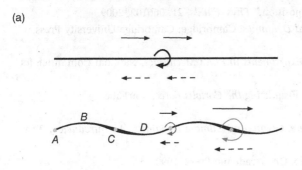

(b) U_1, Vorticity

(c) P

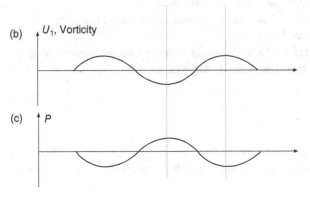

Figure 5.35 Mechanism of vortex roll-up; migration and accumulation of vortices at the inflection points A and C (dots) (a); velocity/vorticity (b) and pressure (c) distributions along the free shear layer

clockwise direction with respect to the point between two vortices. This corotating vortex pair eventually merge into a single vortex and this pairing process continues. It is to be noted that the shape of the rolled-up vortex formed by Kelvin–Helmholtz instability differs between two slip-flow conditions: one with the same flow direction but moving at different speeds, and the other moving in the opposite direction. The rolled-up vortices for the latter are often referred to as cat's eyes.

References

1 Saffman, P.G., *Vortex Dynamics*, Cambridge: Cambridge University Press, 1993.

2 Lugt, H.J., *Vortex Flow in Nature and Technology*, 1st ed. New York: Wiley, 1983.

3 Lighthill, J., *An Informal Introduction to Theoretical Fluid Mechanics*, Oxford: Oxford University Press, 1988.

4 Wu, J.Z., H.Y. Ma, and M.D. Zhou, *Vorticity and Vortex Dynamics*, 2006th ed. Berlin: Springer-Verlag, 2006.

5 Shapiro, A.H., "Vorticity," *Illustrated Experiments in Fluid Mechanics*, National Committee for Fluid Mechanics Films, 1980.

6 Talyor, E.S., "Secondary Flow," *Illustrated Experiments in Fluid Mechanics*, National Committee for Fluid Mechanics Films, 1980.

7 Zhou, Y., "Rayleigh-Taylor and Richtmyer–Meshkov instability induced flow, turbulence, and mixing. I," *LLNL-JRNL-700063*, Lawrence Livermore National Laboratory, 2016.

8 Ogino, Y., N. Ohnishi, S. Taguchi, and K. Sawada, "Baroclinic Vortex Influence on Wave drag Reduction Induced by Pulse Energy Deposition," *Phys. Fluids*, 21, 066102, 2009.

9 Batchelor, G.K., *An Introduction to Fluid Dynamics*, Cambridge: Cambridge University Press, 1964.

10 Fultz, D., "Rotating Flows," *Illustrated Experiments in Fluid Mechanics*, National Committee for Fluid Mechanics Films, 1980.

11 MIT Open Course Ware, *Experiment V: Visualizing the Coriolis Force*, YouTube.

12 UCLA Spinlab, *Coriolis Effect*, YouTube.

13 Marshall, J., and R.A. Plumb, *Atmosphere, Ocean, and Climate Dynamics: An Introductory Text*, San Diego, CA: Academic Press, 2007.

14 Kundu, P.K., *Fluid Mechanics*, San Diego, CA: Academic Press, 1990.

15 Hoskins, H.J., M.E. McIntyre, and A.W. Robertson, "On the Use and Significance of Isentropic Potential Vorticity Maps," *Quart. J. R. Met. Soc.*, 111, 877–946, 1985.

16 Mollo-Christensen, E.L., "Flow Instabilities," *Illustrated Experiments in Fluid Mechanics*, National Committee for Fluid Mechanics Films, 1980.

17 Cohen, R.C.Z., P.W. Cleary, and B.R. Mason, "Simulations of Dolphin Kick Swimming Using Smoothed Particle Hydrodynamics," *Hum. Mov. Sci.*, 31, 604–619, 2012.

Problems

5.1 In vortex dynamics, the curl of the Lamb acceleration vector $\vec{\xi} = \nabla \times \vec{L}$ (where $\vec{L} = \vec{\omega} \times \vec{v}$) represents the rate of change of the vorticity vector, while the vortex is convected, stretched, and tilted by local velocity components, and isotropically expanded by compressibility effect. Physically explain the curl of the Lamb acceleration vector in a curved open channel flow.

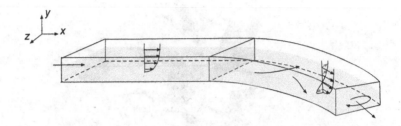

a) Explain what χ defined as

$$\chi(\vec{n}, \vec{x}, t) = \vec{\xi} \cdot \vec{n} = (\nabla \times \vec{L}) \cdot \vec{n} = \lim_{A \to 0} \frac{1}{A} \oint_c \vec{L} \cdot d\vec{s}$$

implicates physically.
b) Discuss $\chi(\vec{n}, \vec{x}, t)$ in the planar Couette and Posieuille flows.
c) Discuss how $\chi > 0$ explains the secondary flow associated with the vortex tilting in the curved channel flow.

5.2 Physically explain the curl of the Lamb acceleration vector $\vec{\xi} = \nabla \times \vec{L}$ in the boundary layer over a cambered airfoil at a small angle of attack.

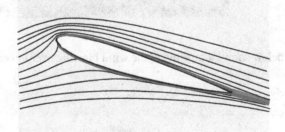

a) Sketch the streamwise distribution of $\vec{L}$ in the boundary layer.
b) Explain what χ in the boundary-layer plane implicates physically.
c) Discuss how $\chi < 0$ explains the lift generation over an airfoil.

5.3 Physically explain the curl of the Lamb acceleration vector $\vec{\xi} = \nabla \times \vec{L}$ in the drainage vortex.

5.4 Birds fly forward by flapping their wings. Explain how birds produce thrust force from the perspective of vortex dynamics.

5.5 Dolphins propel with their undulatory body motions [17]. Explain how dolphins produce thrust force from the perspective of vortex dynamics.

5.6 The photo below shows footprints of the wind blowing over towering rocks on the shore.

a) What vortex dynamics are involved in this phenomenon?

b) Explain why this phenomenon is important for submarines. (Hint: reconnaissance issues.)

5.7 The figure below shows instantaneous eddy structures in a turbulent boundary layer over a flat plate. Explain why the turbulent eddies are tilted approximately 45° with respect to the plate.

5.8 The figures below show the velocity magnitudes plotted on the streamlines in the cerebral blood vessels. Explain how secondary flows are developed in the cerebral blood vessels.

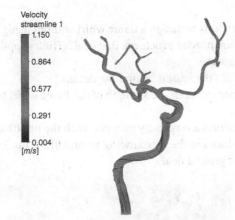

5.9 In an open curved channel flow, the secondary flow downstream can be controlled with an anisotropic porous screen placed upstream. Neglect the end-wall viscous effects.

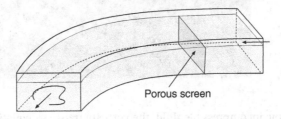

a) How would you distribute porosity to produce a counter-clockwise rotating secondary flow?

b) How would you distribute porosity to produce a clockwise rotating secondary flow?

c) How would you distribute porosity to produce a pair of counter-rotating secondary flows?

5.10 Ben Hur is one of the most famous Hollywood films of the 1950s. What vortex dynamics can we learn from this movie clip of *Chariot Race* (see below)? (Hints: (i) motions of the horse's legs (ii) horses are tied to each other with ropes and covers.)

5.11 A student in a lab wants to design a flame whirl without using any rotating parts.
  a) What supplies are needed to achieve that goal? (Hint: supplies can be used to control the buoyancy and swirl.)
  b) Physically explain the reason behind the design.
  c) With what parameters can the strength of the flame whirl be controlled?

5.12 The photo below shows a very dusty tornado. With the negative pressure built in the vortex core, the tornado draws in the surrounding air and lifts it up. Explain why most of the air is sucked in from the ground field.

5.13 In a barotropic incompressible fluid, the vorticity transport equation in the rotating reference frame reads

$$\frac{D(\vec{\omega}_r + 2\vec{\Omega})}{Dt} = \left((\vec{\omega}_r + 2\vec{\Omega}) \cdot \nabla\right) \vec{v}_r$$

  a) If there is no relative vorticity in the beginning, show that this equation proves the Taylor–Produman theorem:

$$\frac{\partial \vec{v}_r}{\partial z} = 0$$

  b) Physically interpret the Taylor–Produman theorem with respect to the Taylor columns shown in Figure 5.24 (see Example 5.2).
  c) If a small circular column of solid (e.g. hockey puck) is put at the bottom of the rotating cylindrical container, describe how it affects the flow in the rotating water.

5.14 In a laboratory, a student wants to design a flow system that produces a small-scale tornado. The student is provided with a cylindrical tank of diameter D, motor, hose of diameter d, variable-speed water pump, and table.[17]

a) Discuss how to design the system to create a potential swirl and whirlpool.

b) Define the Rossby number R_o.

c) For the *cyclostrophic* and *geostrophic* cases, discuss how the strength of the whirlpool in the rotating system can be changed with the Rossby number R_o.

d) Physically interpret the results of the present experiment with the Coriolis force that can be considered as an inertial force acting in the rotating system.

5.15 In a rapidly rotating cylindrical container filled with water, any vortical disturbances generated in the water hardly mix with each other, and they are formed like curtains (see below).

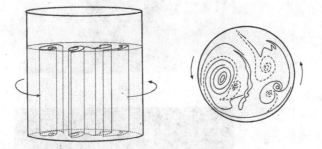

a) Physically explain the reason.

b) If the container does not rotate, what will happen to the vortices?

5.16 A scientist-minded student residing in an high-rise apartment sets up an experimental facility to demonstrate the rotational effect of the earth (see below). A large cylindrical tank is partially filled with water and a vertical pipe is attached to the drainage hole at the bottom of the tank.

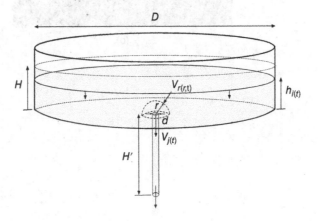

a) Work out the problem as done in Example 5.3.

b) Discuss how the experiment conducted in the high-rise apartment changed the results.

---

17 GFD Lab III: *Radial inflow*, http://marshallplumb.mit.edu/experiments/gfd-iii-radial-inflow.

5.17 The two photos below, (a) milk falling in a cup of coffee and (b) the *crab nebula*, show the same physics. Note that the crab nebula is a supernova remnant and pulsar wind nebula in the constellation of *Taurus*. Discuss the physics involved.

(a)

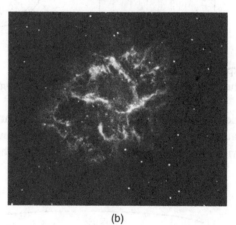

(b)

6

External Viscous Flows

6.1 The Rayleigh Problem

6.1.1 Molecular Diffusion of Tangential Momentum

Let us suppose a stationary fluid of density ρ and viscosity μ is bounded at $y = 0$ by a plate in a semi-infinite domain (Figure 6.1). If the plate is suddenly pulled at a constant speed, a tangential viscous force exerted by the plate does work on the fluid in two ways: (i) straining the fluid and viscously dissipating the mechanical energy at a rate into heat, and (ii) locally accelerating the fluid and changing the kinetic energy at a rate.

The net horizontal viscous force and the transient inertial force of the fluid are balanced by the momentum principle:

$$\rho \, (dA \, dy) \frac{\partial u}{\partial t} = \left(\tau_{yx}(x, y + dy) - \tau_{yx}(x, y) \right) \, dA \tag{6.1}$$

where τ_{yx} represents the viscous stress acting in the x direction on the surface facing the y direction and $dA \, (= dxdz)$ is the wetted area. With the Taylor series, the equation of motion reads

$$\frac{\partial u}{\partial t} = v \frac{\partial^2 u}{\partial y^2} \tag{6.2}$$

where $v = \mu/\rho$ is the kinematic viscosity of the fluid (or hydrodynamic diffusivity). Equation (6.2) can be solved with the initial condition and no-slip condition at the boundary:

$$u(y, 0) = 0; \quad u(0, t > 0) = U_0 \quad \text{and} \quad u(\infty, t > 0) = 0 \tag{6.3}$$

This is a classical molecular diffusion equation for the transport of heat, momentum, vorticity, concentration, etc., where the kinematic viscosity $v = \mu/\rho$ is the diffusivity coefficient. In a microscopic point of view, the tangential momentum supplied by the plate is transversely transported over time by frictional molecular interactions, i.e. the random walks of the molecules (or Brownian motion).

6.1.2 Molecular Diffusion Speed

Molecular diffusion occurs in a continuous transitional region of which the thickness increases with diffusivity and time. It is, however, to be noted that this diffusion problem does not have a fixed natural length or time scale, though its solution depends on space or time. Therefore, it is necessary to construct a scale using space, time, or other dimensional quantities that can represent the problem (e.g. kinematic viscosity v).

Introduction to Fluid Dynamics: Understanding Fundamental Physics, First Edition. Young J. Moon.
© 2022 John Wiley & Sons, Inc. Published 2022 by John Wiley & Sons, Inc.
Companion website: www.wiley.com/go/Moon/IntroductiontoFluidDynamics

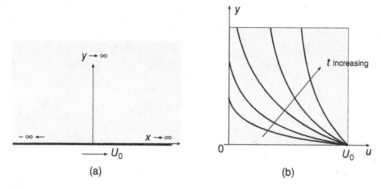

Figure 6.1 Rayleigh problem (a); velocity $u(t)$ in time (b)

The molecular diffusion speed is proportional to the hydrodynamic diffusivity (or kinematic viscosity, $v = \mu/\rho$). For instance, the more viscous the fluid is, the faster the molecular diffusion occurs. In contrast, the molecular diffusion speed is inversely proportional to the fluid density because the diffusion process is based on the Brownian motion of the fluid molecules. The greater the number of molecules, the more molecular interactions occur; thus, diffusion is more slowly processed.

The molecular diffusion speed can be represented by the ratio v/δ, where δ is the reference length scale. To show this, we take an order analysis on the diffusion equation, Eq. (6.2):

$$\left(\frac{1}{t}\right) \sim v\left(\frac{1}{\delta^2}\right) \tag{6.4}$$

From Eq. (6.4), a thickness of transition via molecular diffusion can be estimated as follows:

$$\delta \sim \sqrt{v\,t} \tag{6.5}$$

where δ is referred to as the penetration depth.

If we define the diffusion time scale t as δ/u_d, where u_d is the molecular diffusion speed, Eq. (6.5) reads

$$u_d \sim \frac{v}{\delta} \tag{6.6}$$

which shows that u_d is proportional to the kinematic viscosity v per unit penetration depth.[1] We can find a similar example in the heat conduction of a homogeneous solid. The heat is transferred down the gradient of temperature, and its rate is proportional to the thermal diffusivity, $\alpha = (k/c_p)/\rho$, where k is the conductivity, c_p heat capacity, and ρ density. This also indicates that the diffusion process is based on molecular interactions through the Brownian motion.

---

1 In general, a molecular diffusion speed by friction can be expressed as follows:

$$u_d \sim \frac{v}{\delta} \sim \sqrt{\frac{v}{t}} = \sqrt{\frac{\mu/t}{\rho}} = \sqrt{\frac{K_f}{\rho}} \tag{6.7}$$

where μ/t represents a measure of frictional stiffness, analogous to dilatational stiffness (or bulk modulus, K_d) in compressible flows. More discussion on this continues in Chapter 8 *Compressible Flows*.

Table 6.1   Properties of air and water

	$\rho \; [kg/m^3]$	$\mu \; [Ns/m^2]$	$v \; [m^2/s]$
Air	1.225	1.8×10^{-5}	1.47×10^{-5}
Water	1000	1.3×10^{-3}	1.3×10^{-6}

Example 6.1 *Diffusivity (or kinematic viscosity) of air and water*
Molecular diffusion occurs faster in the air than in water because air density is approximately 1000 times lower, although its viscosity is approximately 100 times higher (Table 6.1). Thus, for the same flow speed, the boundary layer thickness in water is 10 times thinner than that in the air.

6.1.3   Self-Similarity

The solution of the Rayleigh problem can be found by scaling the governing equations, based on the concept of dimensional analysis and scaling laws. Since the thickness of transitional region (or transient boundary layer) is proportional to the kinematic viscosity v and time t to the power of one half, and there exists no characteristic length scale, the velocity profile only grows in time, and its profile at each instant appears to be self-similar.

A new self-similar variable η can then be defined as follows:

$$\eta = y/\delta = y/\sqrt{vt} \tag{6.8}$$

where two independent variables, y and t, are reduced to a single variable η by scaling the length scale with $\sqrt{vt}$, that is, another reference length scale.

We can find the self-similarity solution of Eq. (6.2) with initial and boundary conditions Eq. (6.3). By introducing a self-similar variable η and a nondimensional velocity f defined as follows:

$$\eta = \frac{y}{\sqrt{vt}}, \quad f(\eta) = \frac{u}{U_0} \tag{6.9}$$

the partial differential equation, Eq. (6.2), can be reduced to an ordinary differential equation

$$f'' + \frac{1}{2}\eta f' = 0 \tag{6.10}$$

with new boundary conditions

$$f(0) = 1, \quad f(\infty) = 0 \tag{6.11}$$

The solution to the above problem can be written in terms of the complementary error function as follows:

$$u = U_0 \mathrm{erfc}\left(\frac{y}{\sqrt{4vt}}\right) \tag{6.12}$$

and the force per unit area exerted on the plate reads

$$F = -\mu \left(\frac{\partial u}{\partial y}\right)_{y=0} = \rho \sqrt{\frac{vU_0^2}{\pi t}} \tag{6.13}$$

6.2 Laminar Boundary Layer

6.2.1 Creation, Convection, and Diffusion of Vorticity

6.2.1.1 Boundary Layer Thickness

A flow over a flat plate is shear-resisted close to the surface of the plate by frictional viscous forces. In consequence, the flow develops a boundary layer, where the shear-resisting momentum (or vorticity) created at the leading-edge of the plate is not only diffused in the transverse direction through molecular interactions but also convected in the streamwise direction with the fluid above.

As shown in Figure 6.2, the diffusion of momentum defect in the transverse direction is bounded by the convecting fluid. This boundary is the vertical distance that the viscous shear forces imposed from the wall can ever reach at any x. This is called the boundary layer, whose thickness $\delta(x)$ is defined as the vertical distance from the wall at which the streamwise velocity is 99% of the free stream velocity, i.e. $\delta(x) = y(x)$ at 0.99 U_∞.

The velocity profiles in the spatially growing boundary layer are *self-similar* as in the Rayleigh problem. Thus, an analogy can be taken between the self-similarity of a transiently growing viscous layer and the self-similarity of a spatially growing boundary layer over a flat plate of infinite length. Note that the boundary layer thickness grows in space as follows:

$$\delta \sim \sqrt{\nu t} = \sqrt{\frac{\nu x}{U_\infty}} \tag{6.14}$$

since $t \sim x/U_\infty$ in the flow over the flat plate. For a given ν, the faster the fluid (or the shorter the distance of travel), the thinner the thickness.

With Eq. (6.14), we can also explain that if the fluid is accelerated over a convex surface of the body, the boundary layer thickness will become thinner, but if the fluid is decelerated over a concave surface, the boundary layer thickness will be thicker. In creeping flows, the boundary layer thickness will far exceed the characteristic length of the body, since either ν is very large or U_∞ is very small for a given length scale x.

6.2.2 Reynolds Number

In many external flows with high speed or low kinematic viscosity, the outer potential flow spatially dominates the viscous boundary layer, which is often confined to a thin region adjacent to the solid surfaces, e.g. flows over the aircrafts, rockets, automobiles, submarines. The question is what physical quantities determine the boundary layer thickness δ.

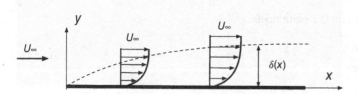

Figure 6.2 Laminar boundary layer and boundary layer thickness $\delta(x)$

If we nondimensionalize Eq. (6.14) by dividing it with the length scale x, it reads

$$\delta/x \sim \sqrt{\frac{v}{U_\infty\, x}} = \frac{1}{\sqrt{Re_x}} \tag{6.15}$$

where $Re_x = U_\infty\, x/v$ is the Reynolds number. Note also that the Reynolds number can be defined by the ratio of the convection speed of the fluid (inertial effect) u_c to the viscous (or molecular) diffusion speed of the frictional momentum defect u_d

$$Re_x = \frac{u_c}{u_d} = \frac{U_\infty}{v/x} = \frac{\rho\, U_\infty\, x}{\mu} \tag{6.16}$$

where the Reynolds number represents a measure of dominance between these two speeds.

In high Reynolds number flows, the molecular transport of the momentum defect from the wall is confined in a thin layer. At low Reynolds numbers, molecular diffusion can penetrate deep into the fluid and thicken the boundary layer. When a solar-powered aircraft, e.g. *Pathfinder*, is at landing, for example, a typical Reynolds number for a flow over the wing is approximately $Re_l = 3 \times 10^5$, where l is the chord length of the wing. Therefore, the boundary layer thickness over the wing is about $l/\sqrt{Re_l} \approx l/550$, and δ is approximately 4.36 mm for $l = 2.4$ m. Note that Eq. (6.15) is only valid for laminar boundary layers.

6.2.3 Displacement Thicknesses and Momentum Thickness

Due to viscous effects, mass flux in the boundary layer is defected so that the boundary layer behaves as if the body is displaced upward by δ^*, which is called the displacement thickness (Figure 6.3). It can be interpreted as an extra thickness that would be added to the body to account for extra blockage by the boundary layer. The concept of the displacement thickness is used in the design of ducts, intakes of air-breathing engines, wind tunnels, etc. by first assuming a frictionless flow and then enlarging the passage walls by the displacement thickness so as to allow the same mass flow rate.

The streamlines outside the boundary layer are displaced by the presence of the boundary layer. Equating mass flux across the two sections A and B yields

$$U_\infty\, \delta^* = U_\infty\, H - \int_0^H u\, dy \tag{6.17}$$

where H is an arbitrary distance larger than δ. Thus, the displacement thickness is defined as follows:

$$\delta^* = \int_0^H \left(1 - \frac{u}{U_\infty}\right) dy \tag{6.18}$$

Figure 6.3 Streamline displacement and displacement thickness δ^*

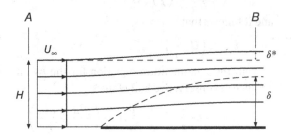

Similarly, a momentum thickness θ is defined such that $\rho\, U_\infty^2\, \theta$ is the momentum loss due to the presence of the boundary layer. The momentum principle states that

$$\rho\, U_\infty^2\, \theta = \rho\, U_\infty^2\, H - \int_0^{H+\delta^*} \rho\, u^2\, dy \tag{6.19}$$

From δ^* defined in Eq. (6.18), Eq. (6.19) is written as follows:

$$\rho\, U_\infty^2\, \theta = \int_0^H \rho\, (U_\infty^2 - u^2)\, dy - \rho\, U_\infty^2 \int_0^H \left(1 - \frac{u}{U_\infty}\right) dy \tag{6.20}$$

For an incompressible flow, the momentum thickness is then defined as follows:

$$\theta = \int_0^H \frac{u}{U_\infty} \left(1 - \frac{u}{U_\infty}\right) dy \tag{6.21}$$

which is the normal distance to a reference plane representing the lower edge of a hypothetical inviscid fluid of uniform velocity U_∞ that has the same momentum flux as occurring in the real fluid with the boundary layer.

6.2.4 Prandtl's Boundary Layer Equation

The concept of the boundary layer is based on two assumptions. In a thin layer adjacent to the body, the velocity gradient normal to the wall $\partial u/\partial y$ is very large, and a very small viscosity of fluid exerts an essential influence in so far as the shearing stress $\tau = \mu\, \partial u/\partial y$ may assume large values. In the remaining region, no such large velocity gradients exist and the influence of viscosity is unimportant. In this region, the flow is frictionless and irrotational.

The thickness of the boundary layer generally increases with viscosity or decreases as the Reynolds number increases:

$$\delta \sim \sqrt{\nu} \tag{6.22}$$

and it is assumed that this thickness is very small compared to the dimension of the body l:

$$\delta \ll l \tag{6.23}$$

In this case, the solution obtained from the boundary layer equations are asymptotic and apply to very large Reynolds numbers.

Let a characteristic velocity scale of u be U and l the streamwise distance over which u changes appreciably. The pressure is made dimensionless with $\rho\, U^2$ and time with l/U. From the continuity equation:

$$\frac{\partial u}{\partial x} + \frac{\partial v}{\partial y} = 0 \tag{6.24}$$

the orders of magnitude of the terms can be expressed as follows:

$$O\left(\frac{U}{l}\right) \sim O\left(\frac{v}{\delta}\right) \tag{6.25}$$

and it follows that

$$v \sim O\left(U \cdot \frac{\delta}{l}\right) \tag{6.26}$$

From the x-momentum equation for the two-dimensional, steady, incompressible flow in the absence of conservative body forces:

$$u\frac{\partial u}{\partial x} + v\frac{\partial u}{\partial y} = -\frac{1}{\rho}\frac{\partial p}{\partial x} + \frac{\mu}{\rho}\left(\frac{\partial^2 u}{\partial x^2} + \frac{\partial^2 u}{\partial y^2}\right) \tag{6.27}$$

and by dividing by U^2/l, the orders of magnitude of the terms are written as follows:

$$O(1) + O(1) \sim O(1) + O\left(\frac{1}{Re}\right) + O\left(\frac{1}{Re} \cdot \frac{l^2}{\delta^2}\right) \tag{6.28}$$

For large Re, the terms of $O(1/Re)$ approach zero, while the last term approaches order one, since $\delta/l \ll 1$. It then follows that

$$\frac{\delta}{l} \sim O\left(\frac{1}{\sqrt{Re}}\right) \tag{6.29}$$

and with the continuity equation,

$$\frac{v}{U} \sim O\left(\frac{1}{\sqrt{Re}}\right) \tag{6.30}$$

Thus, the x-momentum equation can be simplified as follows:

$$u\frac{\partial u}{\partial x} + v\frac{\partial u}{\partial y} = -\frac{1}{\rho}\frac{\partial p}{\partial x} + \frac{\mu}{\rho}\frac{\partial^2 u}{\partial y^2} \tag{6.31}$$

Meanwhile, from the y-momentum equation:

$$u\frac{\partial v}{\partial x} + v\frac{\partial v}{\partial y} = -\frac{1}{\rho}\frac{\partial p}{\partial y} + \frac{\mu}{\rho}\left(\frac{\partial^2 v}{\partial x^2} + \frac{\partial^2 v}{\partial y^2}\right) \tag{6.32}$$

and with division by U^2/l, the orders of magnitude of the terms read

$$O\left(\frac{\delta}{l}\right) + O\left(\frac{\delta}{l}\right) \sim \underbrace{O(?)}_{\equiv\, O(\delta/l)} + O\left(\frac{1}{Re} \cdot \frac{\delta}{l}\right) + O\left(\frac{\delta}{l}\right) \tag{6.33}$$

but $O(\delta/l) \sim O(1/\sqrt{Re})$. Therefore, it can be concluded from the y-momentum equation that for large Re:

$$\frac{\partial p}{\partial y} = 0 \tag{6.34}$$

Since viscous forces are exerted via frictional molecular interferences, pressure cannot change unless the fluid is completely bounded by solid walls. In external boundary layers, pressure does not change across the boundary layer, as proven by Eq. (6.34).

6.2.5 Blasius's Self-Similarity Solution

For a two-dimensional, steady, laminar boundary layer on a flat plate, the continuity and momentum equations read

$$\frac{\partial u}{\partial x} + \frac{\partial v}{\partial y} = 0 \tag{6.35}$$

$$u\frac{\partial u}{\partial x} + v\frac{\partial u}{\partial y} = \nu\frac{\partial^2 u}{\partial y^2} \tag{6.36}$$

which are subject to the following boundary conditions:

$$
\begin{aligned}
u &= U_\infty & \text{at} \quad x &= 0, & \text{all } y \\
u &= v = 0 & \text{at} \quad y &= 0, & 0 \leq x \leq L \\
u &\to U_\infty & \text{at} \quad y &\to \infty, & 0 \leq x \leq L
\end{aligned}
\tag{6.37}
$$

To reduce the continuity and momentum equations into a single equation, a stream function ψ defined by

$$u = \frac{\partial \psi}{\partial y} \quad \text{and} \quad v = -\frac{\partial \psi}{\partial x} \tag{6.38}$$

is introduced. The stream function that satisfies the continuity equation expresses the momentum equation as follows:

$$\frac{\partial \psi}{\partial y} \frac{\partial^2 \psi}{\partial x \, \partial y} - \frac{\partial \psi}{\partial x} \frac{\partial^2 \psi}{\partial y^2} = \nu \frac{\partial^3 \psi}{\partial y^3} \tag{6.39}$$

With the self-similar variables defined as follows:

$$\eta = \frac{y}{\delta(x)} = y \sqrt{\frac{U_\infty}{\nu x}} \tag{6.40}$$

$$f(\eta) = \frac{\psi}{\delta(x) U_\infty} = \frac{\psi}{\sqrt{\nu x U_\infty}} \tag{6.41}$$

where $f(\eta)$ is the dimensionless stream function, we can evaluate the derivatives in the boundary layer equation Eq. (6.39) as follows:

$$\frac{\partial \psi}{\partial y} = U_\infty f' \tag{6.42}$$

$$\frac{\partial \psi}{\partial x} = -\frac{1}{2} \sqrt{\frac{\nu U_\infty}{x}} \left(\eta f' - f \right) \tag{6.43}$$

$$\frac{\partial^2 \psi}{\partial x \, \partial y} = -\frac{U_\infty}{2x} \eta f'' \tag{6.44}$$

$$\frac{\partial^2 \psi}{\partial y^2} = U_\infty \sqrt{\frac{U_\infty}{\nu x}} f'' \tag{6.45}$$

$$\frac{\partial^3 \psi}{\partial y^3} = \frac{U_\infty^2}{\nu x} f''' \tag{6.46}$$

Substituting these expressions into Eq. (6.39) yields an ordinary differential equation for the dimensionless stream function f:

$$f''' + \frac{1}{2} f f'' = 0 \tag{6.47}$$

which is subject to the boundary conditions

$$f(0) = f'(0) = 0 \tag{6.48}$$

$$f'(\eta) \to 1 \quad \text{as} \quad \eta \to \infty \tag{6.49}$$

A numerical solution of this Blasius laminar boundary layer equation presented in Figure 6.4 shows the scaled velocities, u/U_∞ and $v/(\nu x/U_\infty)^{1/2}$. With the Blasius solution, we can also obtain the wall shear stress τ_w, skin-friction coefficient c_f, drag force acting on a length l of the plate F, boundary layer thickness δ, displacement thickness δ^*, and momentum thickness θ as follows:

$$\tau_w = \mu \left(\frac{\partial u}{\partial y} \right)_{y=0} = 0.332 \sqrt{\frac{\rho \mu U_\infty^3}{x}} \tag{6.50}$$

$$c_f = \frac{\tau_w}{1/2 \, \rho \, U_\infty^2} = 0.664 \sqrt{\frac{\mu}{\rho \, U_\infty x}} = \frac{0.664}{\sqrt{Re_x}} \tag{6.51}$$

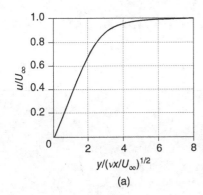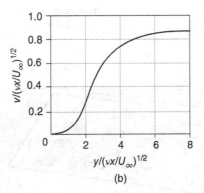

Figure 6.4 Blasius laminar boundary layer; the scaled-velocities: u (a) and v (b)

$$F = 2 \int_0^l \tau_w \, dx = 1.328 \sqrt{\rho \mu U_\infty^3} \tag{6.52}$$

$$\delta = 5.0 \sqrt{\frac{vx}{U_\infty}} = 5.0 \, \frac{x}{Re_x^{1/2}} \tag{6.53}$$

$$\delta^* = \int_0^\infty \left(1 - \frac{u}{U_\infty}\right) dy = 1.72 \sqrt{\frac{vx}{U_\infty}} \tag{6.54}$$

$$\theta = \int_0^\infty \frac{u}{U_\infty} \left(1 - \frac{u}{U_\infty}\right) dy = 0.665 \sqrt{\frac{vx}{U_\infty}} \tag{6.55}$$

It is important to be reminded that the distribution of the wall viscous stress $\tau_w = \mu \, \partial u / \partial y |_w$ along the surface of the flat plate is not only a function of x but also a function of the state of the boundary layer: laminar, transitional, or turbulent.

6.3 Stability and Transition

The boundary layer over the flat plate remains laminar up to $Re_x = 5 \times 10^5$. After this critical Reynolds number, the boundary layer becomes unstable in the streamwise direction, and the Tollmien–Schlichting (TS) waves start to appear. It is to be noted that in wall-bounded parallel shear flows, viscous effects destabilize the flow. Therefore, as the Reynolds number increases further, the boundary layer starts to become unstable in the spanwise direction, and three-dimensional vortex breakdown occurs. Later, turbulent spots are created, and the boundary layer finally becomes fully turbulent at $Re_x > 1 \times 10^7$ [1]. Figure 6.5 schematically illustrates the transition process to turbulence, indicating that the onset of turbulence is associated with the instability of the fluid.

6.3.1 Orr–Sommerfeld Equation

Let us consider a two-dimensional incompressible mean flow superimposed with a two-dimensional disturbance. We simplify the problem by assuming that the flow is parallel, i.e. $\vec{v} = (U(y), 0, 0)$ so that we can decompose the incompressible flow field as follows:

$$u = U + u'$$

$$v = v'$$

$$p = P + p' \tag{6.56}$$

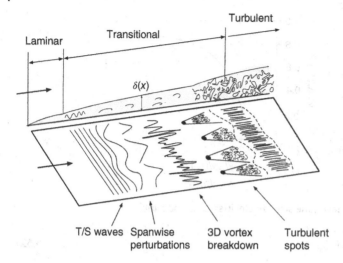

Figure 6.5 Boundary layer transition from laminar to turbulent ($5 \times 10^5 < Re_x < 1 \times 10^7$, flat plate)

where the perturbed variables ϕ' are functions of time and space, i.e. $\phi' = \phi'(x, y, t)$. Note that this parallel flow approximation can describe the fully developed flows in the channel or pipe as well as the flow in the boundary layer. The latter can also be regarded as a parallel flow with good approximation, since flow development in the x-direction is much slower than that in the y-direction.

Now, by substituting Eq. (6.56) into the two-dimensional incompressible Navier–Stokes equations and neglecting the nonlinear terms of the perturbed quantities, we obtain the following momentum equations in the x and y directions:

$$\frac{\partial u'}{\partial t} + U \frac{\partial u'}{\partial x} + u' \frac{\partial U}{\partial x} = -\frac{1}{\rho} \frac{\partial P}{\partial x} - \frac{1}{\rho} \frac{\partial p'}{\partial x} + \nu \left(\frac{\partial^2 U}{\partial x^2} + \nabla^2 u' \right) \tag{6.57}$$

$$\frac{\partial v'}{\partial t} + U \frac{\partial v'}{\partial x} = -\frac{1}{\rho} \frac{\partial P}{\partial y} - \frac{1}{\rho} \frac{\partial p'}{\partial y} + \nu \nabla^2 v' \tag{6.58}$$

and the continuity equation:

$$\frac{\partial u'}{\partial x} + \frac{\partial v'}{\partial y} = 0 \tag{6.59}$$

Since the mean base flow satisfies

$$0 = -\frac{1}{\rho} \frac{\partial P}{\partial x} + \nu \frac{\partial^2 U}{\partial x^2} \quad \text{and} \quad \frac{\partial P}{\partial y} = 0 \tag{6.60}$$

subtracting Eq. (6.60) from Eqs. (6.57) and (6.58) yields the perturbed momentum equations:

$$\frac{\partial u'}{\partial t} + U \frac{\partial u'}{\partial x} + u' \frac{\partial U}{\partial x} = -\frac{1}{\rho} \frac{\partial p'}{\partial x} + \nu \nabla^2 u' \tag{6.61}$$

$$\frac{\partial v'}{\partial t} + U \frac{\partial v'}{\partial x} = -\frac{1}{\rho} \frac{\partial p'}{\partial y} + \nu \nabla^2 v' \tag{6.62}$$

along with the perturbed continuity equation, Eq. (6.59).

It is important to note that in laminar shear flows, any disturbance created by instability propagates in the x-direction as a wave, whereas the perturbation is two-dimensional. We then introduce a stream function $\psi = \psi(x, y, t)$ such that

$$\psi(x, y, t) = \phi(y) \, e^{i(\alpha x - \beta t)} = \phi(y) \, e^{i\alpha(x - ct)} \tag{6.63}$$

where $\phi(y)$ is the amplitude of the disturbance stream function and $\alpha = 2\pi/\lambda$ where λ is the wave length. Note that $\beta = \beta_r + i\,\beta_i$ is a complex number, where β_r is the angular velocity or $2\pi f$ (f is the frequency) and β_i is called the amplification factor. If $\beta_i < 0$, the disturbances will be damped and the mean flow is stable. If $\beta_i > 0$, then the flow becomes unstable. The ratio β/α is denoted by ξ where it is also complex, i.e. $\xi = \xi_r + i\,\xi_i$. The parameter ξ_r is the phase velocity, and ξ_i is the degree of damping or amplification, depending on its sign. The limiting case, $\xi_i = 0$, is called neutral stability.

From Eq. (6.63), we can obtain the velocity perturbations, u' and v', where

$$u' = \frac{\partial \psi}{\partial y} = \phi'(y)e^{i(\alpha x - \beta t)}, \quad \text{and} \quad v' = -\frac{\partial \psi}{\partial x} = -i\,\alpha\,\phi(y)\,e^{i(\alpha x - \beta t)} \tag{6.64}$$

Introducing Eq. (6.64) into Eqs. (6.59), (6.61), and (6.62) and by eliminating the pressure terms, we obtain the fourth-order ordinary differential equation for $\phi(y)$:

$$(U - \xi)(\phi'' - \alpha^2 \phi) - U'' \phi = -\frac{i}{\alpha Re}(\phi'''' - 2\alpha^2 \phi'' + \alpha^4 \phi) \tag{6.65}$$

where $Re = U^* l^*/\nu$, U^* is the maximum velocity of the main flow and l^* is the reference length scale. This stability equation for the disturbance is referred to as the *Orr–Sommerfeld* equation. The boundary conditions of Eq. (6.65) for the boundary layer flow are as follows:

$$u' = v' = 0 \quad (\phi = 0, \quad \phi' = 0) \quad \text{at } y = 0 \text{ and } y = \infty \tag{6.66}$$

The problem of stability is now reduced to an eigenvalue problem Eq. (6.65) with the boundary condition Eq. (6.66). If the mean flow $U(y)$ is specified, there exists a functional relation between the parameters, α, Re, and ξ ($= \xi_r + i\,\xi_i$), which the eigenfunction $\phi(y)$ must satisfy. Thus, if we choose the Reynolds number Re and the disturbance wavelength λ ($= 2\pi/\alpha$), Eqs. (6.65) and (6.66) will give one eigenfunction $\phi(y)$ and one complex eigenvalue ξ.

An α–Re diagram can be drawn for the given laminar flow $U(y)$. Note that the locus of $\xi_i = 0$ corresponds to the curve of neutral stability. An example is shown in Figure 6.6 where neutral stability curves for two velocity profiles are shown. For a given velocity profile, the interior of each curve represents a region of instability, whereas the exterior is a region of stability. The critical Reynolds number of stability Re_{cr} can also be defined on the curve of neutral stability, below which the oscillations decay. Above that value, some are amplified [2].

The critical Reynolds number calculated from stability theory will in general be less than the Reynolds number observed at the *point of transition*. Note, however, that it has been experimentally found that the transition takes place over a finite distance. We often distinguish between these two

Figure 6.6 Curves of neutral stability; two-dimensional boundary layer with two-dimensional disturbances. Source: Republished with permission of Springer Nature from Onset of Turbulence (Stability Theory), Hermann Schlichting (Deceased), Klaus Gersten, pp 415–496, Jan 1, 2017, permission conveyed through Copyright Clearance Center, Inc [1]

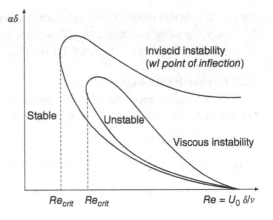

values by referring to the experimental one as the point of transition and the theoretical one as the *point of instability*.

6.3.2 Inviscid Stability

6.3.2.1 Rayleigh Equation

Based on the fact that disturbances are majorly governed by inviscid dynamics, the stability analysis of the laminar flow can be simplified at high Reynolds numbers. As $Re \to \infty$, Eq. (6.65) reduces to

$$(U - \xi)(\phi'' - \alpha^2\phi) - U''\phi = 0 \tag{6.67}$$

which is called the *Rayleigh equation*. Since this inviscid stability equation is of the second order, we need two boundary conditions:

$$\phi = 0 \quad \text{at } y = 0 \text{ and } y = \infty \tag{6.68}$$

representing that there are no normal velocities at the two boundaries. Equations (6.67) and (6.68) define again an eigenvalue problem, with ϕ as the eigenfunction and ξ as the eigenvalue.

6.3.2.2 Rayleigh's Theorem (Point of Inflection)

If we consider an unstable mode for which $\xi_i > 0$ and thus $U - \xi \neq 0$, Eq. (6.67) is rewritten as follows:

$$\phi'' - \alpha^2\phi - \frac{U''}{(U - c)}\phi = 0 \tag{6.69}$$

By multiplying this equation by a conjugate ϕ^* and integrating it from y_1 to y_2 by parts, the first term in Eq. (6.69) is written as follows:

$$\int_{y_1}^{y_2} |\phi^*\phi''dy = \phi^*\phi' \Big|_{y_1}^{y_2} - \int_{y_1}^{y_2} \phi^*\phi'dy = -\int_{y_1}^{y_2} |\phi'|^2dy \tag{6.70}$$

The Rayleigh equation then reads

$$\int_{y_1}^{y_2} \left(|\phi'|^2 + \alpha^2|\phi|^2\right) dy + \int_{y_1}^{y_2} \frac{U''}{(U - \xi)} |\phi|^2dy = 0 \tag{6.71}$$

If we multiply the numerator and denominator by $U - \xi^*$, we can find the imaginary part of Eq. (6.71) as follows:

$$\xi_i \int_{y_1}^{y_2} \frac{U''}{|U - \xi|^2} |\phi|^2dy = 0 \tag{6.72}$$

For $\xi_i > 0$ (unstable mode), Eq. (6.72) is satisfied only if U'' changes sign at least once in the interval, $y_1 < y < y_2$. In other words, the base flow velocity must have at least one *point of inflection* where $U'' = 0$ within the interval (Figure 6.7) [3].[2]

6.3.2.3 Fjortoft's Theorem

Fjortoft, a Swedish meteorologist, discovered a stronger necessary condition for the instability of inviscid parallel flows: somewhere in the flow,

$$U''(U - U_1) < 0$$

where U_1 is the velocity at the point of inflection.

---

2 This is a necessary condition for inviscid instability, not a sufficient condition. The existence of a point of inflection does not guarantee a nonzero ξ_i.

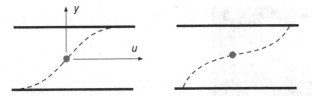

Figure 6.7 Inviscid instability conditions

This instability condition can be proved by multiplying again the numerator and denominator of the second term in Eq. (6.71) by $U - \xi^*$ and taking the real part:

$$\int_{y_1}^{y_2} \frac{U''(U - \xi_r)}{|U - \xi|^2} |\phi|^2 dy = - \int_{y_1}^{y_2} \left(|\phi'|^2 + \alpha^2 |\phi|^2 \right) dy \; < 0 \qquad (6.73)$$

Meanwhile, if the flow is unstable so that $\xi_i > 0$ and the point of inflection exists according to the Rayleigh criterion, then it follows that

$$(\xi_r - U_1) \int_{y_1}^{y_2} \frac{U''}{|U - \xi|^2} |\phi|^2 dy = 0 \qquad (6.74)$$

If we add Eq. (6.73) and Eq. (6.74), we obtain

$$\int_{y_1}^{y_2} \frac{U''(U - U_1)}{|U - \xi|^2} |\phi|^2 dy \; < 0 \qquad (6.75)$$

so that $U''(U - U_1)$ must be negative somewhere in the flow.

6.4 Fundamentals of Turbulent Flow

6.4.1 General Characteristics of Turbulent Flow

Turbulent flows are more common than laminar flows in reality. As shown in Figure 6.8, they are drastically different from flows in laminar state. Turbulent flows can be characterized by randomness, diffusiveness and effective mixing, three-dimensional vorticity fluctuations (or vortex stretching), large Reynolds numbers, dissipation, etc. One of the most important concepts in turbulence is the energy cascade; energy can be transferred through different scales.

Figure 6.8 shows that turbulent flows have various length scales. Large-scale turbulences can be close to the geometric scale (e.g. diameter of the crater), but the smaller ones can be identified by breaking up the scales. The smallest scale is the Kolmogorov scale based on the viscous dissipation rate of energy. Note that the higher the Reynolds number, the smaller the eddy sizes. Thus, turbulence is a property of the flows, not of the fluids.

6.4.2 Turbulent Fluctuations and Time-Mean Flow Profiles

A turbulent boundary layer over a flat plate is presented in Figure 6.9. With the second invariant of the velocity gradient tensor Q, the figure clearly shows the most common feature in the turbulent boundary layer, i.e. instantaneous turbulent eddies including hairpin vortex. In this case, the instability is triggered at the leading-edge of the plate which has a finite thickness, and the boundary layer quickly transits to turbulence over the plate surface.

Figure 6.8 Energy cascade of turbulent jet in volcanic eruption; small-scale turbulences embedded in large-scale turbulent motions. Source: Yosh Ginsu/Unsplash

Figure 6.9 Turbulent boundary layer on a flat plate; large eddy simulation at $Re_l = 6 \times 10^5$ (l: plate length)

It is interesting to note that the turbulent boundary layer can broadly be divided into two regions: the first region is the viscous sublayer which occupies less than 1% of the total turbulent boundary layer thickness. The momentum transfer in this region is dominated by viscous shear. Outside of the viscous sublayer is a full turbulent region, where turbulent eddies transfer energy from largest to smallest through a cascade process.

In Figure 6.10, two velocity profiles in the laminar and turbulent boundary layers are compared, where the instantaneous velocity profiles in the turbulent boundary layer are taken at a number of instances. The turbulent boundary layer on the right shows that velocity increases sharply in the sublayer near the wall. The turbulent velocity profile in the boundary layer is difficult to be determined since the flow is no longer steady. However, its average profile is consistent over time. Note also that the turbulent boundary layer thickness is expressed as a function of the Reynolds number:

$$\delta \approx 0.37 \, \frac{x}{Re_x^{1/5}} \tag{6.76}$$

6.4.3 The Reynolds Equations

6.4.3.1 The Reynolds Decomposition

For incompressible flows, the instantaneous flow quantities can be decomposed into an average and fluctuation as follows:

$$u_i = \overline{u}_i + u_i' \tag{6.77}$$

$$p = \overline{p} + p' \tag{6.78}$$

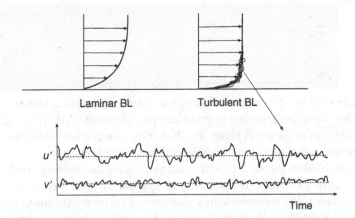

Figure 6.10 Laminar boundary layer vs. turbulent boundary layer (instantaneous snap shots and time-averaged) (top); velocity fluctuations, $u'(t)$ and $v'(t)$, in the turbulent boundary layer (bottom)

The basic idea of this decomposition is that we substitute Eqs. (6.77) and (6.78) into the equations of motion and subtract the mean equation.

The continuity equations for the mean and turbulent fluctuation then read as follows:

$$\frac{\partial \bar{u}_i}{\partial x_i} = 0 \tag{6.79}$$

$$\frac{\partial u'_i}{\partial x_i} = 0 \tag{6.80}$$

where both equations are divergence free.

By applying the same procedure to the Navier–Stokes equations, we can obtain for the mean and turbulent fluctuation:

$$\frac{\partial \bar{u}_i}{\partial t} + \frac{\partial \bar{u}_i \bar{u}_j}{\partial x_j} = -\frac{1}{\rho_0}\frac{\partial \bar{p}}{\partial x_i} + \nu \frac{\partial^2 \bar{u}_i}{\partial x_i^2} - \frac{\partial \overline{u'_i u'_j}}{\partial x_j} \tag{6.81}$$

$$\frac{\partial u'_i}{\partial t} + \bar{u}_j \frac{\partial u'_i}{\partial x_j} + u'_j \frac{\partial \bar{u}_i}{\partial x_j} + \frac{\partial u'_i u'_j}{\partial x_j} - \frac{\partial \overline{u'_i u'_j}}{\partial x_j} = -\frac{1}{\rho_0}\frac{\partial p'}{\partial x_i} + \nu \frac{\partial^2 u'_i}{\partial x_j^2} \tag{6.82}$$

where $-\rho_0 \overline{u'_i u'_j}$, the *Reynolds stress* (or *turbulent stress*), represents a momentum transfer by turbulent motion. Equation (6.81), referred to as the *Reynolds-averaged* Navier–Stokes equations for the mean flow, can be written as follows:

$$\frac{\partial \bar{u}_i}{\partial t} + \frac{\partial \bar{u}_i \bar{u}_j}{\partial x_j} = -\frac{1}{\rho_0}\frac{\partial \bar{p}}{\partial x_i} + \frac{1}{\rho_0}\frac{\partial}{\partial x_i}\left(\bar{\tau}_{ij} - \rho_0 \overline{u'_i u'_j}\right) \tag{6.83}$$

where $\bar{\tau}_{ij} = 2\,\mu\,\bar{\epsilon}_{ij}$.

6.4.4 The Reynolds Stress

As the second-order Cartesian stress tensor σ_{ij} was defined in Chapter 3, the Reynolds stress tensor $\tau'_{ij} = -\rho_0 \overline{u'_i u'_j}$ can be defined over the three mutually orthogonal surfaces of an infinitesimal cubic element at a point (x, y, z):

$$[\tau'] = \rho_0 \begin{bmatrix} -\overline{u'^2} & -\overline{u'v'} & -\overline{u'w'} \\ -\overline{v'u'} & -\overline{v'^2} & -\overline{v'w'} \\ -\overline{w'u'} & -\overline{w'v'} & -\overline{w'^2} \end{bmatrix} \tag{6.84}$$

where $-\rho_0\,\overline{u_i'u_j'}$ represents one normal and two tangential turbulent stress vectors that act on each orthogonal surface; the first index i denotes the surface element on which the stress vector is acting, and the second index j is the direction of the stress vector in action. Note that the Reynolds stress is symmetric, i.e. $\tau_{ij}' = \tau_{ji}'$ so that only six components are independent (Figure 6.11).

The Reynolds stresses are visualized in Figure 6.12 with a hairpin vortex that develops (with ejection) and bursts (with sweep) in a turbulent boundary layer. Suppose we have a shear flow where $d\overline{u}/dy > 0$. On a $y = \text{constant}$ plane specified with a unit normal vector $\vec{n}$ outward to the surface (i.e. $\vec{n} > 0$), $\tau_{yx}' = -\rho_0\,\overline{v'u'}$ is the Reynolds stress exerted in the positive x-direction when a fluid particles moves towards the surface with v' (< 0) (sweep).

If the unit normal vector $\vec{n}$ is pointing in the opposite direction to the surface (i.e. $\vec{n} < 0$), the Reynolds stress acting on this surface becomes negative for the same motion while v' is defined as positive (ejection). This Reynolds stress associated with the turbulent motion of the fluid can be interpreted as a transfer of x-momentum in the direction of y. By symmetry, $\tau_{yx}' = \tau_{xy}'$, implying

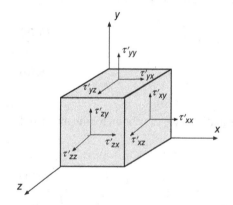

Figure 6.11 Reynolds stresses τ_{ij}' on the six surfaces of a cubic fluid element

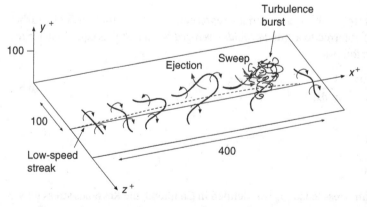

Figure 6.12 Growth and burst of a hairpin vortex in turbulent boundary layer; Reynolds stresses associated with ejection and sweep

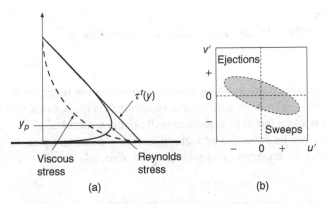

Figure 6.13 Viscous (or molecular) stress and the Reynolds stress in turbulent boundary layer (a); correlations between u' and v' clearly indicate ejection and sweep (b)

that on a x-constant plane, there must exist a transport of y-momentum in the x-direction, exerting a stress in the y-direction. Following the index notation, this is $-\rho_0 \, \overline{u'v'}$.

With the Reynolds stress, we can now compute the total stress $\tau_{ij}^t = \overline{\tau}_{ij} + \tau_{ij}' = \overline{\tau}_{ij} - \rho_0 \, \overline{u_i'u_j'}$, where $\overline{\tau}_{ij}$ is the molecular (or viscous) stress. As shown in Figure 6.13a, the total stress in a fully developed turbulent channel (or pipe) flow consists of the molecular (or viscous) stress and the Reynolds stress. The Reynolds stresses dominate in the region between δ and y_p, whereas in the region between 0 (wall) and y_p, viscous stresses prevail. As also shown in Figure 6.13b, u' and v' are well correlated throughout the channel (or pipe), showing an ellipsoid of the scattered data of the u' and v' pair at different times (i.e. anisotropic turbulence of probability density function). If the turbulent flow were completely isotropic, the scattered data would show a spherical symmetry, meaning that $\overline{u_i'u_j'} = 0$ and $\overline{u'^2} = \overline{v'^2} = \overline{w'^2}$; in this case, $u' = \pm v'$.

6.4.5 Eddy Viscosity and Mixing Length Theory

6.4.5.1 The Boussinesque Approximation
For a Newtonian fluid, the deviatoric viscous stress tensor τ_{ij} can directly be related to the strain rate tensor ϵ_{ij}:

$$\overline{\tau}_{ij} = 2 \, \mu \, \overline{\epsilon}_{ij} = \rho_0 \, \nu \left(\frac{\partial \overline{u}_i}{\partial x_j} + \frac{\partial \overline{u}_j}{\partial x_i} \right) \tag{6.85}$$

and this is referred to as the constitutive relation.

Similarly, the turbulent stress τ_{ij}' can be expressed as follows:

$$\tau_{ij}' = -\rho_0 \, \overline{u_i'u_j'} = \rho_0 \, \nu_t \left(\frac{\partial \overline{u}_i}{\partial x_j} + \frac{\partial \overline{u}_j}{\partial x_i} \right) \tag{6.86}$$

by employing the concept of *eddy viscosity* (or turbulent viscosity) proposed by Boussinesque [4]. Therefore, the total stress tensor τ_{ij}^t can be written as follows:

$$\tau_{ij}^t = \overline{\tau}_{ij} + \tau_{ij}' = \rho_0 \, (\nu + \nu_t) \left(\frac{\partial \overline{u}_i}{\partial x_j} + \frac{\partial \overline{u}_j}{\partial x_i} \right) \tag{6.87}$$

It is important to note that ν_t is not a property of fluids as the kinematic viscosity ν is. It rather depends on the mean velocity; to be more specific, it is approximately proportional to the square of the mean velocity.

6.4.5.2 Prandtl's Mixing Length Theory

Suppose we have a simple laminar shear flow of a gas. In this case, the viscous stress is

$$\bar{\tau}_{xy} = \bar{\tau}_{yx} = \rho_0 \, v \, \frac{\partial \bar{u}}{\partial y} \tag{6.88}$$

where v is a property of the fluid. The kinetic theory states that the momentum is transported through collision between the molecules. To be more specific, the momentum is transported via process of molecular diffusion, which is also referred to as random walks of the molecules.

In this case, the kinematic viscosity v can be estimated by a length scale (e.g. mean free path λ) and the root-mean-square (rms) speed of the molecular motion (e.g. speed of sound c)

$$v \sim c \, \lambda \tag{6.89}$$

For air, for example, $\lambda = 6 \times 10^{-8} \, m$ and $c = 340 \, m/s$, so that $v \approx 2 \times 10^{-5} \, m^2/s$.

If the flow is turbulent, one can speculate that the diffusivity may be qualitatively similar to that of a laminar flow and may simply be represented by an analogy with Eq. (6.89). The eddy viscosity can then be written as follows:

$$v_t \sim u' \, l' \tag{6.90}$$

where u' is a scale of the fluctuating velocity and l' is the scale of mixing due to turbulent motion, which is called Prandtl's mixing length (Figure 6.14).

According to Prandtl's mixing length theory, u' represents the velocity of a lump of fluid particles that move bodily, crossing a transverse length in the streamwise and lateral directions. The fluctuating velocity of a turbulent eddy can be approximated by Taylor's expansion:

$$u' = \bar{u}(y) - \bar{u}(y - l') \approx l' \, \frac{d\bar{u}}{dy} \tag{6.91}$$

Now, it follows that

$$|u'| \approx l' \left| \frac{d\bar{u}}{dy} \right| \tag{6.92}$$

and the eddy viscosity reads

$$v_t \sim l'^2 \left| \frac{d\bar{u}}{dy} \right| \tag{6.93}$$

Although the eddy viscosity v_t is explicitly expressed by Eq. (6.93), the mixing length l' is still unknown and differs from case to case.

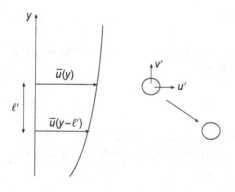

Figure 6.14 Prandtl's mixing length theory

6.5 Turbulence Production and Energy Cascade

6.5.1 Kinetic Energy of the Mean Flow

If the Reynolds-averaged Navier–Stokes equations Eq. (6.83) is multiplied by $\bar{u}_i$, we obtain the kinetic energy equation of the mean flow per unit mass

$$\frac{\partial(\bar{u}_i^2/2)}{\partial t} + \frac{\partial(\bar{u}_i^2/2)\bar{u}_j}{\partial x_j} = \underbrace{\frac{1}{\rho_0}\frac{\partial}{\partial x_j}\left(-\bar{p}\,\bar{u}_i\,\delta_{ij} + \bar{u}_i\,\bar{\tau}_{ij} - \rho_0\,\bar{u}_i\,\overline{u_i' u_j'}\right)}_{①}$$

$$\underbrace{-\frac{1}{\rho_0}\,\bar{\tau}_{ij}\,\frac{\partial \bar{u}_i}{\partial x_j}}_{②} + \underbrace{\overline{u_i' u_j'}\,\frac{\partial \bar{u}_i}{\partial x_j}}_{③} \qquad (6.94)$$

where $\bar{\tau}_{ij} = 2\,\mu\,\bar{e}_{ij}$.

It is to be noted that the term ① takes the form of divergence. If Eq. (6.94) is integrated over a volumetric space, it can be transformed into a surface integral by the divergence theorem. This term only transports or redistributes the kinetic energy of the mean flow from one point to another without creating or dissipating it.[3] The term ② is the rate of viscous dissipation of kinetic energy in the mean flow. Through the deformation work done on the fluid at a rate against intermolecular frictions, it converts the mean kinetic energy into heat at a rate. Note that this term reappears in the thermal energy equation with reversed signs.

The term ③ represents the rate of deformation work done on the fluid via interactions between the mean flow and the Reynolds stresses:

$$\overline{u_i' u_j'}\,\frac{\partial \bar{u}_i}{\partial x_j} = -\frac{1}{2}\,\nu_t\left(\frac{\partial \bar{u}_i}{\partial x_j} + \frac{\partial \bar{u}_j}{\partial x_i}\right)^2 < 0 \qquad (6.95)$$

Similarly to the deformation work done by molecular interactions, the term ③ represents loss of mean kinetic energy to the production of turbulent fluctuations, i.e. the Reynolds stress. This term is often called the turbulent energy production (or shear production of turbulence) [5]. Note that it also reappears in the turbulent kinetic energy equation with the sign reversed, while acting as a sink for the mean kinetic energy.

In a fully turbulent flow at high Reynolds numbers, the viscous transport term $(1/\rho_0)\,\partial(\bar{u}_i\,\bar{\tau}_{ij})/\partial x_j$ and the viscous dissipation term $(\bar{\tau}_{ij}/\rho_0)\,\partial\bar{u}_i/\partial x_j$ are negligible, based on the scaling using the Reynolds number. These terms are in the order of $\nu(\bar{u}/L)^2$, whereas the Reynolds stress is in the order of $u_{rms}'^2(\bar{u}/L)$ (u_{rms}' is the rms value of the turbulent fluctuation). In high Reynolds number flows, experiments and numerical simulations have shown that u_{rms}' and $\bar{u}$ are of the same order. Thus, the ratio between the viscous transport term (or viscous dissipation term) and the Reynolds stress is $O(1/Re)$, which becomes small when Re is high.

6.5.2 Kinetic Energy of Turbulence

To trace the rate of deformation work done by Reynolds stress, we must obtain the kinetic energy of turbulence. Multiplying the equation of turbulent fluctuation Eq. (6.82) by u_i' and applying the

---

3 See Section 3.6.4 *Total energy equation in conservative form*, Chapter 3 *Differential Equations of Motions*.

Reynolds-averaging procedure yields

$$\frac{\partial(\overline{u_i'^2/2})}{\partial t} + \frac{\partial(\overline{u_i'^2/2})\overline{u}_j}{\partial x_j} = \underbrace{\frac{1}{\rho_0}\frac{\partial}{\partial x_j}\left(-\overline{p'\,u_i'}\,\delta_{ij} - \rho_0\,\overline{(u_i'^2/2)u_j'} + 2\mu\,\overline{u_i'\,\epsilon_{ij}'}\right)}_{\textcircled{1}'}$$

$$\underbrace{-\frac{1}{\rho_0}\,\overline{\tau_{ij}'\frac{\partial u_i'}{\partial x_j}}}_{\textcircled{2}'} \underbrace{-\,\overline{u_i'u_j'}\frac{\partial \overline{u}_i}{\partial x_j}}_{\textcircled{3}'} \tag{6.96}$$

where the rate of strain produced by turbulent fluctuating velocities is defined as follows:

$$\epsilon_{ij}' = \frac{1}{2}\left(\frac{\partial u_i'}{\partial x_j} + \frac{\partial u_j'}{\partial x_i}\right) \tag{6.97}$$

It is to be noted that the term $\textcircled{1}'$ is in the flux divergence form and thus simply transports or redistributes the turbulent kinetic energy. The first two terms are the transport of turbulent kinetic energy by turbulent fluctuations, but the third term is the transport of turbulent kinetic energy by viscosity.

The term $\textcircled{2}'$ is the rate of viscous dissipation of turbulent kinetic energy:

$$\frac{1}{\rho_0}\,\overline{\tau_{ij}'\frac{\partial u_i'}{\partial x_j}} = \nu\,\overline{\left(\frac{\partial u_i'}{\partial x_j}\right)^2} = 2\nu\,\overline{\epsilon_{ij}'\epsilon_{ij}'} > 0 \tag{6.98}$$

This term is not negligible in the turbulent kinetic energy equation, though its analogue $\textcircled{2}$ is negligible in the kinetic energy of the mean flow. It is always positive, indicating that this term, a sink in the turbulent kinetic energy, viscously dissipates the turbulent kinetic energy into heat at a rate in the thermal energy equation. The order of this term is, in fact, comparable to that of the turbulent energy production term $\textcircled{3}'$, implying that the turbulent kinetic energy is in an equilibrium state because the transport term does not create nor destroy the turbulent kinetic energy.

The term $\textcircled{3}'$ is the rate of deformation work, identical to $\textcircled{3}$ in Eq. (6.94), but its sign is reversed. This implies that this term represents a loss of mean kinetic energy and a gain of turbulent kinetic energy. Therefore, this term, appearing as a source in the turbulent kinetic energy equation, represents the rate of production of turbulent kinetic energy by the interaction between the Reynolds stress and the mean shear [6].

6.5.3 Energy Cascade Process

6.5.3.1 Vortex Stretching in 3D Straining Field

In a three-dimensional straining field of mean shear flow, turbulent eddies can be exposed to vortex stretching. When the direction of a vorticity vector of an eddy is aligned to the positive principal axis of strain rate, the vortex is stretched at a rate along the positive principal axis; it is also contracted at a rate along the negative principal axis. This vortex stretching results in a positive shear production $-\overline{u'v'}d\overline{u}/dy$ with $u' > 0$ and $v' < 0$. As a result, the eddy becomes smaller in size (or thinner), while its vortical strength (or rotational kinetic energy) increases. A good example is the hairpin vortex which grows out of the unstable low-speed streaks. As the flow convects, it stretches at an angle of

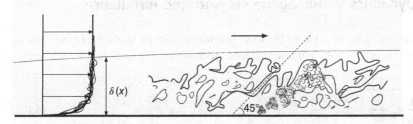

Figure 6.15 Vortex stretching along the principal axes of strain rate in three-dimensional straining field

45° along the positive principal axis and finally bursts, accompanying ejections and sweeps of the neighboring fluid (Figure 6.15).

6.5.3.2 Kolmogorov's Microscales

This process of energy extraction from the large-scale eddies to small scales is essentially inviscid and occurs in a broad spectrum of scales. The so-called *energy cascade* continues until the energy of the smallest eddies is viscously dissipated at a rate into heat [7, 8]. Conversely speaking, the viscosity determines the scales at which turbulent energy is effectively dissipated at a rate into heat, according to $2\nu \; \overline{\epsilon'_{ij}\epsilon'_{ij}}$.

As shown in Figure 6.15, a turbulent boundary layer clearly exhibits a variety of scales of eddies. The largest ones in the boundary layer continuously break into smaller eddies through the process of energy cascade. The smallest eddies are long and thin vortex filaments, all tangled up like spaghetti. It is interesting to note that molecular diffusion occurs on the large surfaces of these smallest eddies, which thereby produces effective mixing in a turbulent flow.

It can be estimated that the rate of turbulent energy production ③′ is in the order $0(u'^3/l')$, where u' and l' are the fluctuating velocity scale and the length scale of the largest scale of eddies; l' is known as the integral scale. This production rate of turbulent energy must equal the rate of viscous dissipation ②′ (denoted by ε) so that

$$u'^3/l' \sim \varepsilon \sim \nu \, (\upsilon/\eta)^2 \tag{6.99}$$

where the smallest length scale η and velocity scale υ are known as Kolmogorov's microscales. However, the Reynolds number based on the smallest scales is in the order of unity, i.e. $Re = \upsilon \, \eta/\nu \sim 0(1)$.

Therefore, Kolmogorov's microscales υ and η can be estimated as follows:

$$\eta \sim (\nu^3/\varepsilon)^{1/4} \quad \text{or} \quad \eta \sim l' Re^{-3/4} \tag{6.100}$$

$$\upsilon \sim (\nu \, \varepsilon)^{1/4} \quad \text{or} \quad \upsilon \sim u' Re^{-1/4} \tag{6.101}$$

where the Reynolds number is defined as $Re = u'l'/\nu$.

For a free shear layer, Kolmogorov's length scale η is about $6 \times 10^{-2} \; mm$ if the Reynolds number is 10^3 and $l' = 1 \; cm$. Another example is a household mixer of 10 Watt in 1 kg of water. The energy dissipation rate ε is 10 $W/kg = 10 \; m^2/s^3$ and Kolmogorov's length scale η can be estimated as $2 \times 10^{-2} \; mm$.[4] Kolmogrov's length scale is the finest structure in turbulent flow, which becomes smaller as the Reynolds number increases.

---

4 Landahl and Mollo-Christianson [9].

6.6 Spectral Dynamics in Homogeneous Isotropic Turbulence

A practical approximation to homogeneous isotropic turbulence can be found for turbulence far downstream of a uniform grid of bars. For homogeneous turbulence, the turbulent kinetic energy equation becomes

$$\frac{\partial(\overline{u_i'^2/2})}{\partial t} = \frac{1}{\rho_0}\overline{\tau_{ij}'\frac{\partial u_i'}{\partial x_j}} - \overline{u_i'u_j'}\frac{\partial \overline{u}_i}{\partial x_j} \tag{6.102}$$

Note that unless production equals dissipation, the turbulent kinetic energy varies with time, and the flow is not stationary.

6.6.1 Spatial Correlations

To understand the spatial structure of turbulence, a two-point spatial correlation tensor is defined as follows:

$$R_{ij}(\vec{r}) = \overline{u_i'(\vec{x}_1)\,u_j'(\vec{x}_2)} \tag{6.103}$$

where $\vec{r} = \vec{x}_2 - \vec{x}_1$ is the separation vector. It has two properties: $R_{ij}(\vec{r}) = R_{ji}(-\vec{r})$ and $R_{ij}(\vec{r}) \to 0$ as $\vec{r} \to \infty$.

If the separation vector is chosen along the x-axis, it follows that

  (i) $R_{11}(r) = \overline{u'(x)\,u'(x+r)}$ is a longitudinal correlation,

  (ii) $R_{22}(r) = \overline{v'(x)\,v'(x+r)}$ and $R_{33}(r) = \overline{w'(x)\,w'(x+r)}$ are transversal correlations, and

(iii) $R_{12}(r) = \overline{u'(x)\,v'(x+r)}$ is a cross-correlation.

We can now describe the spatial structure of homogeneous turbulence by defining the spectral tensor $\Phi_{ij}(\vec{\kappa})$

$$\Phi_{ij}(\vec{\kappa}) = \frac{1}{8\pi^3}\iiint_{-\infty}^{\infty} R_{ij}(\vec{r})\,e^{-i\vec{\kappa}\cdot\vec{r}}\,d\vec{r} \tag{6.104}$$

where $\vec{\kappa} = (2\pi/\lambda^2)\vec{\lambda}$ is the wavenumber vector and $\vec{\lambda}$ is the wavelength vector. Note that $R_{ij}(\vec{r})$ and $\Phi_{ij}(\vec{\kappa})$ are a Fourier transform pair:

$$R_{ij}(\vec{r}) = \frac{1}{8\pi^3}\iiint_{-\infty}^{\infty} \Phi_{ij}(\vec{\kappa})\,e^{-i\vec{\kappa}\cdot\vec{r}}\,d\vec{\kappa} \tag{6.105}$$

6.6.2 Turbulent Energy Spectrum

A turbulent kinetic energy e' is defined as follows:

$$e' = \frac{1}{2}\overline{u_i'^2} = \frac{1}{2}R_{ii}(0) \tag{6.106}$$

To relate e' to $R_{ii}(0)$, we need to eliminate the direction-dependency of $\Phi_{ii}(\vec{\kappa})$. One legitimate way is to define the turbulent energy spectrum $E(\kappa)$

$$E(\kappa) = \frac{1}{2}\iint_{\kappa}\Phi_{ii}(\vec{\kappa})\,d\Omega \tag{6.107}$$

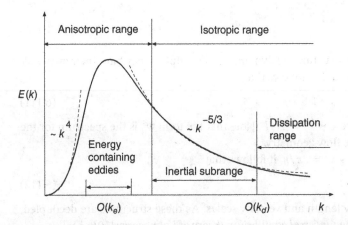

Figure 6.16 Energy spectrum; inertial subrange scales with $\kappa^{-5/3}$ between $\kappa_e = 2\pi/l$ and $\kappa_d = 2\pi/\eta$. Source: Republished with permission of Springer Nature from Correlation Function and Spectrum, Frans T. M. Nieuwstadt, Bendiks J. Boersma, Jerry Westerweel, pp 183–213, Jan 1, 2016, permission conveyed through Copyright Clearance Center, Inc

where $\kappa = |\vec{\kappa}|$ and Ω is a spherical surface in $\vec{\kappa}$ space. Then, the turbulent kinetic energy e' reads

$$e' = \frac{1}{2} R_{ii}(0) = \frac{1}{2} \int_0^\infty E(\kappa) \, d\kappa \tag{6.108}$$

The turbulent energy spectrum $E(\kappa)$ shows the distribution of turbulent energy over various wavenumbers κ. For instance, $E(\kappa)$ having a maximum at $\kappa_e \sim 1/l'$ (l' is the integral scale) represents large and energetic turbulent eddies. In contrast, $E(\kappa)$ decays at wavenumbers close to $\kappa_d \sim 1/\eta$ (η is Kolmogorov's microscale), where the eddies are viscously dissipated at a rate. The region in between, which is called the *inertial subrange*, corresponds to the energy cascade process (Figure 6.16).

6.6.2.1 One-Dimensional Spectrum
If the separation vector $\vec{r}$ only varies in one-direction, e.g. $\vec{r} = (x, 0, 0)$, it follows that

$$R_{ij}(x, 0, 0) = \int_{-\infty}^\infty e^{i\kappa_1 x} \left(\iint_{-\infty}^\infty \Phi_{ij}(\kappa_1, \kappa_2, \kappa_3) \, d\kappa_2 \kappa_3 \right) d\kappa_1$$

$$= \int_{-\infty}^\infty e^{i\kappa_1 x} \, \mathsf{P}_{ij}(\kappa_1) \, d\kappa_1 \tag{6.109}$$

where $\mathsf{P}_{ij}(\kappa_1)$ is the one-dimensional spectrum and $R_{ij}(x, 0, 0)$ is the one-dimensional correlation function.

It is to be noted that $\mathsf{P}_{11}(\kappa)$, describing the spatial structure of turbulent eddies along a line, is different from $E(\kappa)$ due to the *aliasing effect*. One noticeable difference is that $\mathsf{P}_{11}(\kappa)$ is maximum at a small κ, where $E(0) \approx 0$. Now, with this one-dimensional spectrum, the characteristic length scales of the macroscale structure of turbulence can be defined as follows:

$$\overline{u'^2} \, l'_l = \int_0^\infty R_{11}(x, 0, 0) \, dx, \qquad \overline{u'^2} \, l'_t = \int_0^\infty R_{22}(x, 0, 0) \, dx \tag{6.110}$$

where l'_l is the longitudinal length scale and l'_t is the transversal length scale.

6.6.3 Kolmogorov's $\kappa^{-5/3}$ Law

6.6.3.1 Inertial Subrange

Since a macrostructure and a microstructure are dynamically decoupled, they have their own scaling parameters. For the macrostructure, it follows that

$$P_{11}(\kappa) = l'u'^2 \, \Phi_e(\kappa l') \tag{6.111}$$

which is valid for the region, where $\kappa \sim \kappa_e = 2\pi/l'$. Note that the term Φ_d is the spectrum for the large eddies, which depends on the flow geometry.

For the microstructure where $\kappa \sim \kappa_d = 2\pi/\eta$, it follows that

$$P_{11}(\kappa) = \eta \, v'^2 \, \Phi_d(\kappa \eta) \tag{6.112}$$

where η and v are the Kolmogorov length and velocity scales. As these structures are decoupled, Φ_d is universal; this is known as the *universal equilibrium theory of Kolmogorov* [10].

A matching condition between the two equations Eqs. (6.111) and (6.112) leads us to conclude that

$$\lim_{\kappa l' \to \infty} l'u'^2 \, \Phi_e(\kappa l') = \lim_{\kappa \eta \to 0} \eta v^2 \, \Phi_e(\kappa \eta) \tag{6.113}$$

Eliminating u' and v in Eq. (6.113) by applying the relation $\varepsilon = u'^3/l' = v^3/\eta$ leads to

$$\lim_{\kappa l' \to \infty} \frac{\Phi_e(\kappa l')}{(\kappa l')^{-5/3}} = \lim_{\kappa \eta \to 0} \frac{\Phi_e(\kappa \eta)}{(\kappa \eta)^{-5/3}} = \alpha_1 \tag{6.114}$$

where α_1 is a universal constant, called Kolmogorov's constant.

If we use the relation $\varepsilon = u'^3/l' = v^3/\eta$, it follows that

$$P_{11}(\kappa) = \alpha_1 \, \varepsilon^{2/3} \kappa^{-5/3} \tag{6.115}$$

where α is approximately 0.26, and that

$$E(\kappa) \sim u'^2 l' = \alpha_2 \, \varepsilon^{2/3} \kappa^{-5/3} \tag{6.116}$$

where α_2 is approximately 1.6. Note that the inertial subrange is clearly indicated with a slope of $-5/3$ [10].

6.7 Boundary Layer Separation

6.7.1 Creeping Flows

Let us suppose we have a highly viscous fluid (e.g. honey, oil) flowing over a circular cylinder at a very low speed (Figure 6.17). At low Reynolds numbers (e.g. $Re < O(1)$), the molecular diffusion speed far exceeds the convection speed of the flow. Thus, the viscous force dominates the inertial force. In this case, the viscous forces exerted at the surface of the body pull the outer flow tangent to the surface, creating rotation in the field. Therefore, the so-called *creeping flow* is attached all the way to the end of the surface of the body.

As shown in the figure, the streamlines appear to be almost the same as those of the potential flow, which hypothetically results in zero drag. However, the drag in this highly viscous flow is not

Figure 6.17 Creeping flow around a circular cylinder at very low Reynolds numbers

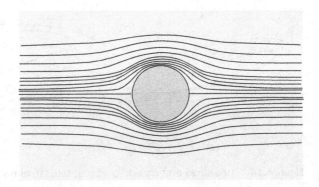

zero because flow attachment is built by the viscous shear forces. In creeping flows, the inertial force is negligible compared to the pressure and viscous forces. In this case, the drag force can be obtained by solving the Navier–Stokes equations reduced to $\nabla p = \mu \, \nabla^2 \, \vec{v}$, with the continuity equation $\nabla \cdot \vec{v} = 0$. In the regime of $Re_D \leq 1$, the drag force on the sphere is found to be $\mathcal{D}_{sp} = 3\pi\mu DU$, which is called the Stokes law [11].

6.7.2 Moderate to High Reynolds Number Flows

From moderate-to-high Reynolds numbers, a boundary layer develops at the body surface where viscous stresses are confined. As explained before, the thickness of the boundary layer depends on the Reynolds number. Outside the boundary layer, the fluid moves with inertia and changes its momentum at a rate with the forces that only act on the center of mass of the fluid particles, for instance, pressure forces and gravitational body forces.[5] The flow outside the boundary layer is often referred to as an inviscid or irrotational flow, whereas the boundary layer is viscous and rotational. Note that all fluids have viscosity no matter how small they are. Therefore, an inviscid flow means that the influence of net viscous force is outreached.

6.7.2.1 Favorable and Unfavorable Pressure Gradients

Over a convex surface of a body (e.g. fore of the circular cylinder), the flow is accelerated with a favorable pressure gradient. Since the diffusion time scale becomes smaller, the boundary layer becomes thinner, while newly creating vortices. As a result, the boundary layer tends to be attached to the surface. Over a concave surface, however, the flow is decelerated with an unfavorable pressure gradient. In this case, the boundary layer becomes thicker while destroying the vortices, as the diffusion time scale becomes larger [12].

If the flow deceleration exceeds a certain limit (e.g. aft the circular cylinder where the space in the transverse direction increases too rapidly), the convecting vortices in the boundary layer are all destroyed at some point (called a separation point) and the flow direction starts to be reversed with entrainment of the neighboring fluid. The mathematical definition of the separation point is the location where the vertical gradient of velocity is zero, i.e. $\partial v_t / \partial n|_{\mathrm{wall}} = 0$ (v_t: tangential velocity).

An example of the flow over an airfoil is illustrated in Figure 6.18. The figure on the left shows that the bounder layers over and underneath the airfoil are attached to the surfaces, and a lift is the

---

5 A nonconservative body force such as the Coriolis force does not act on the center of mass of the fluid particles. See Chapter 5 *Vortex Dynamics*.

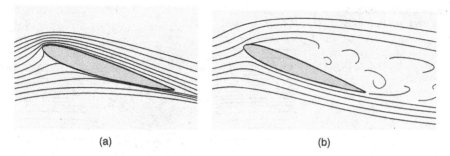

(a) (b)

Figure 6.18 Streamlines over an airfoil; without stall (a), with stall ($\alpha > 15° - 20°$) (b)

Figure 6.19 Reduction of pressure drag due to separation of boundary layer

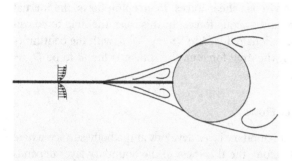

vertical force acting on the airfoil. If the angle of attack exceeds 15°–20°, the boundary layer separates over the upper surface of the airfoil as in figure on the right. This boundary layer separation close to the leading-edge of the airfoil is called *stall*. When a stall occurs, drag increases drastically and lift drops significantly, resulting in failure of pressure recovery. In internal viscous flows, flow separation leads to a loss of mechanical energy and an increase of energy supplied to the system.[6]

6.7.2.2   Creation of Separation for Drag Reduction

A fluid approaching a circular cylinder gradually decelerates and finally stagnates at the nose of the cylinder. If a plate is attached to the cylinder as in Figure 6.19, a boundary layer develops over the plate surface but will separate off the surface on the way to the stagnation point. Unless the Reynolds number of the flow is substantially low, the only way to adjust the flow over the plate surface is for the boundary layer to separate from the surface, destroying all the vortices in the boundary layer. If then, the flow path of the major stream is lifted up and the space after the separation point is filled with the recirculating flow, which would not have existed if there were no plate attached to the cylinder. This early departure from the centerline results in lower stagnation pressure, which reduces the pressure drag of the cylinder [13].

6.8   Drag Force

A body immersed in a flow experiences a force acting onto it. The component of the force parallel to the incoming flow is called drag force and the other perpendicular to it is called lift force. These two forces are directly related to the viscous and pressure forces acting on the body surfaces.

---

6 More discussion continues in Chapter 7 *Interval Viscous Flows*.

There are two different types of drag forces acting on the body; one is the skin friction drag and the other is the pressure drag. The pressure drag results from an asymmetric pressure distribution with respect to the axis perpendicular to the incoming flow and is usually associated with the lack of pressure recovery caused by separation of the boundary layer. The dominance between the two drag forces is typically determined by the shape of the body and also by the state of the boundary layer, as to whether it is laminar or turbulent.

6.8.1 Skin Friction Drag

6.8.1.1 Skin Friction Coefficient (Flat Plate)

A flow over a flat plate is frictionally resisted by viscous forces acting on the surface. In order to find the viscous force acting on the surface, it is necessary to know the distribution of the wall shear stress $\tau_w = \mu\, \partial u / \partial n|_w$ along the surface, which is a function of x and also the state of the boundary layer: laminar, transitional, or turbulent.

As discussed in Section 6.3, a boundary layer developed over the plate remains laminar up to $Re_x = 5 \times 10^5$. Afterward, it experiences a transition process and finally becomes turbulent at $Re_x > 1 \times 10^7$. Thus, the wall shear stress τ_w is determined not only by the fluid's viscosity but also by the velocity profile, especially close to the wall. In laminar boundary layers, it is solely determined by the convection and molecular diffusion of vortices created at the wall. However, if the boundary layer becomes turbulent, the velocity profile is influenced not only by interactions between the molecules but also by fluctuating turbulent eddies that must undertake the energy cascade process in the boundary layer.

The skin friction coefficient for the flat plate (Figure 6.20) is expressed with the Reynolds number as follows:

$$C_f = \frac{0.664}{\sqrt{Re_x}} \quad \text{(laminar, } 5 \times 10^5 < Re_x)$$

$$C_f = \frac{0.0371}{Re_x^{1/5}} \quad \text{(transitional, } 5 \times 10^5 < Re_x < 1 \times 10^7)$$

$$C_f = \frac{0.227}{(\log Re_x)^{2.58}} \quad \text{(turbulent, } 1 \times 10^7 < Re_x < 1 \times 10^9) \tag{6.117}$$

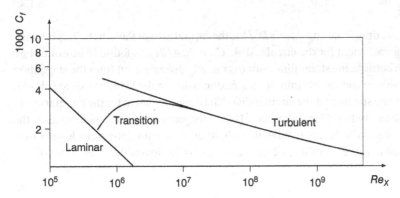

Figure 6.20 Skin friction coefficient C_f vs. Reynolds number Re_x (flat plate); laminar (Blausius), turbulent (Prandtl–Schlichting)

where the skin friction coefficient is defined by

$$C_f = \frac{\tau_w(x)}{\frac{1}{2}\rho U^2} \tag{6.118}$$

The drag coefficient is defined by

$$C_D = \frac{F_D}{1/2\,\rho\,U^2 A_w} = \frac{2\int \tau_w(x)\,dA}{1/2\,\rho\,U^2 A_w} \tag{6.119}$$

where A_w is the wetted area. Note that the skin friction coefficient for the laminar case can be obtained by Blasius's self-similarity solution, but those for transition and turbulent boundary layers are determined by correlating the experimental data.

6.8.2 Pressure Drag

6.8.2.1 Inertial Resistances

Although a fluid is a deformable substance, it exerts a force back to the solid when being accelerated. Transport vehicles, such as aircrafts, automobiles, are thus resisted by the fluid, and this resistance force is considered as drag to the fluid. The drag force depends on the fluid's density and viscosity, as well as the shape of the solid body and the rate at which the velocity is forced to change (i.e. acceleration). This implicates that drag force owes to viscous and pressure forces.

For example, if one waves a hand in the air, the motion does not get interfered much. But if the waving of the hand is done in water, it would be much more difficult. If such actions were to be conducted in a very viscous oil, it would be even harder. The drag force varies with the fluid's density and viscosity and also depends on how the palm faces the fluid during motions or at what rate the fluid is forced to change speed. In many circumstances, the pressure force acting against the motion is the primary resisting force.

6.8.2.2 Drag Coefficients (Circular Disk, Sphere, and Circular Cylinder)

Flow over a circular disk, sphere, or circular cylinder is a good example of how the characteristics of skin friction drag and pressure drag depend on the Reynolds number. The drag coefficients for these three cases are presented in Figure 6.21.

In the regime of Stokes' law ($Re_D \leq 1$), the drag coefficient for the sphere is

$$C_D = \frac{D_{sp}}{\rho\,U^2/2 \cdot A_f} = \frac{24}{Re} \tag{6.120}$$

where $D_{sp} = 3\pi\mu DU$ is the drag force and $A_f = \pi D^2/4$ is the frontal area of the sphere. It is interesting to note that the drag coefficient for the circular disk, $C_D = 20.4/Re_D$, is found to be quite close to the Stokes law. This is because the streamline patterns radially diverging out from the stagnation point on the windward side or converging into the stagnation point on the leeward side are similar with each other between the sphere and the circular disk. The drag coefficient for the circular cylinder is, however, lower than predicted by Stokes' law. Due to its geometric two-dimensionality, the streamlines over the circular cylinder are parallel to each other. As a result, pressure changes over the surface are not as great as those over the sphere or circular disk, i.e. loss of the three-dimensional effect.

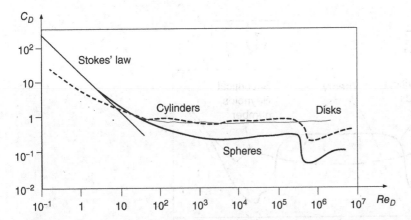

Figure 6.21 Drag coefficient C_D vs. Reynolds number Re_D; circular disk, sphere, and circular cylinder

After Stokes' flow regime, the drag coefficient reflects the various characteristics of drag related to body geometry. Since the circular disk has sharp edges, the separation point remains fixed and C_D becomes independent of the Reynolds number, regardless of whether the boundary layer over the disk frontal surface is laminar or turbulent. Note also that $C_D \approx 1$ indicates that no noticeable pressure recovery occurs on the back surface of the circular disk.

The sphere also shows a nearly constant drag coefficient at $10^3 < Re_D < 5 \times 10^5$, because the separation point nearly stays at the same location until the boundary layer over the surface transits to turbulence. The drag coefficient of the sphere is, however, nearly one half lower than that of the circular disk, due to partial recovery of the pressure on the leeward side of the sphere. This owes to the hemisphere (windward side) that changes the characteristics of the wake. The more streamlined the shape of the body, the smaller the drag coefficient.

The drag coefficient of the circular cylinder is a mixture of those of the circular disk and sphere. Up to transition, C_D is nearly the same as that of the circular disk ($C_D \approx 1$). Although the wake characteristics of the cylinder is similar to that of the sphere, the vortex shedding pattern is symmetric about the centerline due to geometric two-dimensionality. Therefore, the effect of partial recovery of pressure aft the cylinder disappears in a time-averaged sense. Nevertheless, a sudden drop of drag coefficient and the rest of the profile in the turbulent boundary layer regime are resemblant to the sphere, although the magnitude of the drag coefficient is higher.

The drag coefficients for the circular cylinder and sphere can be correlated as for the Reynolds number less than 2×10^5 (laminar);

$$C_D \approx 1 + 10 \, Re_D^{-0.67} \qquad \text{(cylinder)}$$

$$C_D \approx \frac{24}{Re_D} + \frac{6}{1 + \sqrt{Re_D}} + 0.4 \qquad \text{(sphere)} \qquad (6.121)$$

It is also to be noted that the surface roughness lowers the Reynolds number at the transition Re_{tr} from 3×10^5 to 8×10^5, for example.

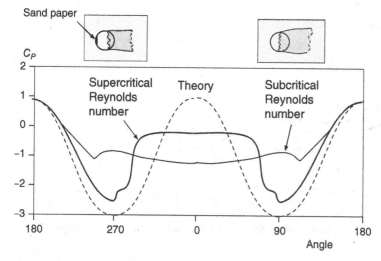

Figure 6.22 Drag reduction by separation delay (golf ball effect); turbulent boundary layer (supercritical) (left) vs. laminar boundary layer (subcritical) (right); critical Reynolds number ($Re_{cr} = 3 \times 10^5$)

6.8.2.3 Golf Ball Effect

A sudden drop of C_D indicates a shift of the separation point when transition occurs at $3 \times 10^5 < Re_D < 4 \times 10^5$. The lower drag at the turbulent boundary layer regime can be explained by the so-called golf ball effect, as shown in Figure 6.22. With the delay of the separation point to the point aft the sphere, a pressure recovery is partially achieved and thus, the pressure difference between the fore and aft of the cylinder surface is reduced. This phenomenon is often called the golf ball effect since the surface of the golf ball is intentionally roughed with dimples for longer travel distances. In experiments, we trigger the boundary layer with some artificial means such as sand papers and wires, or sometimes with acoustic waves.

6.8.2.4 Drag Reduction

Fuel efficiency of transport vehicles such as automobiles, commercial aircrafts, cargo ships is one of the most challenging topics in fluid dynamics research. As far as fuel consumption is concerned, the reduction of drag force is a primary issue. Figure 6.23 shows the drag coefficients for various automobiles, such as the passenger car, microbus, bus, truck, and trailer. A general strategy for reducing drag is to change the shape of the bluff body to a slender one and control the boundary layer to delay the separation. However, technical implementation of this strategy will require a significant amount of research effort.

Example 6.2 *Nature-inspired aerodynamics*

If we look at a dolphin from above (Figure 6.24a), we observe that the general shape is very much the same as that of an airfoil. In fact, early wing designs were based on anatomical studies on dolphins, trout, and tuna in the late eighteenth century. In dolphins, the point of maximum thickness occurs at around 45% of its length in order to push the point of flow separation backward and minimize pressure drag. This design has inspired the shape of boat hulls and submarines (Figure 6.24b).

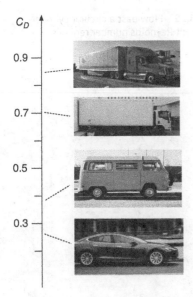

Figure 6.23 Drag coefficients C_D for sedan, microbus, truck, and trailer

(a)   (b)

Figure 6.24 Drag reduction by shape modification; dolphin of streamlined teardrop-shape (a) and submarine based on dolphin biomimetic design (b). Source: (a) NOAA/Wikimedia Commons/CC BY 2.0. (b) Darren Halstead/Unsplash

6.9 Bluff Body Aerodynamics

6.9.1 Karman's Vortex Shedding

Bluff bodies aerodynamics is strongly dependent on the Reynolds number (Figure 6.25). The wake behind a circular cylinder is stationary at $Re_D < 40$ due to dominant viscous effects. At $Re_D > 40$, however, the so-called *Karman vortex* shedding occurs; any measure of asymmetry of vortices within the two boundary layers over and under the cylinder surface will create a nonzero circulation around the cylinder. As a result, the vortex shedding produces oscillatory lift and drag forces exerted on the cylinder. The Karman vortex shedding has its origin in the instability of the flow.

The frequency of oscillation f is represented by the Strouhal number, a nondimensional frequency, defined as follows:

$$St = \frac{fD}{U} \tag{6.122}$$

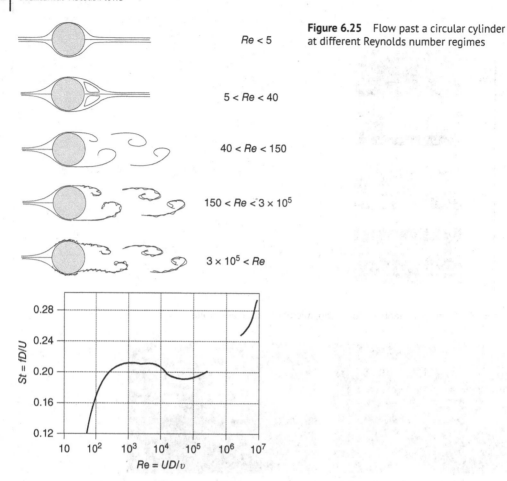

Figure 6.25   Flow past a circular cylinder at different Reynolds number regimes

$Re < 5$

$5 < Re < 40$

$40 < Re < 150$

$150 < Re < 3 \times 10^5$

$3 \times 10^5 < Re$

Figure 6.26   Strouhal number St vs. Reynolds number Re_D (circular cylinder)

where D is the diameter of the cylinder, and U is the free stream velocity. Note that the vortex shedding frequency increases from $Re_D = 40$ to 600, approaching an asymptotic value of the Strouhal number close to 0.21 (Figure 6.26).

The separated boundary layer and the wake behind the cylinder show a great deal of unsteady flow characteristics, which strongly depend on the Reynolds number. The two-dimensionality of the flow remains only until $Re_D = 150$ at which St corresponds to 0.18. Between 150 and 300, transition occurs at the free shear layers emitted from the cylinder surface and the *spanwise coherence* breaks into a finite scale; the wake immediately becomes turbulent. As shown in Figure 6.25, the boundary layer over the cylinder surface remains laminar up to $Re_D = 3 \times 10^5$ and becomes turbulent afterward.

6.9.2   Aerodynamic Sound

The spanwise-coherent vortex shedding can cause serious structural damage to cylindrical bodies, especially when the vortex shedding frequency is resonant to the natural frequency of the structure. This flow-induced vibration is often encountered in industrial problems such as off-shore structures, rod-bundles in nuclear reactors.

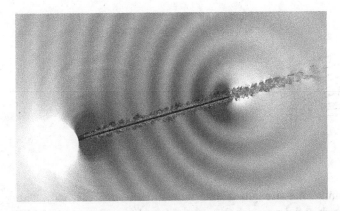

Figure 6.27 Aerodynamic noise by vortex shedding at the trailing-edge of a flat plate; hydrodynamics (large eddy simulation, LES) and acoustics (linearized perturbed compressible equation, LPCE); $M = 0.3, Re_l = 6 \times 10^5$

The vortex shedding also generates aerodynamic sound. When a vortex is shed off the body or when turbulent eddies are passing the sharp edges of the body, the near-field pressure fluctuates and generates sound.[7] Figure 6.27 shows a dipole tone radiating from the trailing-edge of a flat plate, where vortex shedding produces an oscillatory lift force at $St \approx 0.2$. Note that noise at other higher frequencies is also produced by small-scale turbulent eddies scattered at the trailing-edge. The leading-edge of the plate also produces high-frequency noise when the small-scale spanwise vortices created by Kelvin–Helmholtz instability in the free shear layer merge into the larger-scale ones.

Since the near-field pressure is determined by the viscous-inviscid interactions, the aerodynamic sound is a rather perplexing topic, and a scale of resolution is an important issue in physical modeling. Aeroacoustic research covers a variety of noise issues related to airborne sources: jet noise, airframe noise, combustion noise, fan noise, bio-fluid sounds, etc. [14].

6.10 Lift Force

6.10.1 The Origin of Lift

Lift force is much less intuitive than drag force. It is produced by an asymmetry of the forces exerted on the surfaces of a lift-producing object (e.g. a cambered airfoil) in the direction transverse to the incoming flow. This force distribution originates from a circulatory flow around the airfoil to which fluid's viscosity plays an essential role. Without viscosity, there is no circulation created and thus, no lift is produced. The existence of the circulation around the cambered airfoil can be visualized by streamline patterns: (i) upwash and downwash fore and aft the airfoil and (ii) Kutta condition (Figure 6.28).

The Kutta condition, which states that the flow speed must be finite and tangent to the trailing-edge profile, is considered as a necessary condition for lift generation. This condition is simply the result of viscous boundary layers over and underneath the cambered airfoil surface. To be more specific, the strength and distribution of vortices in the boundary layers are the

---

7 More discussion continues in Chapter 8 *Compressible Flows*.

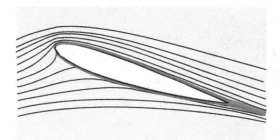

Figure 6.28 Streamlines around a cambered airfoil at small angle of attack

fundamental quantities that determine the strength of the circulation around the body. The Biot–Savart law states that the upwash and downwash flows are referred to as flows induced by vortices in the boundary layers, but the vortices in the boundary layers and the induced flows in the irrotational fields are determined by the mutual interactions between viscous and potential effects.

The origin of lift has been explained by the vortices generated in the boundary layers, which are asymmetrically distributed on the upper and lower surfaces of the airfoil. The main force representing the lift is the surface pressure force, as distributed in Figure 6.29a. A higher-speed flow over the airfoil results in low pressure, while a lower-speed flow produces a higher pressure underneath the airfoil. Meanwhile, this flow speed difference results from the curvature effect of the streamlines, which is basically the centrifugal effect of the fluid under the conservation of angular momentum.

A typical distribution of the pressure coefficients over and underneath the airfoil is shown in Figure 6.29b, where the pressure coefficient is defined by

$$C_p = \frac{p - p_\infty}{1/2 \, \rho \, U_\infty^2} \tag{6.123}$$

Note that the lift force on the airfoil is positive, regarding the signs of the C_p curves.

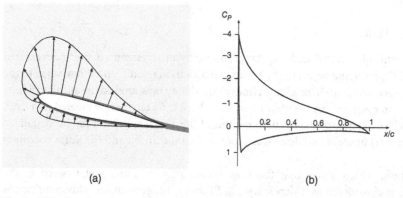

(a) (b)

Figure 6.29 Local surface pressure represented by vectors (a); pressure coefficient C_p distributions at the upper and lower surfaces of the airfoil at small angle of attack (b)

6.10.2 The Kutta–Joukowski Theorem

If there is a net rotational motion inside a closed circuit, it must result in a circulatory flow along the loop. As shown in Figure 6.30, the y-momentum equation shows

$$\underbrace{-\Delta_{34}\, p \cdot A_x - L}_{L_1} = \underbrace{\Delta_{12}(\rho\, u)v \cdot A_y}_{L_2} + \underbrace{\Delta_{34}(\rho\, v)v \cdot A_x}_{L_3} \tag{6.124}$$

where $\Delta_{ab}(\ ) \cdot A_{ab} = \int (\ )\, dA|_b - \int (\ )\, dA|_a$ and L (> 0) is the resulting lift force on the body. The lift force is then expressed as follows:

$$-L = \underbrace{\Delta_{12}(\rho\, u)v \cdot A_y}_{L_2} + \underbrace{\Delta_{34}\, p \cdot A_x}_{L_1} + \underbrace{\Delta_{34}(\rho\, v)v \cdot A_x}_{L_3} \tag{6.125}$$

If $u = U_\infty + u'$ and $v = U_\infty \tan\alpha$ with u', $v \ll U_\infty$, then Eq. (6.125) can be approximated as follows:

$$-L = \underbrace{\Delta_{12}(\rho\, U_\infty)\, v \cdot A_y}_{L_2} + \underbrace{\Delta_{34}\, p \cdot A_x}_{L_1} \tag{6.126}$$

by letting $O(v^2)$ and $O(u'v) \to 0$ (i.e. $L_3 \to 0$).

If $A_y \to \infty$, then, $\Delta_{34}\, p \to 0$, since p_3 and p_4 both approach the ambient pressure p_∞. Therefore, the lift force is written as follows:

$$L = -\rho\, U_\infty\, (\Delta_{12}\, v \cdot A_y) \tag{6.127}$$

or per unit span, it reads

$$L/s = -\rho\, U_\infty \oint \vec{v} \cdot d\vec{y} = -\rho\, U_\infty\, \Gamma \tag{6.128}$$

This is known as the Kutta–Joukowski theorem for lift.

We have shown that the circulatory flow around a solid body results in a lift force on the body. It is linearly proportional to the circulation Γ that is also a function of the flow speed U_∞, the chord length of the airfoil c, and its angle of attack α, i.e. $\Gamma \sim U_\infty\, c \sin\alpha$. Note that this is only valid within $\alpha < 15°$–$20°$. If the angle of attack exceeds this critical angle, a stall occurs, which drastically drops the lift force.

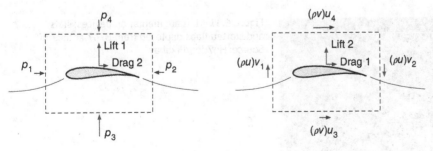

Figure 6.30 Lift generated by circulating flow around the cambered airfoil; Lift = Lift 1 + Lift 2; Lift 1 by pressure difference; Lift 2 by momentum flux difference (both in the vertical direction)

The lift coefficient is defined as follows:

$$C_L = \frac{L}{1/2\,\rho\,U_\infty^2 A_p} \tag{6.129}$$

where A_p is the planform area of the surface.

[**Notes**] The momentum balance in the horizontal direction shows a drag force D (> 0) exerted on the cambered airfoil

$$-\Delta_{12} p \cdot A_y - D = \Delta_{12}(\rho\,u)u \cdot A_y + \Delta_{34}(\rho\,v)u \cdot A_x \tag{6.130}$$

or

$$-D = \Delta_{12} p \cdot A_y + \Delta_{12}(\rho\,u)u \cdot A_y + \Delta_{34}(\rho\,v)u \cdot A_x \tag{6.131}$$

As the closed loop expands, $\Delta_{12} p \to 0$ and $\Delta_{34}(\rho\,v)u \to 0$, and it becomes

$$D = -\Delta_{12}(\rho\,u)u \cdot A_y \tag{6.132}$$

where the drag is mostly associated with the momentum defect by wake.

6.10.3 Aircraft Wings

6.10.3.1 Lift-augmented Devices

As shown in Kutta–Joukowsky's theorem, lift force is proportional to the second power of the free stream velocity. During takeoff and landing, the aircraft does not have enough lift since the speed of the aircraft is low. Therefore, lift-augmented devices such as slats and flaps are used to prevent stalls even at high angles of attack (Figure 6.31). When the slats are deployed, a gap flow between the slat and the main wing forms a wall jet over the main wing via the Coanda effect. The wall jet will then delay the boundary layer separation, injecting a high-momentum fluid near the wing surface. Similarly, the deployed flaps will delay the boundary layer separation over the flap surface.

Figure 6.32 shows a three-element airfoil of McDonald Douglas (30P30N) in high-lift configuration with slat and flap deployed at 30°. The instantaneous field of vorticity computed by large eddy simulation (LES) at $M = 0.17$ and $Re_c = 1.7 \times 10^6$ clearly shows well-attached boundary layers on the slat, main wing, and flap surfaces (Figure 6.32a). The Coanda effect utilized by the slat and flap is well demonstrated in this figure. Furthermore, the flow field is essentially unsteady, although it appears to be stable as far as the overall force balance is concerned. The local flow field is unstable and creates pressure fluctuations, especially, in the cove region of the slat. To be more

Figure 6.31 Lift augmentation devices; slats and slotted flaps deployed from B747 at take-off. Source: Ho7dog/Pixabay

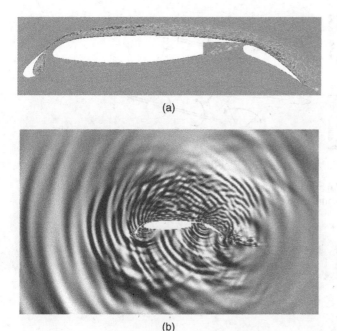

(a)

(b)

Figure 6.32 McDonald Douglas three-element airfoil in high-lift configuration (30P30N) with slat and flap deployed at 30°; instantaneous field of vorticity (LES, a) and acoustic field (LPCE, b) at $M = 0.17$ and $Re_c = 1.7 \times 10^6$

specific, the free shear layer emitted from the leading-edge of the slat becomes unstable, interacting with the slat's trailing-edge, which then causes a feed-back mechanism for noise generation.

Figure 6.32b shows an instantaneous acoustic field produced by the MD 30P30N three-element airfoil in high-lift configuration. The acoustic field was obtained by solving the linearized perturbed compressible equation (LPCE) with the noise source Dp/Dt acquired from the flow field computed by LES [15]. The figure clearly visualizes the noise radiating from the gap between the slat and the main wing and from the gap between the main wing and the flap. In this case, the trailing-edge of the slat is found to be the major noise source.

6.10.3.2 Trailing Vortices

Aircraft wings have a finite span, and due to the end effect, the flow at the wingtips rolls into a spiral vortex (or called a trailing vortex) (Figure 6.33a). The end-effect is based on the fact that air is being spilled over the wingtip by the pressure difference between the lower and upper surfaces of the wing while the aircraft is moving forward. A sink is then created at the wingtip while producing a three-dimensional spiral vortex.

It is important to understand that two counter-rotating trailing vortices are directly coupled with the flows induced by the Biot–Savart law–upwash flows on the side trails of the wingtips (outside the trailing vortex) and downwash flows behind the aircraft (inside the trailing vortex) (Figure 6.33b). In aviation, small aircrafts are prohibited from following large commercial aircrafts at close range due to the downwash effect imposed by trailing vortices. In contrast, migrating birds fly with a V-shaped fleet formation to take advantage of the upwash flows on the side trails; birds adjust their positions to make the best of the subtle effects of vortex dynamics (Figure 6.34).

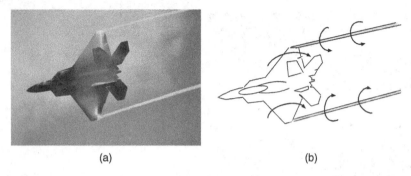

(a)   (b)

Figure 6.33 Wingtip trailing vortices (a) and the induced flow fields; upwash on the side trails and downwash behind the aircraft (b). Source: Inu Etc/Unsplash

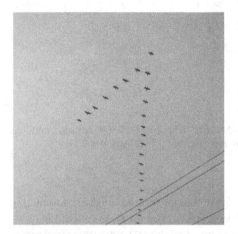

Figure 6.34 Migration of snow goose with V-shape formation; lift-support by upwash on the wing side trails. Source: Luke Jernejcic/Unsplash

6.10.3.3 Induced Drag

Another drawback of downwash flow is induced drag. The downwash flow behind the aircraft reduces the effective angle of attack of the wing, redirecting the incoming flow. As a result, the lift force vector is tilted in the clockwise direction and its component parallel to the incoming flow then acts as an induced drag. Figure 6.35 shows the total drag against the flight speed of the aircraft, where the total drag is the sum of the induced drag and the parasitic drag (skin friction drag, pressure drag, interference drag, and wave drag). Note that wave drag is a part of pressure drag caused by formation of shock waves around a body. It is clearly shown in the figure that the induced drag is an important issue for aircrafts in low-speed conditions under the speed for minimum drag.

To minimize the induced drag, a ratio of the span to the chord can be increased. Figure 6.36 shows a seagull, a typical long rider which tends to glide with the aspect ratio of about 30, and a high altitude reconnaissance aircraft, Lockheed's U2 spy-plane with the aspect ratio of 23. On the other hand, birds with small aspect ratios (e.g. swallows, sparrows) have more maneuvering capabilities than just gliding.

Another important development for reducing induced drag is winglets. As shown in Figure 6.37, the wingtips are extended to attenuate the pressure difference across the wing surfaces and thereby to reduce the strength and size of the wingtip vortex. Various forms of winglets are in use today and

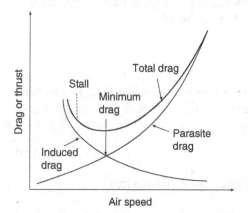

Figure 6.35 Total drag vs. air speed; total drag = induced drag + parasitic drag (skin friction drag, pressure drag, interference drag, wave drag, etc.)

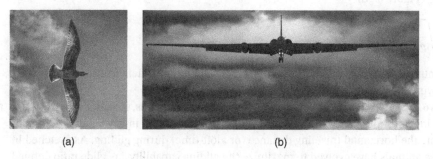

Figure 6.36 High aspect-ratio wings; seagull (a) and Lockheed's U2 spy-plane (b). Source: (a) Pascalmwiemers/Pixabay. (b) Seguir/Pixabay

Figure 6.37 Reduction of fuel consumption and air pollution by winglets. Source: Pgn/Pixabay

known to reduce fuel consumption by more than 10% and noise footprint by 6%. Winglets save billions of dollars in fuel costs and also reduce carbon dioxide emissions.

6.10.4 Gliding Aerodynamics

For aircrafts or flying animals, landing is one of the most important elements in flight dynamics. Safe landing requires low landing-speed to reduce the impact. In steady-state flight conditions, lift must equal weight:

$$W = L = C_L A_p \frac{1}{2}\rho\, V^2 \tag{6.133}$$

where $W = mg$ is the weight and V is the forwarding speed. It then follows that

$$V = \sqrt{\frac{2\,W}{\rho\, C_L\, A_p}} \tag{6.134}$$

For a given weight, the minimum speed can be obtained by increasing the lift coefficient C_L or the planform area A_p (Figure 6.38):

$$V_{\min} = \sqrt{\frac{2\,W}{\rho\, C_{L\max}\, A_{p\max}}} \tag{6.135}$$

where the minimum speed is also called *stall speed*. During landing, we often hear sound generated by motors deploying the flaps.

In nature, Eq. (6.135) holds for many flying animals such as frogs, lizards, snakes, fishes, etc. [16]. In this case, the low minimum speed is important not only for avoiding injury when landing but also for increasing the horizontal traveling distance (or aloft-time) during gliding. As sketched in Figure 6.39, these animals have evolved to maximize the gliding capability, i.e. glide ratio defined as $L/D : 1$. The glide ratio is based on the following relations for gliding at an angle of θ:

$$L\cos\theta + D\sin\theta = W \tag{6.136}$$

$$L\sin\theta = D\cos\theta \tag{6.137}$$

$$\tan\theta = H/S = D/L \tag{6.138}$$

where H and S are the vertical and horizontal distances during gliding. The first two equations are the equations of vertical and horizontal motions with no accelerations in each direction. The typical glide ratios for frog, snake, and dragon lizard (*Draco*) are 1, 1.875, and 2.5, respectively. Thus, for $H = 10\ m$, the horizontal traveling distances of these animals are approximately 10 m, 18.75 m, and 25 m, respectively.

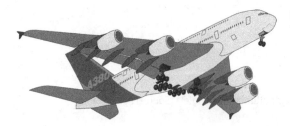

Figure 6.38 Airbus 380 with the fully deployed flaps in landing

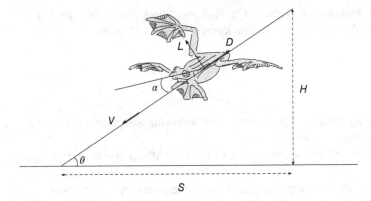

Figure 6.39 Frog gliding with webbed hands and feet

References

1 Schlichting, H., *Boundary Layer Theory*, 4th ed. New York: McGraw-Hill, 1979.

2 Mollo-Christensen, E.L., "Flow Instabilities," *Illustrated Experiments in Fluid Mechanics*, National Committee for Fluid Mechanics Films, 1980.

3 Drazin, P.G., *Introduction to Hydrodynamic Stability*, Cambridge: Cambridge University Press, 1981.

4 Nieuwstadt, F.T.M., J. Westerweel, and B.J. Boersma, *Turbulence: Introduction to Theory and Applications of Turbulent Flows*, 1st ed. Berlin: Springer-Verlag, 2018.

5 Bailly, C., and G. Comte-Bellot, *Turbulence*, 1st ed. Berlin: Springer-Verlag, 2015.

6 Kundu, P.K., *Fluid Mechanics*, San Diego, CA: Academic Press, 1990.

7 Davidson, P, *Turbulence: An Introduction for Scientists and Engineers*, Oxford: Oxford University Press, 2004.

8 Stewart, R.W., "Turbulence," *Illustrated Experiments in Fluid Mechanics*, National Committee for Fluid Mechanics Films, 1980.

9 Landahl, M.T., and E. Mollo-Christianson, *Turbulence and Random Processes in Fluid Mechanics*, Cambridge University Press, 1986.

10 Tennekes, H., and J.L. Lumley, *A First Course in Turbulence*, 7th ed. Cambridge: MIT Press, 1981.

11 Batchelor, G.K., *An Introduction to Fluid Dynamics*, Cambridge: Cambridge University Press, 1964.

12 Lighthill, J., *An Informal Introduction to Theoretical Fluid Mechanics*, Oxford: Oxford University Press, 1988.

13 Abernathy, F., "Fundamentals of Boundary Layers," *Illustrated Experiments in Fluid Mechanics*, National Committee for Fluid Mechanics Films, 1980.

14 Williams, J.E.F., and J. Lighthill, "Aerodynamic Generation of Sound," *Illustrated Experiments in Fluid Mechanics*, National Committee for Fluid Mechanics Films, 1980.

15 Seo, J.H., and Y.J. Moon, "Linearized Perturbed Compressible Equations for Low Mach Number Aeroacoustics," *J. Comput. Phys.*, 281(2), 702–719, 2006.

16 Dudley, R., G. Byrnes, S.P. Yanoviak, B. Borrell, R.M. Brown, and J.A. McGuir, "Gliding and the Functional Origins of Flight: Biomechanical Novelty or Necessity?," *Annu. Rev. Ecol. Evol. Syst.*, 38, 179–201, 2007.

17 Ajanic, E., M. Feroskhan, S. Mintchev, F. Noca, and D. Floreano, "Bioinspired Wing and Tail Morphing Extends Drone Flight Capabilities," *Science Robotics*, 5(47), 1–12, 2020.

Problems

6.1 For a laminar boundary layer on a flat plate, answer the following questions regarding creation, convection, and diffusion of vortices.
a) Physically explain why the convection speed of the flow affects the boundary layer thickness.
b) Physically explain why the viscosity of the fluid affects the boundary layer thickness.
c) Physically explain why the density of the fluid affects the boundary layer thickness.

6.2 We can learn the concept of self-similarity from the two pictures shown below: Russian nesting dolls (*Matryoshka*) and trailing-vortices shed from the tips of an aircraft. With these two pictures, physically explain the concept of self-similarity, and give examples to which this concept can be applied.

(a)

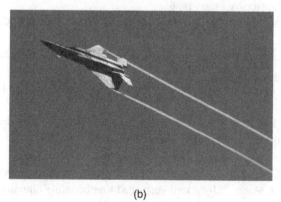

(b)

6.3 The Rayleigh problem discussed in Section 6.1 can be casted into a problem of vorticity diffusion.

a) Derive the vorticity diffusion equation from Eq. (6.2).
b) Transform initial and boundary conditions given for velocity into those for vorticity.
c) Physically interpret the diffusion equation of vorticity and the corresponding initial condition.

6.4 Consider a two-dimensional laminar jet shown in the figure. Since the jet velocity profiles are self-similar, a self-similar solution is sought by setting the new variables: $\eta = 1/3\, x^a y/\sqrt{v}$ and $\Psi = \sqrt{v}\, x^\beta f(\eta)$. An important physical invariant property of this jet flow is that the x-momentum flux $J = \rho \int_\infty^\infty u^2\, dy$ is conserved, i.e. independent of x.

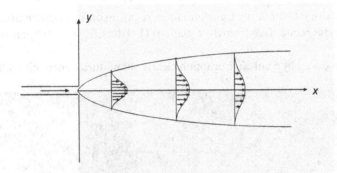

a) State a proper form of the simplified Navier–Stokes equations.
b) State the boundary conditions in terms of the physical coordinate variables.
c) Transform each term in the simplified Navier–Stokes equations in terms of self-similar variables.
d) Find α and β.
e) Derive an ordinary differential equation with the boundary conditions in terms of self-similar variables.

6.5 Consider the flow field of the far downstream wake produced by a two dimensional flat plate of width W. In the region far downstream (far wake zone), the static pressure in the wake may be considered constant and equal to atmospheric pressure. Assume that the flow is steady, laminar, and of constant flow properties. The free stream velocity is U_∞, and the velocity in the far wake is u. Also, the velocity defect u_1 is defined as $U_\infty - u$.

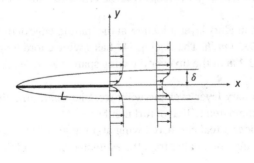

a) Show that the drag D on the plate's both sides is $D = 2\,\rho \int_0^\infty u(U_\infty - u)\, dy$ by taking a proper control volume.

b) From the boundary layer equations and with the order of magnitude analysis, show that the governing equation for the velocity defect u_1 in the far wake, where u_1 is very small compared to U_∞, is derived as follows:

$$U_\infty \frac{\partial u_1}{\partial x} = \nu \frac{\partial^2 u_1}{\partial y^2}$$

and state the boundary conditions.

c) Show that the drag is also expressed as follows:

$$D = 2\,\rho W \int_0^\infty u\,u_1\,dy = 2\,\rho W\,U_\infty \int_0^\infty u_1\,dy$$

d) By use of self-similarity transformation, set up an ordinary differential equation with the boundary conditions from the equation (1). (Hints: use $\eta = Ax^m y$ and $u_1(x,y) = Bx^m f(\eta)$.)

6.6 A jet produced by a volcanic eruption clearly shows turbulences on multiple scales.

a) Identify them in the picture from largest to smallest.
b) Calculate the wavenumber of each length scale obtained from (a) and find the nondimensional wavenumber with Prandtl's mixing length l'.
c) Estimate the Kolmogorov length scale η of the volcanic turbulent jet.

6.7 Estimate the boundary layer thickness at the trailing-edge of the wing of Boeing 747 cruising at a speed of 933 km/h. The Boeing 747 has a wing chord length that varies from 14.84 m at the root to 3.7 m at the tip over the wing span of 26.72 m. Assume that the wing is a flat plate.

a) If the boundary layer over the wing were laminar, how the boundary layer thickness would be compared to that of turbulent case.
b) If we consider a real cambered wing at an angle of attack of 5°, is the boundary layer thickness larger or smaller than the estimated one? Explain the reason.

6.8 Explain the physical basis of the Boussinesque approximation when defining turbulent stresses.

6.9 Physically interpret the Reynolds stress tensor τ'_{ij}, comparing with the viscous (or molecular) stress tensor $\overline{\tau}_{ij}$ for the mean shear flow.

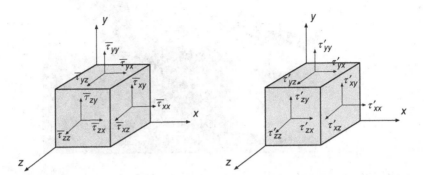

a) A viscous stress tensor is symmetric, i.e. $\overline{\tau}_{yx} = \overline{\tau}_{xy}$. Explain the physical reason.
b) The Reynolds stress tensor is also symmetric, i.e. $\tau'_{yx} = \tau'_{xy}$. Explain the physical reason.

6.10 Show that the term ③ in the kinetic energy of the mean flow and the term ②′ in the kinetic energy of turbulence (see Section 6.5 *Turbulence production and energy cascade*) can, respectively, be expressed as follows:

$$③: \quad \overline{u'_i u'_j} \, \frac{\partial \overline{u}_i}{\partial x_j} = -\frac{1}{2} \, \nu_t \left(\frac{\partial \overline{u}_i}{\partial x_j} + \frac{\partial \overline{u}_j}{\partial x_i} \right)^2$$

$$②': \quad \frac{1}{\rho_0} \overline{\tau'_{ij} \frac{\partial u'_i}{\partial x_j}} = \nu \overline{\left(\frac{\partial u'_i}{\partial x_j} \right)^2} = 2\nu \, \overline{e'_{ij} e'_{ij}}$$

and physically interpret the two terms.

6.11 In the kinetic energy of the mean flow and turbulence, answer the followings:
a) physically explain the flux divergence terms:

$$① : \quad \frac{1}{\rho_0} \frac{\partial}{\partial x_j} \left(-\overline{p} \, \overline{u}_i \, \delta_{ij} + \overline{u}_i \, \overline{\tau}_{ij} - \rho_0 \, \overline{u}_i \, \overline{u'_i u'_j} \right)$$

$$①' : \quad \frac{1}{\rho_0} \frac{\partial}{\partial x_j} \left(-\overline{p' u'_i} \, \delta_{ij} - \rho_0 \, \overline{(u'^2_i/2)u'_j} + 2\mu \, \overline{u'_i e'_{ij}} \right)$$

b) Discuss how the relations between these terms and the kinetic energy of the mean flow and turbulence is related to the divergence of Bernoulli's function $B(t)$ in potential flows.

6.12 In three-dimensional straining field, vortex stretching is the primary mechanism for turbulent energy cascade. Physically explain the reason.

6.13 Flow separation can be explained by creation and destruction of vortices in the boundary layer. Physically explain how vortices can be created or destroyed in the boundary layer.

6.14 Humpback whales have special patterns of bumps at the leading-edge of their fins, called *tubercles*. Discuss the effect of tubercles and give examples where the tubercles can be used for industrial applications.

6.15 A falcon's wings and tail morphology are the most important elements in aerodynamic maneuverability [17]. The picture below shows a Falcon in aggressive flying mode. It is found that the falcon with fully stretched wings and tail increases the lift and drag coefficients by 68% and 50 – 70%, respectively at an angle of attack between 20 and 35°. In cruise mode, wings are half-stretched and tail is completely folded.

a) In an aerodynamics point of view, what is the role of an unfolded tail and stretched wings?

b) Explain why falcons fly in fast cruise mode with the tail completely folded and the wings half-stretched.

6.16 If a facial tissue is dropped in the air, it falls slowly, gently dancing around. If we drop the facial tissue again but with a notebook underneath it, the tissue falls almost at the same speed as the notebook. Let's assume the followings for a facial tissue:

o Mass per unit area of the facial tissue: $m' = 15 \ (g/m^2)$.

o Areal size: $A = 10 \times 20 \ (cm^2)$.

o Mass: $m = 15 \times 10^{-3} \times 0.02 = 3 \times 10^{-4} \ (kg)$

a) Assuming that the facial tissue is a rigid body, derive an equation of motion.

b) Find the time history of $U(t)$ and $h(t)$ and plot them versus time. Compare the solution with the terminal speed obtained by $U_t = \sqrt{2\,mg/\rho A\, C_D}$.

c) Physically explain the drastic change of tissue falling when the notebook is put underneath it.

6.17 When a frog jumps off the branch of a tree, it first parachutes and glides afterward. Parachuting performance is an important parameter for aerodynamics of flying frogs. Let us assume the followings for a flying frog:

- Mass: $m = 25$ g
- Hands and feet: a half ellipsoid (hollow) with $b/a = 1/5$, where $a = 1$ and 2 cm for hands and feet, respectively (hollow side faces the ground)
- Body: an ellipsoid (filled) with $b/a = 1/3$, where $a = 10$ cm (convex side faces the ground)
- Arms, legs, and flaps: neglected (for simplicity)
- Drag coefficient C_D: 1.42 (hollow hemisphere)
                      1.0 (circular disk)
                      1.18 (rectangular plate of aspect ratio = 3)

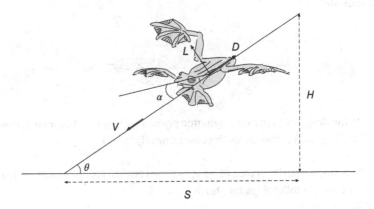

a) Derive an equation of motion for free fall with $\alpha = 90°$ and express $v(t)$ and $h(t)$, in terms of the given parameters.

b) Plot $h(t)$ and $v(t)$ against time.

c) Find the time and velocity at the momentum of touch-down for $H = 10$ m.

d) Determine the drag force of the frog at the terminal speed.

6.18 (Problem 17 is continued.) If the frog turns its body and decreases the angle of attack α as it falls, a lift force is created and the frog moves forward and glides at an angle of θ, as sketched in Problem 6.17. Let us suppose a model flying-frog is available. An experiment for gliding and parachuting performances of flying frog is conducted by dropping the model frog with different angles of attack α.

a) Draw free-body diagrams of forces (L and D) for $\alpha = 90°, 60°, 30°$ and $0°$.

b) Sketch L, D, L/D (= S/H), and θ vs. α between $0°$ and $90°$.

c) Discuss the results of (b), with respect to the gliding ratio, L/D (= S/H), of the flying frog.

6.19 (Problem 17 is continued.) The gliding performance of the flying frog can be defined by measuring the minimum speed V_{min}

$$V_{min} = \sqrt{\frac{2\,W}{\rho\,C_{Lmax}\,A_p}} = f(\alpha) \qquad\qquad (6.139)$$

where the maximum lift coefficient C_{Lmax} and planform area A_p are determined by shapes of the hands and feet of the frog and its body posture at an angle of attack α. The lower the minimum speed V_{min} is, the larger the aloft time becomes – thus, the farther the frog can move forward. In a wind tunnel test with a flow speed of 4.5 m/s, lift forces are measured: $L = 15, 58, 75, 53, 20\,mN$ for $\alpha = 90°, 60°, 40°, 20°, 0°$.

a) Sketch L vs. α.

b) Sketch V_{min} vs. α.

6.20 (Problem 17 is continued.) To examine the webbing effect of hands and feet, a parachuting performance test ($\alpha = 90°$) is conducted with the following conditions: (i) without webbing of hands and feet, (ii) without webbing of feet, (iii) webbing of hands, and (iv) with webbing of hands and feet.

a) Plot the drag forces per unit dynamic pressure, $D/(\rho\,U^2/2)$, for each case from (i) to (iv).

b) Plot the terminal speeds for the cases from (i) to (iv).

6.21 The parachuting performance of a flying lizard (Draco) is compared to that of a flying frog. Let's assume the followings for Draco:

o Mass: $m = 40\,g$

o Membranes that may be extended to create wings (*patagia*): a half ellipsoid (hollow) with $b/a = 1/5$, where $a = 10\,cm$

o Arms, legs, and flaps: neglected (for simplicity)

o Drag coefficient C_D : 1.42 (hollow hemisphere)

a) Find $U(t)$ and $h(t)$ of Draco and compare them with that of flying frog.

b) Explain why the gliding performance L/D : 1 of a flying lizard is superior to that of a flying frog.

6.22 Flow separation increases drag due to changes of pressure on the surface of the body. Therefore, pressure recovery is an important aerodynamic factor for drag reduction. Another feature of flow separation is unsteadiness.

a) Explain why the steel frames wire the tower.
b) Let us suppose there is wind of speed 10 m/s, and the tower has a diameter of 2 m. If there were no steel frames covering the tower, find the lateral force produced by the wind. What is the frequency of the sound that the tower would produce?

6.23 On a windy day, a clothesline is generating a sound. Note that wind speed is 3 m/s, diameter of the clothesline is 5 mm, and kinematic viscosity of the air is 1.5×10^{-5} m^2/s.

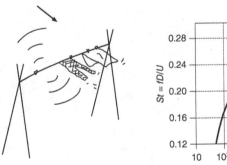
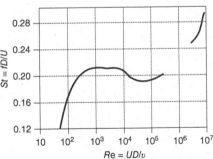

a) Explain what creates sound.
b) Find the frequency of the tonal sound. The figure below shows the measured Strouhal number vs. the Reynolds number. Compare the frequency with that of Problem 6.22(b) and discuss the results.

7

Internal Viscous Flows

7.1 Pipe Flows

7.1.1 Pressure Loss by Frictions

A flow in a pipe or a duct is different from a flow past an object immersed in fluid. In a fully developed pipe flow, the whole fluid is viscously strained at a rate by frictional forces. In this case, the fluid has no freedom to change its momentum at a rate because it is completely bounded by solid walls. Therefore, the only force that can act against the frictional resistance is the pressure difference force.

The pressure difference in a pipe or a duct is often referred to as pressure loss (or head loss). According to the total energy conservation equation, the energy (or flow work) transferred at a rate by the pressure difference force is viscously dissipated at a rate into heat while straining the fluid at a rate against frictions. This mechanical energy loss by frictions is the primary concern in internal flow systems such as industrial pipelines, ventilation ducts in transport vehicles, processing units of viscous liquids, blood vessels in human bodies.

It is interesting to note that, internal flow systems can generally be viewed as a system of hydrodynamic resistance, where the volumetric flow rate Q is determined by the viscosity μ and density ρ of the fluid, the cross-sectional area of the flow passage A, its length L, and the pressure gradient force and/or gravitational body force that balances the frictional force exerted on the fluid (Figure 7.1). The volumetric flow rate Q is defined by

$$Q = \int_A \vec{v}(x, t) \cdot \vec{n} \, dA \qquad (7.1)$$

where $\vec{v}(\vec{x}, t)$ is influenced not only by the specific geometry and flow conditions but also by the state of flow, i.e. laminar or turbulent. The velocity field in the internal flow system is determined by the equation of motion.

7.1.2 Reynolds' Experiment

7.1.2.1 Laminar or Turbulent

Osborne Reynolds (1883) conducted a famous dye injection experiment of pipe flow (Figure 7.2) [1]. The importance of this experiment was to show that there is a dependency of flow state to parameters such as flow speed, fluid density and viscosity, the pipe diameter. The ratio between two velocities, convection speed of the fluid and molecular diffusion speed, is found

Introduction to Fluid Dynamics: Understanding Fundamental Physics, First Edition. Young J. Moon.
© 2022 John Wiley & Sons, Inc. Published 2022 by John Wiley & Sons, Inc.
Companion website: www.wiley.com/go/Moon/IntroductiontoFluidDynamics

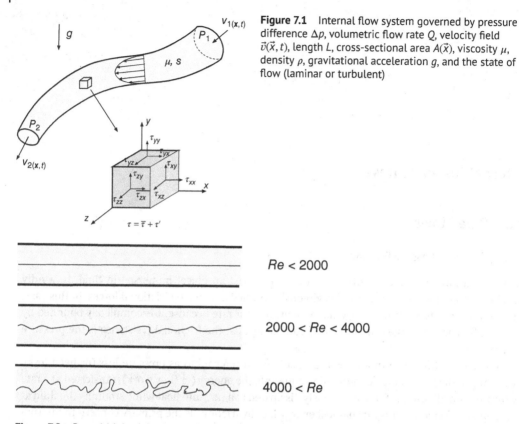

Figure 7.1 Internal flow system governed by pressure difference Δp, volumetric flow rate Q, velocity field $\vec{v}(\vec{x}, t)$, length L, cross-sectional area $A(\vec{x})$, viscosity μ, density ρ, gravitational acceleration g, and the state of flow (laminar or turbulent)

$Re < 2000$

$2000 < Re < 4000$

$4000 < Re$

Figure 7.2 Reynolds' dye injection experiment on pipe flow; laminar ($Re_D < 2000$), transitional ($2300 < Re_D < 4000$), turbulent ($Re_D > 4000$)

to be an important parameter to determine the flow state: laminar or turbulent. This ratio, called the Reynolds number, is defined by

$$Re = \frac{\rho \, UD}{\mu} \tag{7.2}$$

If $Re < 2000$, the flow is laminar. At $Re_{cr} \approx 2300$, the flow starts to oscillate and experiences a transition between $2300 < Re < 4000$. At $Re > 4000$, the flow becomes turbulent. The transition process is influenced by many factors: free-stream disturbances, surface roughness, vibrations, sound waves, etc. The location of transition also varies over time in random ways. Placing an obstruction such as a tripping wire in the boundary layer flow stimulates the naturally occurring processes and hastens the onset of transition.

Turbulence is the random fluctuations of velocities and violent mixing action. Random mixing in turbulent flows can be described as a cross-stream momentum flux as small eddies are thrown. These act as apparent stresses with an order similar to inertial forces. The apparent (or Reynolds) stresses are much larger than viscous (or molecular) stresses, since $\tau' \sim O(\rho \, \overline{u'v'})$ but $\overline{\tau} \sim O(\mu \, \partial^2 u / \partial y^2)$; therefore, turbulent flow results in higher mechanical energy loss than laminar flow.

7.1.3 Entrance Flow

As a fluid enters a pipe through a bell mouth-like entrance, the boundary layer starts to grow. Once the viscous shear force from the pipe wall reaches the centerline, a fully developed flow region

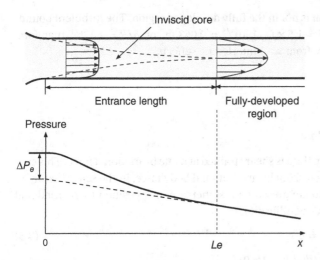

Figure 7.3 Pipe flow from entrance to a fully developed region; Δp_e for flow acceleration in the entrance region

develops. From this point, the flow inside the pipe is completely under the influence of viscous forces, i.e. all the fluid elements are angularly strained and rotated at a rate without any acceleration in the streamwise direction, i.e. $\partial u/\partial x = 0$. The entrance length is the distance from entrance to the location at which the fully developed region begins (Figure 7.3).

The centerline velocity increases as x increases up to the fully developed region, which means that the fluid in the inviscid core region accelerates due to mass conservation. Therefore, the pressure varies axisymmetrically in the entrance. In the fully developed region, however, the pressure varies only in the streamwise direction as there is no velocity change in the streamwise direction. In this region, the pressure force is balanced only with the viscous shear force, regardless of whether the flow is laminar or turbulent.

The entrance length L_e is experimentally found to be a function of the Reynolds number:

$$\frac{L_e}{D} = 0.06\, Re \qquad \text{(laminar)}$$

$$\frac{L_e}{D} = 4.4\, Re^{1/6} \qquad \text{(turbulent)} \tag{7.3}$$

Some typical numbers are listed; $Le/D = 140$ at $Re = 2300$ (laminar), $Le/D = 20$ at $Re = 10,000$, and $Le/D = 30$ at $Re = 100,000$ (turbulent).

Example 7.1 A wind tunnel of 3 m diameter takes 300 m^3/s volumetric flow rate of air. The test section is located 4 m from the entrance. Estimate the boundary layer thickness at the test section.

In this case, the Reynolds number is

$$Re = \frac{\rho\, UD}{\mu} = \frac{1.2 \times 42.4 \times 3}{1.49 \times 10^{-5}} = 8.5 \times 10^6 \tag{7.4}$$

The flow is found fully turbulent and the entrance length is therefore estimated as

$$Le/D = 4.4\, Re^{1/6} = 62.9$$

Since $Le = 190\ m > 4\ m$, the test section is not in the fully developed region. The turbulent boundary layer grows as $\delta \sim x^{4/5}$ and so $\delta = 1.5 \times (4/190)^{4/5} = 0.068\ m$, or $\delta/R = 1.13\%$; therefore, the test section is considered rather safe from wall-interference effects.

7.2 Poiseuille Flow

7.2.1 A Fully Developed Laminar Flow

In a fully developed pipe flow, the entire fluid is shear strained at a rate by frictional force. This case is equivalent to hydrostatic pressure increasing by gravitational body force, in the sense that there is no acceleration of the flow. The upstream pressure must increase in opposition to the frictional shear force in order to be in a static force equilibrium:

$$\Sigma F_x = -dp\ (\pi R^2) - \tau_w\ (2\pi R)\ dx = 0 \tag{7.5}$$

where $dp = p_2 - p_1\ (< 0)$ and $\tau_w = -\mu\ (du/dr)_w\ (> 0)$.

This reads

$$dp/dx = -2\ \tau_w/R = \text{const.} \tag{7.6}$$

Since the flow is fully developed, there is no change in velocity along the pipe. Thus, the wall shear stress τ_w is constant and so is the pressure gradient dp/dx. It is important to be reminded that Eq. (7.6) holds true regardless of whether flow is laminar or turbulent.

If Eq. (7.6) is integrated from 0 to x, we have an expression of pressure in terms of x:

$$p(x) = p_0 - 2\ \frac{\tau_w}{R}\ x \tag{7.7}$$

where p_0 is the reference pressure at $x = 0$. For example, a pressure drop over a distance L can be expressed as follows:

$$|\Delta p| = 4\ (L/D)\ \tau_w \tag{7.8}$$

which shows that the pressure drop linearly increases not only with the wall shear stress, τ_w but also with the aspect ratio of the pipe L/D. This equation is valid for all Reynolds numbers as long as the flow is fully developed.

The force balance equation is valid for any radial distance r so that

$$-dp\ (\pi r^2) = \tau\ (2\pi r\ dx) \tag{7.9}$$

Then, the viscous stress reads

$$\tau\ (r) = -\frac{r}{2}\left(\frac{dp}{dx}\right) \tag{7.10}$$

where the viscous stress is found to be a linear function of r, since dp/dx is constant.

If the flow is laminar, the shear stress can simply be expressed as $\tau = -\mu\ (du/dr)$ since the flow is parallel to the wall at any instant. It then follows that

$$-\mu\ \frac{du}{dr} = -\frac{r}{2}\left(\frac{dp}{dx}\right) \tag{7.11}$$

and the velocity profile is expressed as follows:

$$u(r) = -\frac{1}{4\mu}\left(\frac{dp}{dx}\right)(R^2 - r^2) \tag{7.12}$$

The laminar pipe flow with a parabolic velocity distribution is known as the *Hagen–Poiseuille flow*. Note that the velocity magnitude is linearly proportional to the pressure gradient dp/dx and inversely proportional to the viscosity μ.

Example 7.2 The Poiseuille flow solution can also be obtained by integrating the Navier–Stokes equations. If fluid in the pipe does not accelerate and is incompressible with no body force, then the Navier–Stokes equations in polar coordinates are written as follows:

$$0 = -\frac{\partial p}{\partial z} + \mu \left\{ \frac{1}{r}\frac{\partial}{\partial r}\left(r\frac{\partial u}{\partial r}\right) \right\} \tag{7.13}$$

Since $p = p(z)$ and $u = u(r)$, it is expressed as follows:

$$\frac{dp}{dz} = -\frac{1}{r}\frac{d(r\,\tau)}{dr} \tag{7.14}$$

where $\tau = -\mu\,du/dr$.

If both sides are integrated from $r = 0$ to R

$$\left(\frac{dp}{dz}\right)\int_0^R r\,dr = -\int_0^{R\tau_w} d\,(r\,\tau) \tag{7.15}$$

or

$$\frac{1}{2}R^2\left(\frac{dp}{dz}\right) = -R\,\tau_w \tag{7.16}$$

then finally it reads

$$|\Delta p| = 4\,(L/D)\,\tau_w \tag{7.17}$$

7.2.2 Pressure Loss and Friction Factor

The volumetric flow rate through the pipe is obtained by integrating $u(r)$ over the cross-sectional area of the pipe:

$$Q = \int_0^R u(r)\,2\pi r\,dr = \frac{\pi R^4}{8\,\mu\,L}\,|\Delta p| \tag{7.18}$$

where $|\Delta p| = 4\,(L/D)\,\tau_w$. This relation shows that the volumetric flow rate Q is proportional to $|\Delta p|$ and R to the power of 4 and inversely proportional to μ and L.[1]

Equation (7.18) shows that the averaged velocity of the Poiseuille flow is found to be twice the centerline velocity:

$$\bar{u} = \frac{Q}{\pi R^2} = \frac{R^2}{8\,\mu}\frac{|\Delta p|}{L} = \frac{1}{2}\,u(0) = \frac{1}{2}\,u_{max} \tag{7.19}$$

and that it is proportional to the pressure gradient of the pipe, as well as to the radius to the power of 2. Also, it is inversely proportional to the fluid viscosity. From Eq. (7.19), the pressure drop (or loss) can be written as follows:

$$|\Delta p| = \frac{64}{Re_D} \cdot \frac{L}{D} \cdot \frac{\rho\,\bar{u}^2}{2} \tag{7.20}$$

---

1 A hydrodynamic resistance R_h can be defined as follows:

$$R_h = \frac{|\Delta p|}{Q} = \frac{8\,\mu\,L}{\pi R^4} = f(\mu,\,L,\,R)$$

which is the ratio of the pressure difference to the volumetric flow rate.

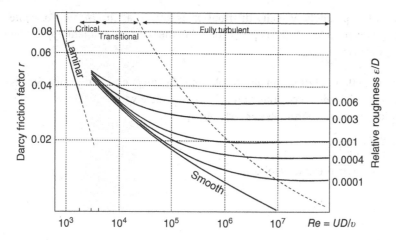

Figure 7.4 Moody's chart (pipe flow)

If we express the pressure loss in the pipe as a head loss

$$gh_f = \frac{|\Delta p|}{\rho} \tag{7.21}$$

the head loss coefficient κ can be defined as follows:

$$\kappa = \frac{gh_f}{\bar{u}^2/2} = \frac{64}{Re} \cdot \frac{L}{D} \tag{7.22}$$

where Darcy–Weisbach's friction factor f is defined as follows:

$$f = \frac{\kappa}{L/D} = \frac{64}{Re_D} \tag{7.23}$$

This relation of f vs. Re_D, called Moody's chart, is presented in Figure 7.4. In the laminar flow regime ($Re_D < 2000$), this follows exactly Eq. (7.23) derived from the Poiseuille flow solution. It is interesting to note that the loss coefficient κ is only a function of L/D and Re_D.

7.2.2.1 A Fully Developed Turbulent Flow

If the flow in pipe transits from laminar to turbulent, the field of Poiseuille flow drastically changes, reflecting turbulent characteristics such as random fluctuations of the physical quantities (e.g. momentum, heat), three-dimensional nature of vortex dynamics, and a turbulent energy cascade process. Figure 7.5 shows turbulent pipe flow computed by an incompressible large-eddy simulation at $Re_D = 5000$. The figure, represented by the second invariant of the velocity gradient tensor Q, clearly shows the random nature of the turbulent flow field.

The friction factor for a turbulent pipe flow is given by Colebrook-White

$$\frac{1}{\sqrt{f}} = -2.0 \log \left(\frac{\epsilon/D}{3.7} + \frac{2.51}{Re_D \sqrt{f}} \right) \tag{7.24}$$

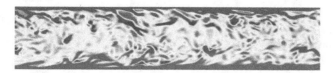

Figure 7.5 Turbulent pipe flow; large eddy simulation at $Re_D = 5000$ (or $Re_\tau = 175$)

where ϵ/D is the relative roughness, and it is an implicit function of f.

7.2.3 Total Energy Equation

If we take a control surface at the inner surface of the pipe, the mechanical energy equation in integral form reads

$$0 = \int_{CV} \left(-\vec{v} \cdot \nabla p + \vec{v} \cdot \nabla [\tau] \right) dV \tag{7.25}$$

i.e. the rate of flow work done on the fluid by pressure force is the same as the rate of viscous shear work.

According to the divergence theorem,

$$\int_{CV} \nabla \cdot \vec{v} \, [\tau] \, dV = \oint_{CS} \vec{v} \, [\tau] \cdot \vec{n} \, dA = 0 \tag{7.26}$$

It then follows that the rate of viscous shear work is identical to the rate of viscous dissipation:

$$-\int_{CV} \vec{v} \cdot \nabla [\tau] \, dV = \int_{CV} [\tau] \cdot [\nabla \vec{v}] \, dV = \Phi \quad (> 0) \tag{7.27}$$

where the viscous dissipation rate can also be expressed as follows:

$$\Phi = \int_{CV} 2 \, \mu \, \epsilon_{ij} \epsilon_{ij} \, dV \tag{7.28}$$

Thus, it is concluded that the rate of flow work is identical to the rate of viscous dissipation:

$$-\int_{CV} \vec{v} \cdot \nabla p \, dV = \Phi \tag{7.29}$$

Meanwhile, the thermal energy equation

$$\int_{CV} \rho \, (\vec{v} \cdot \nabla) \, e \, dV = \int_{CV} [\tau] \cdot [\nabla \vec{v}] \, dV \tag{7.30}$$

shows that the rate of viscous dissipation equals the rate of increase of internal energy. If we combine Eqs. (7.25), (7.27), and (7.30), it is shown that the convective rate of change of internal energy equals the rate of flow work:

$$\int_{CV} \rho \, (\vec{v} \cdot \nabla) \, e \, dV = \int_{CV} - \vec{v} \cdot \nabla p \, dV \tag{7.31}$$

If we rearrange Eq. (7.31) by adding the terms, $\nabla \cdot \rho \, \vec{v}$ and $p \, \nabla \cdot \vec{v}$ (both are zero for incompressible flow), it reads

$$\int_{CV} (\rho \, \vec{v} \cdot \nabla \, e + e \, \nabla \cdot \rho \, \vec{v}) \, dV = \int_{CV} (-\vec{v} \cdot \nabla p - p \, \nabla \cdot \vec{v}) \, dV \tag{7.32}$$

or

$$\int_{CV} \nabla \cdot (\rho \, e \, \vec{v}) \, dV = \int_{CV} - \nabla \cdot (p \, \vec{v}) \, dV \tag{7.33}$$

Using the divergence theorem, it finally reads

$$\int_{CS} (\rho \, e \, \vec{v}) \cdot \vec{n} \, dA = \int_{CS} - (p \, \vec{v}) \cdot \vec{n} \, dA \tag{7.34}$$

i.e. a net rate of flow work done on the fluid in the pipe equals the net internal energy flux across the inlet and outlet boundaries [2].

In inviscid (or ideal) flow in the nozzle (or diffuser), the energy supplied to the fluid to build the pressure difference would have done flow work on the fluid to change the kinetic energy. In a fully developed viscous flow, however, the energy input is completely wasted by viscous dissipations while shear-straining the fluid at a rate. Therefore, this is considered as energy loss and the dissipated energy is converted into heat and transferred to its surroundings.

Example 7.3 Water flowing in a pipe of constant diameter experiences a pressure drop of 6895 kPa due to friction alone. There is no heat transfer. Calculate the temperature change.

Water, being nearly incompressible, suffers no change in velocity. The density of water is 1000 kg/m^3 and specific heat $c_v = 4187\ J/kg\ °K$. The rate of flow work done by pressure difference force equals the rate of change of internal energy of the water:

$$\rho\ (e_2 - e_1)\ Q = \rho\ c_v\ (T_2 - T_1)\ Q = -\Delta p\ \ Q \tag{7.35}$$

where $\Delta p = p_2 - p_1 = -6895\ kPa$. Thus, the temperature increase is

$$T_2 - T_1 = -\frac{\Delta p}{\rho\ c_v} = 6895 \times 10^3 / (10^3 \times 4187) = 1.65\ °K \tag{7.36}$$

7.3 Laminar Channel Flows

7.3.1 Plane Couette Flow

The Couette flow subject to a pressure gradient can be described by the Navier–Stokes equations as follows:

$$\frac{dp}{dx} = \mu\ \frac{d^2 u}{dy^2} \tag{7.37}$$

If we integrate Eq. (7.37) with two boundary conditions, i.e. $u(0) = 0$ and $u(h) = U_0$, the velocity profile is given as follows:

$$u(y) = \frac{U_0}{h}\ y - \frac{1}{2\mu}\left(\frac{dp}{dx}\right)\ y(h - y) \tag{7.38}$$

Then, the wall shear stresses at $y = 0$ and $y = h$ read, respectively,

$$\tau_w|_0 = \mu\frac{U_0}{h} - \frac{h}{2}\left(\frac{dp}{dx}\right), \qquad \tau_w|_h = \mu\frac{U_0}{h} + \frac{h}{2}\left(\frac{dp}{dx}\right) \tag{7.39}$$

The volumetric flow rate per unit depth is

$$Q = \int_0^h u(y)\ dy = \frac{1}{2}\ U_0\ h - \frac{h^3}{12\ \mu}\left(\frac{dp}{dx}\right) \tag{7.40}$$

from which we can obtain the average velocity:

$$\bar{u} = Q/h = \frac{1}{2}\ U_0 - \frac{h^2}{12\ \mu}\left(\frac{dp}{dx}\right) \tag{7.41}$$

Example 7.4 *Viscous fluid driven by a vertical belt*
A continuous belt pulls a film of liquid (thickness: h) from a chemical bath at a constant velocity U_0 (Figure 7.6). The bath is filled with polymer liquid of density ρ and viscosity μ.

(a) Set the governing equation of motion for the polymer film.
(b) Define the boundary conditions at the belt surface and at the free surface of the polymer film.
(c) Find the nondimensionalized velocity profile in the film.

Figure 7.6 Polymer liquid driven by a vertical belt; h (polymer film thickness)

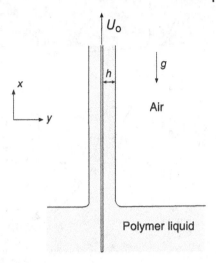

Air

Polymer liquid

(d) Discuss the nondimensional variables in the solution.

(a) The governing equation can be written as follows:

$$+\rho\, g = \mu\, \frac{d^2 u}{dy^2} \tag{7.42}$$

where the gravitational body force acts as a force to drain down the polymer film while the frictional force exerted by the belt prevents the film falling off.

(b) The boundary condition at the belt surface is the nonslip condition, i.e. $u(y) = U_0$ at $y = 0$. The boundary condition at the free surface of the film is the free viscous-stress (or traction) vector condition, i.e. $du/dy = 0$ at $y = h$, where h is the thickness of the polymer film.

(c) Integrating Eq. (7.42) twice and applying the boundary conditions yields

$$u(y) = U_0 + \frac{\rho\, g}{\mu}\, y\, \left(\frac{y}{2} - h\right) \tag{7.43}$$

This solution can be nondimensionalized as

$$\bar{u}\,(\bar{y}) = 1 + Re_h\,\bar{y}\,\left(\frac{\bar{y}}{2} - 1\right) \tag{7.44}$$

where the Reynolds number Re_h is defined as follows:

$$Re_h = \frac{\rho\, g h^2}{\mu\, U_0} = \frac{\rho\,(g\, h/U_0)\, h}{\mu} \tag{7.45}$$

(d) Note that $g\, h/U_0$ is the velocity represented by a ratio between the specific gravitational energy that can be converted into the specific kinetic energy and the reference velocity U_0; therefore, the higher the Reynolds number is, the larger the down-draining effect becomes; conversely, the lower the Reynolds number is, the larger the frictional effect becomes (Figure 7.7).

7.3.2 Plane Poiseuille Flow

The Navier–Stokes equations for a steady flow in a channel with two parallel flat walls separated by h is written as follows:

$$\frac{dp}{dx} = \mu\, \frac{d^2 u}{dy^2} \tag{7.46}$$

with the boundary conditions $u = 0$ for $y = 0$ and $y = h$.

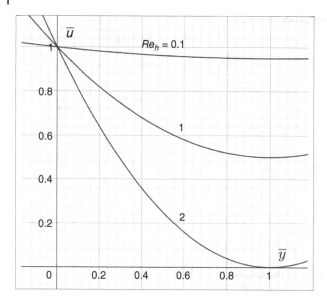

Figure 7.7 $\bar{u}(\bar{y})$ at $Re_h = 0.1$, 1, and 2; $Re_h = \rho\,(g\,h/U_0)\,h/\mu$ (*h*: polymer film thickness)

Since $\partial p/\partial y = 0$, the pressure gradient in the direction of flow is constant. We call this a plane Poiseuille flow, and the solution to Eq. (7.46) reads

$$u(y) = -\frac{1}{2\mu}\left(\frac{dp}{dx}\right) y\,(h-y) \tag{7.47}$$

The maximum speed at the centerline is

$$u_{max} = u(h/2) = -\frac{h^2}{8\mu}\left(\frac{dp}{dx}\right) \tag{7.48}$$

and the volumetric flow rate per unit depth is

$$Q = \int_0^h u(y)\,dy = -\frac{h^3}{12\,\mu}\left(\frac{dp}{dx}\right) \tag{7.49}$$

It is interesting to note that the mean velocity is found to be

$$\bar{u} = Q/h = \frac{h^2}{12\,\mu}\left(-\frac{dp}{dx}\right) = \frac{2}{3}\,u_{max} \tag{7.50}$$

which is different from that in pipe flow, e.g. $\bar{u} = (1/2)\,u_{max}$.

The vorticity in the channel is expressed as follows:

$$\omega_z(y) = -\frac{h}{2\mu}\frac{dp}{dx}\left(1 - 2\frac{y}{h}\right) \tag{7.51}$$

which is a linear function of *y*, implying that the viscous stress in the field is also the same linear function; thus, the net viscous force is entirely uniform in the flow field (Figure 7.8). In fact, a work done by pressure force on the surface facing the flow is dispensed through the work to shear-deform the fluid at the side surfaces. If the net viscous force per unit volume is expressed in terms of the vorticity vector, it is written as $-\mu\nabla \times \vec{\omega}$. If Eq. (7.51) is substituted into this, we simply obtain dp/dx, which is constant throughout the field. From this, it can be said that the flow work done on the fluid at a rate is viscously dissipated at the same rate against shear work done on the fluid.

Figure 7.8 Plane Poisuille flow; linear distribution of vorticity ω^*

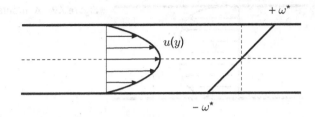

7.3.2.1 Pressure Loss and Friction Factor

From Eq. (7.50), pressure loss then reads

$$\Delta p = \frac{12\,\mu\,\bar{u}\,L}{h^2} = \frac{2\,\tau_w\,L}{h} \tag{7.52}$$

where the wall shear stress is

$$\tau_w = \mu\,\frac{du}{dy}\bigg|_{y=0} = -\frac{h}{2}\left(\frac{dp}{dx}\right) \tag{7.53}$$

The friction factor f is then expressed as follows:

$$f = \frac{48}{Re} \tag{7.54}$$

7.3.3 Hydraulic Diameter

For noncircular conduits, a hydraulic diameter is often used to represent a length scale equivalent to the diameter used for circular pipes. Based on the conservation principle of momentum,

$$-\Delta p = \tau_w(P/A)\Delta x \tag{7.55}$$

the equivalent concept can be utilized, letting

$$\tau_w(P/A)\Delta x = \tau_w\left(\frac{\pi D_{eq}}{\pi D_{eq}^2/4}\right)\Delta x \tag{7.56}$$

and thus, the hydraulic diameter is defined as follows:

$$D_h = \frac{4A}{P} \tag{7.57}$$

where P is the perimeter of the conduit and A is the cross-sectional area. The hydraulic diameter must be differentiated from the geometric equivalent length, $D_{eq} = \sqrt{4A/\pi}$, based on $\pi D_{eq}^2/4 = A$.

The hydraulic diameters for noncircular conduits can be easily calculated. For example, the hydraulic diameter of the circular pipe is $D_h = 4(\pi D^2/4)/(\pi D) = D$, as expected. For a square duct, $D_h = 4a^2/(4a) = a$, and for a rectangular duct, $D_h = 2ab/(a+b)$, etc.

7.4 Turbulent Channel Flows

In this section, we derive the velocity profile in turbulent channel flow, employing the concept of turbulent eddy viscosity proposed by Boussinesque (Figure 7.9).[2]

---

2 See Section 6.4 *Fundamentals of turbulent flow* in Chapter 6 *External Viscous Flows*.

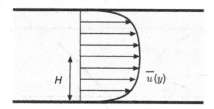

Figure 7.9 A turbulent channel flow (2*H*: channel height)

7.4.1 The Reynolds-Averaged Equations

In channel flows, a gradient of the flow variable in the streamwise and spanwise directions is zero, i.e. $\partial/\partial x = 0$ and $\partial/\partial z = 0$. Thus, the continuity equation reduces to

$$\frac{\partial \bar{v}}{\partial y} = 0 \tag{7.58}$$

With the boundary condition $\bar{v} = 0$ at $y = 0$, integration of Eq. (7.58) yields $\bar{v} = 0$ in the channel, which also implies $\bar{u} = \bar{u}(y)$.

Similarly, the Reynolds-averaged Navier–Stokes equations in the *x*- and *y*-directions reduce to

$$0 = -\frac{1}{\rho_0}\frac{\partial \bar{p}}{\partial x} + v\frac{\partial^2 \bar{u}}{\partial y^2} - \frac{\partial(\overline{u'v'})}{\partial y} \tag{7.59}$$

$$0 = -\frac{1}{\rho_0}\frac{\partial \bar{p}}{\partial y} - \frac{\partial \overline{v'^2}}{\partial y} \tag{7.60}$$

The latter yields a solution of $\bar{p}/\rho_0 = f(x) - \overline{v'^2}$, which indicates that $\partial\bar{p}/\partial x \neq f(y)$. In this case, $\partial\overline{v'^2}/\partial x = 0$, due to homogeneity of turbulence in the *x*-direction. Now, it follows that

$$\bar{p}(x,y) = \overline{p^*}(x) - \rho_0\,\overline{v'^2}(y) \tag{7.61}$$

where $\overline{p^*}(x)$ is the mean pressure in the channel and $\partial\bar{p}/\partial x = d\overline{p^*}/dx$.

7.4.1.1 Frictional Velocity

In a turbulent boundary layer, there exists a very thin layer at the wall, called the *viscous* (or *laminar*) *sublayer*. This layer is known to be a Couette flow. Thus, the velocity profile is linear with distance. Furthermore, the Reynolds number is considered to be close to 1 so that convection speed u_τ is similar to the molecular diffusion speed u_d from the wall. With these two, it follows that

$$\frac{du}{dy} = \frac{u_\tau}{v/u_d} = \frac{u_\tau}{v/u_\tau} \tag{7.62}$$

where u_τ is called *frictional velocity* [3]. The definition of wall shear stress is then expressed as follows:

$$\tau_w = \mu\frac{du}{dy}\Big|_w = \mu\frac{u_\tau}{v/u_\tau} \tag{7.63}$$

and the frictional velocity reads

$$u_\tau = \sqrt{\tau_w/\rho_0} \tag{7.64}$$

7.4.1.2   A General Solution

With the boundary conditions for the channel flow:

$$\text{at} \quad y = H, \quad \frac{\partial \overline{u}}{\partial y} = 0 \quad \text{and} \quad \overline{u'v'} = 0; \quad \text{at} \quad y = 0, \quad \overline{u} = 0 \tag{7.65}$$

we can integrate Eq. (7.59) in the y-direction from 0 to H to which we find that

$$\int_0^H \left(\nu \frac{\partial^2 \overline{u}}{\partial y^2} - \frac{\partial(\overline{u'v'})}{\partial y} \right) dy = -u_\tau^2 = \frac{H}{\rho_0} \frac{\partial \overline{p}}{\partial x} \tag{7.66}$$

We can now write the integral as follows:

$$\nu \frac{\partial \overline{u}}{\partial y} - \overline{u'v'} = \tau_t(y)/\rho_0 = u_\tau^2 \left(1 - \frac{y}{H} \right) \tag{7.67}$$

showing that the total viscous stress $\tau_t(y)$ is a linear function of y.

7.4.2   Wake (or Core) Region

The mean velocity profile of turbulent channel flow can be broken into several layers. In the region in the center, called *wake* (or *core*) region, molecular stress is negligible compared to Reynolds stress. Furthermore, the mixing length l' can be estimated as follows:

$$l' = \beta H \tag{7.68}$$

where H represents the characteristics length scale of turbulence in the core region and β is a constant that can be determined by experimental data. Thus, the eddy viscosity in Eq. (6.93) reads

$$\nu_t \sim \beta^2 H^2 \left| \frac{d\overline{u}}{dy} \right| \tag{7.69}$$

and Eq. (7.67) can be written as follows:

$$\beta^2 H^2 \left| \frac{d\overline{u}}{dy} \right| \frac{d\overline{u}}{dy} = u_\tau^2 \left(1 - \frac{y}{H} \right) \tag{7.70}$$

The solution is

$$\overline{u}(y) = u_0 - \frac{2}{3} \frac{u_\tau}{\beta} \left(1 - \frac{y}{H} \right)^{3/2} \tag{7.71}$$

where u_0 is a constant that represents the velocity at the center plane of the channel. For turbulent pipe flow, the value β is found to be about 0.13.

7.4.3   Wall (or Inner) Region: Law of the Wall

As the flow approaches the wall (or $y \to 0$), the eddy sizes will be different from those in the wake region; in this case, the mixing length becomes limited in size; therefore, we can express the mixing length as a linear function of y:

$$l' = \kappa y \tag{7.72}$$

where κ, called Von Karmann's constant, is found to be about 0.4 by experiment.

By following the same procedure, Eq. (7.67) can be written as follows:

$$\kappa^2 y^2 \left| \frac{d\overline{u}}{dy} \right| \frac{d\overline{u}}{dy} = u_\tau^2 \tag{7.73}$$

assuming that $y/H \ll 1$. The solution of Eq. (7.73) reads

$$\overline{u}(y) = \frac{u_\tau}{\kappa} \, \ln \left(\frac{y}{y_0} \right) \tag{7.74}$$

where y_0 is an integration constant.

7.4.3.1 Viscous Sublayer

It is to be noted that Eq. (7.74) is not valid at $y = 0$. In the region very close to the wall, the eddy viscosity becomes kinematic viscosity, and Eq. (7.67) is written as follows:

$$\tau_w/\rho_0 = \nu \, \frac{\partial \overline{u}}{\partial y} = u_\tau^2 \tag{7.75}$$

If we integrate Eq. (7.75), the solution reads

$$\overline{u}(y) = \frac{u_\tau^2}{\nu} \, y \tag{7.76}$$

which exactly satisfies the boundary condition at the wall.

7.4.3.2 Logarithmic Layer

It is interesting to note the flow in the viscous sublayer is not completely laminar. There exist fluctuations of velocity induced by the flow above, called *logarithmic layer*. To complete the mean velocity profile in the wall region, we must match the viscous sublayer to the logarithmic layer.

If we introduce the dimensionless units for wall distance y^+ and velocity u^+ (also referred to as *wall units*)

$$y^+ = \frac{y}{\nu/u_\tau} \quad \text{and} \quad u^+ = \frac{\overline{u}}{u_\tau} \tag{7.77}$$

Equation (7.76) can be cast into a dimensionless form:

$$\frac{\overline{u}}{u_\tau} = \frac{y}{\nu/u_\tau} \quad \text{or} \quad u^+ = y^+ \tag{7.78}$$

According to the universal representation of the mean velocity in the wall region (Figure 7.10), Eq. (7.78) is only valid for $y^+ < 5$ and Eq. (7.74) is valid for $y^+ > 30$. The value $y^+ = 5$ can be regarded as the thickness of the viscous sublayer δ_s, i.e.

$$\frac{\delta_s}{\nu/u_\tau} = 5 \tag{7.79}$$

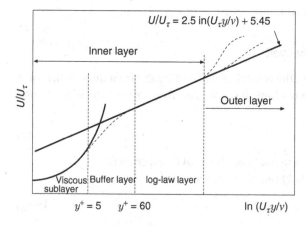

Figure 7.10 Universal representation of the mean velocity in the near-wall region in turbulent boundary layer

Figure 7.11 Typical distributions of the normalized Reynolds stress $-\overline{u'v'}/u_\tau^2$ (solid) and viscous stress $v\,(\partial\overline{u}/\partial y)/u_\tau^2$ (broken) (turbulent pipe flow)

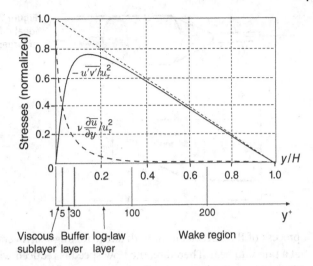

If Eqs. (7.74) and (7.78) are matched at $y^+ = 11$,

$$\overline{u}/u_\tau = u^+ = y^+ = \frac{1}{\kappa}\,\ln\left(\frac{y^+}{y_0^+}\right) \tag{7.80}$$

and thus

$$y^+ = \frac{y_0}{v/u_\tau} = 0.135 \tag{7.81}$$

The mean velocity profile for the wall region is then given by the so-called *law of the wall*

$$u^+ = \begin{cases} y^+ & 0 < y^+ < 5 \\ (1/\kappa)\left(\ln y^+ + 2\right) & y^+ > 30 \end{cases} \tag{7.82}$$

where $\kappa = 0.4$ is the Von Karmann constant. The region between $5 < y^+ < 30$ is referred to as the *buffer layer*, where the molecular viscous stress and the turbulent stress are equally important (Figure 7.11) [4–6].

7.5 Local Losses

In external viscous flows, flow separation results in a drag force (pressure drag) acting on the body. In contrast, flow separation (or mixing) in internal viscous flows causes head (or total pressure) loss in the fluid. This section deals with the head losses in auxiliary components of the pipes, for instance, sudden contractions or enlargements, elbows, tees and laterals, entrances, exits, valves.

7.5.1 Sudden Enlargement

Let us suppose we have an intake, as shown in Figure 7.12. Due to the sudden increase in pipe diameter, pressure is not fully recovered as in an ideal diffuser. In an ideal diffuser (e.g. diverging angle is less than 5°–7°), the inertial force associated with flow deceleration is perfectly balanced by the normal reaction force, i.e. pressure force. If the diverging angle exceeds 5°–7°, the flow is more likely to separate from the body surface. Once the flow is separated, the local pressure changes very little because of mixing in the free shear layer with turbulent eddies. To be more specific, the

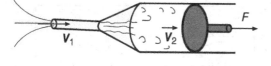

Figure 7.12 Pressure loss in sudden enlargement; ideal diffuser (broken) and intake with sudden enlargement (solid)

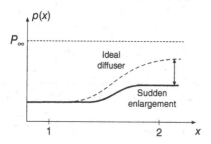

process of flow deceleration with shear mixing is irreversible, as it dissipates the kinetic energy at a rate into heat. Therefore, the flow speed is reduced without full pressure recovery.[3] Note that the pressure distribution in Figure 7.12 shows the piston moving at the same speed. If we apply the same force to the piston, less flow will be sucked into the pipe (per unit time) with sudden enlargement because the pressure distribution (solid line) should shift upwards so that p_2 matches the ideal diffuser. Note that the force applied to the piston F equals $(p_\infty - p_2)A_2$ and the volumetric flow rate is proportional to $p_\infty - p_1$.

7.5.1.1 Local Loss Coefficient

Let us consider two pipes of different diameters connected through a sudden enlargement (Figure 7.13). In this case, pressure does not recover fully as in an ideal diffuser (e.g. $A_2/A_1 > 1$). This is due to the fact that the flow is separated at the corner of the junction. When the flow is separated, it is divided into a core through-flow and a recirculating flow. Toward the reattachment point, pressure is only partially recovered by area increase because of mixing within the shear layer. The shear force in the free shear layer contributes to reduce the core flow speed and therefore pressure does not recover fully as in an ideal diffuser.

The pressure loss in sudden enlargement can be estimated by the control volume analysis. For the control volume taken inside the sudden enlargement,

Continuity equation:

$$u_1 A_1 = u_2 A_2 \tag{7.83}$$

Momentum equation:

$$(p_1 - p_2) A_2 - \bar{\tau}_w A_w = \rho \left(u_2^2 A_2 - u_1^2 A_1 \right) \tag{7.84}$$

Free shear layer

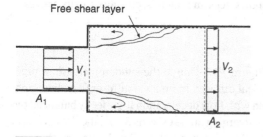

Figure 7.13 Control volume analysis for the sudden enlargement; control surface (broken line)

---

3 The extreme case is the open jet that totally dissipates the kinetic energy without recovering any pressure.

Bernoulli equation:

$$\frac{p_1}{\rho} + \frac{u_1^2}{2} = \frac{p_2}{\rho} + \frac{u_2^2}{2} + gh_f \tag{7.85}$$

or

$$gh_f = \frac{1}{\rho}(p_{01} - p_{02}) \tag{7.86}$$

where $p_0 = p + \rho\, u^2/2$ is the total pressure.

Combining Eq. (7.83) and (7.84) and assuming $\bar{\tau}_w\, A_w \approx 0$ yields

$$p_2 - p_1 = \rho\, u_1^2\; \beta(1 - \beta) \tag{7.87}$$

where $\beta = A_1/A_2$ $(0 < \beta < 1)$. If we substitute Eq. (7.87) into the Bernoulli equation, the local loss coefficient then reads

$$\kappa = \frac{gh_f}{u_1^2/2} = (\beta - 1)^2 \tag{7.88}$$

If $\beta = 1$, then $\kappa = 0$ and $p_1 = p_2$: a trivial case. If $\beta \to 0$, then $\kappa \to 1$, i.e. $gh_f \to u_1^2/2$, while $p_1 \to p_2$; all kinetic energy of the upstream flow is dissipated through viscosity in the form of a jet, without recovering any pressure. Note that the maximum pressure recovery occurs at $\beta = 1/2$, with $\kappa = 1/4$.

7.5.1.2 Energy Dissipation Rate

If we look at the energy balance, the mechanical energy equation shows

$$\int_{CV} \rho\, \vec{v} \cdot \nabla(v^2/2)\; dV = \int_{CV} \left(-\vec{v} \cdot \nabla p + \vec{v} \cdot \nabla[\tau] \right)\; dV \tag{7.89}$$

In order to have the same flow acceleration, an extra pressure difference is required to overcome the viscous force caused by flow separation.

Since

$$\oint_{CS} \vec{v}\,[\tau] \cdot \vec{n}\; dA = \int_{CV} \nabla \cdot \vec{v}\,[\tau]\; dV = 0 \tag{7.90}$$

the energy dissipation rate can be written as follows:

$$\int_{CV} [\tau] \cdot [\nabla \vec{v}]\; dV = -\int_{CV} \vec{v} \cdot \nabla[\tau]\; dV = -\int_{CV} \left\{ \rho\, \vec{v} \cdot \nabla(v^2/2) + \vec{v} \cdot \nabla p \right\}\; dV \tag{7.91}$$

and if we add $(v^2/2)\,\nabla \cdot (\rho\, \vec{v}) = 0$ and $p\,(\nabla \cdot \vec{v}) = 0$, it finally reads

$$\int_{CV} [\tau] \cdot [\nabla \vec{v}]\; dV = -\int_{CS} \left(\rho\, v^2/2 + p \right)\; \vec{v} \cdot \vec{n}\; dA = (p_{01} - p_{02})\, Q \tag{7.92}$$

where $p_0 = p + \rho\, v^2/2$ is the total (or stagnation) pressure and $Q = \int \vec{v} \cdot \vec{n}\; dA$ is the volumetric flow rate.

Furthermore, the thermal energy equation shows that the difference of stagnation pressure results in convective rate of change of the internal energy:

$$\int_{CV} \rho\, (\vec{v} \cdot \nabla)\, e\; dV = \int_{CV} [\tau] \cdot [\nabla \vec{v}]\; dV = (p_{01} - p_{02})\, Q \tag{7.93}$$

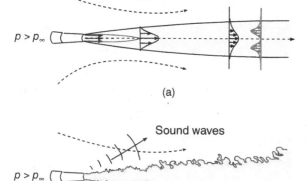

Figure 7.14 Laminar (a) and turbulent (b) jets; velocity vectors (black) and viscous shear stress vectors (gray); sound waves generated by coalescence of vortices in the free shear layer (turbulent jet)

7.5.2 Exits and Jets

The exit flow cannot be the opposite of the entrance flow. As flow exits the pipe, a jet is formed, imparting a momentum at a rate to the ambient fluid at rest (Figure 7.14). The momentum difference between the two starts being attenuated by viscous forces in the free shear layer. To be more specific, the jet flow is being retarded by the ambient fluid and the ambient flow is being entrained by the jet flow, via action and reaction.[4]

In jets, the pressure force cannot react to any deceleration or viscous force in the free shear layer because the jets are unbounded.[5] In this case, the pressure at the exit is the same as the ambient pressure since pressure recovery is not possible to occur; the jet is decelerated only by the shear (molecular and turbulent) stresses distributed across the jet. Furthermore, the jet spreads and eventually dissipates its momentum; in other words, the momentum imposed by the jet will even out far downstream by attenuation.

The local loss coefficient of the jet at $\beta \to 0$ is $\kappa = 1$ with zero pressure recovery. In contrast to the fully developed pipe flow, where the viscous shear work dissipates at a rate the flow work done by pressure force, the viscous shear work in the jet dissipates at a rate the kinetic energy, whereas some of it is used to increase the kinetic energy of the ambient fluid at rest.

The jet is similar to the boundary layer, in a way. In a laminar jet, the momentum is laterally diffused out by viscous stresses distributed across the jet via molecular mixing. A jet width can be defined as a lateral distance at $u = 0.001\ U_c$, where U_c is the jet centerline velocity. The jet spreads linearly in the beginning, but as $\sqrt{x}$ later downstream. The centerline velocity remains almost constant over a distance of about 5–7 jet diameters until the potential core disappears due to the merging of the free shear layers. After this point, the centerline velocity decays as $1/\sqrt{x}$ because the jet is spreading laterally by molecular mixing. A turbulent jet spreads more quickly by turbulent mixing. Note that turbulent mixing involves an energy cascade through vortex stretching in a three-dimensional straining field.

---

4 The wall-bounded shear flow (or boundary layer) or unbounded shear flow are the zone in which attenuation is being processed to even out the momentum difference.
5 The pressure force can only react if the shear flow is bounded by solid walls, e.g. pipes or ducts, or if the shear flow is forced to be parallel or nearly parallel, e.g. at the leading-edge of the boundary layer.

Figure 7.15 Vena Contracta in sudden contraction

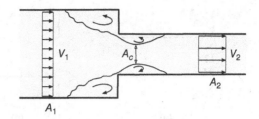

7.5.3 Sudden Contraction

In the sudden contraction of a pipe, flow separates before the contraction and also at the junction corner as sketched in Figure 7.15. The second bubble blocks the flow passage so that the cross-sectional area of the flow shrinks to less than that of the smaller pipe. The location where the cross-sectional area is minimum is called *Vena Contracta*. In this case, the flow does not accelerate as well as in an ideal nozzle. To have the same flow rate, a higher pressure difference would be required to compensate the head loss caused by viscous resistance forces.

7.5.3.1 Local Loss Coefficient
The loss coefficient for the sudden contraction is

$$\kappa \approx 0.42 \left(1 - \beta'\right) \tag{7.94}$$

where $\beta' = A_2/A_1$ ($0 < \beta' < 1$) and $\kappa = gh_f/(u_2^2/2)$. If $\beta' = 1$, then $\kappa = 0$, which is a trivial case. If $\beta' \to 0$, then $\kappa \to 0.42$. This case corresponds to the entrance with a vertical wall (local loss is solely at the Vena Contracta) (see Figure 7.17b).

Figure 7.16 shows two different types of suddenly contracted pipes: sucking in the ambient fluid through a smaller pipe (a), and ejecting out the fluid to the ambient fluid through a smaller pipe (b). The only difference is the location where the pressure is fixed; the far upstream pressure is fixed to the ambient pressure (c), while the downstream pressure at the exit is fixed to the ambient pressure (d). Since the flow separates through the sudden contraction, it becomes more difficult to transport the same flow rate with the same energy input. Therefore, a larger pressure difference is required to meet the same flow rate; we need an additional energy input (or flow work) that can act against shear force work done at a rate along the free shear layer.

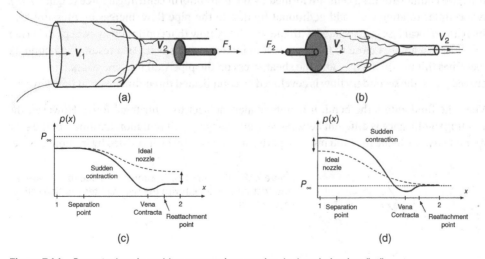

Figure 7.16 Pressure loss in sudden contraction; suction (a,c) and ejection (b,d)

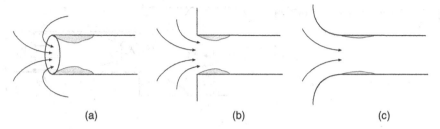

(a)   (b)   (c)

Figure 7.17 Loss coefficients for different types of entrances; $\kappa = 0.8 \sim 1$ (a), $\kappa = 0.4 \sim 0.5$ (b), $\kappa \approx 0.05$ (c)

7.5.3.2 Entrances

The entrance flow is often encountered when the outside fluid at rest is sucked into the pipe by imposition of a negative pressure downstream. In this case, a nozzle-like stream tube is formed with pressure lowered along the streamline from the ambient pressure to the pressure at the entrance. The Bernoulli principle can be applied since energy is conserved. Note that the pressure rapidly decreases toward the entrance due to the contraction of the cross-sectional area of the stream tube.

The local pressure loss, however, occurs at the inlet. The loss coefficient for entrance varies considerably with inlet types (Figure 7.17). The entrance with no inlet guidance may have a local loss coefficient close to $\kappa = 0.8 \sim 1$. This means that there will be a total loss of kinetic energy supplied by the downstream energy input. It can be reduced to $0.4 \sim 0.5$ by introducing an inlet guidance; the simplest are vertical walls. The minimal loss coefficient can be achieved with a bell mouth type entrance. If the corners are rounded, the loss coefficient will be lowered from 0.42 to 0.05, which corresponds to the limiting case, the Bell mouth entrance.

7.5.4 Curved Pipes

7.5.4.1 Secondary Flow

In curved pipes and ducts, the flow is separated by a local adverse pressure gradient in the streamwise direction. This flow separation inevitably entails local losses. It is, however, to be noted that flow separation is not the only cause of local losses in curved pipes and ducts. A secondary flow occurs as two counter-rotating cells are formed by the nonuniform centrifugal force (Figure 7.18). The two counter-rotating cells add additional friction to the pipe flow, increasing pressure loss. This increase of local loss is associated with the secondary flow that continuously sweeps the faster moving fluid into an area close to the pipe wall to increase wall shear. As a result, the boundary layer becomes thinner on average and the shear stress on the pipe wall thus increases.

In curved pipes, the secondary flow is developed by complicated three-dimensional flow physics.

(1) When the fluid enters the bend, it is immediately subject to centrifugal force. However, the resulting significance is different between streams carrying different momentums. For the sake of simplicity, we divide the fluid into two parts: one is strongly shear resisted by frictions (a layer

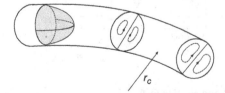

Figure 7.18 Pressure loss in curved pipe due to increase of wall friction associated with the two counter-rotating cells in secondary flow

of low-momentum fluid), and the other is either relatively less sheared, or completely nonsheared (a region of high-momentum fluid).

(2) In the bend, the high-momentum fluid will fully exert the centrifugal force toward the outer surface, increasing the pressure in the radial direction, but the low-momentum fluid cannot meet the same radial increase of pressure unless it allows migration of high-momentum fluid. In consequence, the streamwise flow draws a circle with a smaller radius of curvature. The break of *centrifugal-force-uniformity* will create a secondary flow; the low-momentum fluid must migrate toward the inner side of the bend, while the high-momentum fluid does toward the outer side.

(3) The migrating fluid in the bend must draw a circle about the streamwise component of the vorticity produced by vortex tilting. Note that as soon as the fluid enters into the bend, it tends to move faster at the inner side than at the outer side (due to difference in radius of curvature), in order to conserve the angular momentum (i.e. inertial effect) (see Section 6.5 *Vortex tilting*).

Example 7.5   *Secondary flows in the cerebral vessels*

A blood flow in the cerebral vessels is one good example of internal viscous flows. The vessels have complex branches, different cross-sectional areas, different radius of curvature, sudden enlargement (e.g. aneurysm), sudden contraction (e.g. stenosis), etc. Figure 7.19 shows pressure and streamlines with velocity magnitude in brain vessels. The secondary spiral flows are clearly identifiable in many locations of the curved vessels. More interesting part is the flow in the aneurysm, where a jet enters in and turns its direction by forming a swirl. This three-dimensional complex flow is, in fact, pulsating along with the cardiac cycle, which risks a rupture of aneurysm [7].

7.5.5   Total Head Loss

7.5.5.1   Pipes in Series

If the pipes are arranged in series with fittings, valves, exits, and entrances, it is straightforward to find the total head loss of the system, i.e.

$$h_t = h_f + h_l = \sum_i \left[f_i \left(\frac{L}{D} \right)_i + \sum_j \kappa_j \right] \frac{\bar{u}_i^2}{2g} \tag{7.95}$$

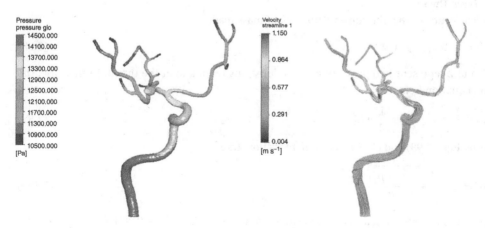

Figure 7.19   Pressure and streamlines with velocity vector magnitude in brain vessels (at systole phase)

where h_f is the frictional loss of the pipe, h_l is the local loss, and $Q_i = A_i \bar{u}_i = Q$. Thus, the total pressure loss Δp equals $\rho g h_t$.

If the flow rate of the system Q is known, the total head loss of the system can easily be found by Eq. (7.95). If the total head loss of the system h_t is known, and it is desired to find Q, we must use an iterative solution procedure since f_i is a function of Q (or Reynolds number) [8].

7.5.5.2 Pipes in Parallel

If the pipes are arranged in parallel with fittings, valves, exits, and entrances, the system finds the total head loss and flow rates as follows:

$$h_t = h_{ti} \quad \text{and} \quad Q = \sum_i Q_i = \sum_i A_i \bar{u}_i \tag{7.96}$$

where h_{ti} in each pipe is

$$h_{ti} = \left[f_i \left(\frac{L}{D} \right)_i + \sum_j \kappa_j \right] \frac{\bar{u}_i^2}{2g} \tag{7.97}$$

If the total head loss is known, it is easy to solve for Q_i in each pipe. Adding them all will find the total flow rate of the system Q. If Q is known, the head loss of each pipe h_{ti} is guessed, and Q_i is calculated and iterated until $Q - \sum_i Q_i = 0$ is satisfied [9].

7.6 Internal Systems

7.6.1 Flow Meters

In internal systems, it is often necessary to measure the flow rate. There are several types of flow meters available today: (i) obstruction type (differential pressure or variable area), (ii) inferential (turbine type), (iii) fluid dynamic (vortex shedding), (iv) ultrasonic, (v) Coriolis force, etc. In this section, we discuss the most conventional obstruction-type flow meters: orifice meter, nozzle meter, and Venturi meter (Figure 7.20).

7.6.1.1 Basic Theory

For an ideal Venturi tube, the conservation law of mass reads

$$\dot{m} = \rho A_1 u_1 = \rho A_2 u_2 \tag{7.98}$$

where 1 and 2 represent two pressure-tab positions: upstream and at the throat. In this case, the Bernoulli equation holds

$$p_1 + \frac{1}{2} \rho u_1^2 = p_2 + \frac{1}{2} \rho u_2^2 \tag{7.99}$$

Combining Eqs. (7.98) and (7.99) yields the mass flow rate

$$\dot{m} = A_2 \sqrt{\frac{2\rho (p_1 - p_2)}{1 - \beta^2}} \tag{7.100}$$

where $\beta = A_2 / A_1$. This inviscid theory indicates that if $\beta \to 1$, $(1 - \beta^2) \to 0$ with $(p_1 - p_2) \to 0$ and $A_2 \to A_1$, thus $\dot{m}$ becomes a finite value. Meanwhile, if $\beta \to 0$, $(1 - \beta^2) \to 1$ with $(p_1 - p_2) \to \infty$ but

Figure 7.20 Flow meters (obstruction type); orifice, flow nozzle, and Venturi tube

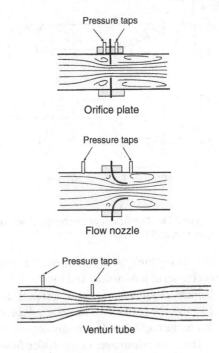

Pressure taps

Orifice plate

Pressure taps

Flow nozzle

Pressure taps

Venturi tube

$A_2 \to 0$, thus $\dot{m}$ approaches zero. In the Venturi tube, the pressure difference occurs as a reaction to the inertial force of the fluid in acceleration.[6]

7.6.1.2 Head Loss

In a real flow meter, the mass flow rate Eq. (7.100) is corrected as follows:

$$\dot{m} = C_d \, A_2 \sqrt{\frac{2\rho \, (p_1 - p_2)}{1 - \beta^2}} \tag{7.101}$$

where C_d is the discharge coefficient that varies from 0 to 1. If $C_d = 1$, it corresponds to the ideal Venturi meter, whereas if $C_d \approx 0$, the mass flow rate of the system is close to 0 due to frictional head loss. This discharge coefficient takes into account the viscous effect in the flow meter when counting the real mass flow rate. Therefore, C_d depends on the type of flow meter, the diameter ratio β, and the Reynolds number Re, where Re is based on the orifice diameter, velocity, and diameter ratio β. The pressure tap positions 1 and 2 are shown in Figure 7.21, and the orifice-type flow meters often use three different types of taps: corner taps, flange taps, and radius taps ($1D$ and $1/2D$).

The relation between $\dot{m}$, Δp, and C_d is quite nonlinear and complicated. As shown in Figure 7.21, the discharge coefficients C_d vs. the diameter (or opening) ratios β and those vs. the Reynolds numbers are all different for orifice meters and nozzle meters [10]. As expected, the discharge coefficients of the flow nozzles are higher than those of the orifice meters. For a given β, C_d is a function

---

6 The relation between the mass flow rate and the pressure difference in the Poiseuille flow is

$$\dot{m} = \frac{\pi R^4}{8 \, \mu \, L} \cdot \rho \, |\Delta p|$$

where the pressure difference occurs as a reaction to the viscous force exerted on the fluid in shear straining.

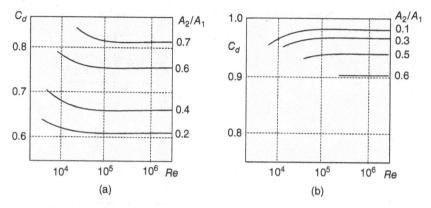

Figure 7.21 Coefficient of discharge C_d vs. Reynolds number Re for various $\beta = A_2/A_1$; orifice meter (a) and flow nozzle (b)

of not only the type of flow meter but also the Reynolds number. The orifice meters show that C_d decreases as β decreases and increases as Re decreases. In contrast, the flow nozzles show that C_d increases as β decreases and decreases as Re decreases. At Reynolds numbers greater than about 10^5–10^6, the coefficients become independent of β. At Reynolds numbers lower than $Re < 10^5$, the discharge coefficient is strongly dependent on β, Re, and the type of flow meter.

The main advantages of the orifice flow meter are as follows: (i) it is simple and robust, (ii) standards are well established and comprehensive, (iii) plates are cheap and may be used on gases, liquids, and wet mixtures (e.g. steam). Its drawbacks are as follows: (i) low dynamic range: maximum to minimum mass flow rates only 4:1 at best, (ii) affected by upstream swirl, (iii) large head loss, and (iv) performance changes with plate damage or buildup of dirt.

7.6.2 Valves

Valves are devices that regulate the flow rate of the internal system by blocking or unblocking the oncoming fluid with opposing walls (Figure 7.22a). There are various kinds of valves in the market: gate, ball, globe, butterfly, etc. As the valve is closed further, the flow fore the valve tends to get stagnated, increasing the pressure and decreasing the velocity of the oncoming fluid via potential interactions. On the other hand, the increase of upstream pressure causes a frictional head loss, forcing the fluid to pass through a narrow passage in the valve (see hydrodynamic resistance, $R = \Delta p/Q$).

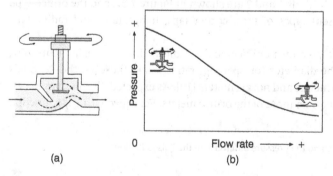

Figure 7.22 Schematic of valve (a) and inverse relation between pressure difference Δp and volumetric flow rate Q for different valve closures (b)

Figure 7.22b shows a plot of Δp vs. Q (or β). As the valve is closed further (i.e. $\beta \rightarrow 0$), the pressure difference across the valve increases, while the flow rate Q and discharge coefficient C_d decrease at the same time. This is due to the increase of frictional head loss. According to the following relation:

$$\Delta p \sim \left(\frac{Q}{C_d}\right)^2 \tag{7.102}$$

where $Q/C_d = f(\beta)$. When the valve is fully closed, the limiting value of the pressure difference approaches a finite value since the discharge coefficient C_d also approaches zero as the flow rate Q does.

If a new valve has been invented, manufacturers must provide a specifications sheet that includes a quantity known as the valve flow-coefficient C_q. The valve flow coefficient C_q is usually provided in a plot as a function of settings of the valve in terms of the number of turns the valve is opened. This information delivers how much friction is produced by the valve in all of its operating positions.

Example 7.6 *Stenosis in the coronary arteries*

One of the major branched vessels in the human body is coronary arteries of the heart. In some critical conditions, abnormalities developed in these arteries such as *stenosis* can threat our lives. From the perspective of internal viscous flow, the stenosis is a valve, which causes pressure loss that accompanies reduction of blood flow rate, i.e. a condition of *ischemia*. Discuss the hemodynamics of stenosis in the branched vessels.

A stenosis is often formed in the coronary arteries of the human heart (Figure 7.23). It is one of the most common heart diseases, which sometimes lead to fatal death. This blockage of the vessel is analogous to the orifice valve in the pipe. Frictional head loss results in a large pressure difference across the stenosis, followed by a significant reduction in flow rate of the blood. When a person is climbing stairs or running, the stenosis creates a lack of oxygen in the body (called *ischemia*), causing severe pain in the chest.

As medical treatments, it is necessary to measure how much the blood flow rate is reduced for a given stenosis. A fractional flow reserve (FFR) is widely regarded as a stenosis-specific index that reflects the effect of the coronary stenosis on the myocardial perfusion. It is defined as the maximum myocardial blood flow in the presence of a stenosis, $Q_{S\text{max}}$, divided by the theoretical maximum myocardial blood flow in the absence of the stenosis, $Q_{N\text{max}}$:

$$\text{FFR} = \frac{Q_{S\text{max}}}{Q_{N\text{max}}} = \frac{(p_d - p_v)/R_{C\text{min}}}{(p_a - p_v)/R_{C\text{min}}} \sim \frac{p_d}{p_a} \tag{7.103}$$

where p_d is the mean pressure distal to the lesion, p_a is the mean pressure proximal to the lesion or mean aortic pressure, p_v is the venous pressure (≈ 0), and $R_{C\text{min}}$ is the hemodynamic resistance of

Figure 7.23 Stenosis in coronary arteries; pressure loss by stenosis causes ischemia

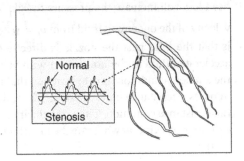

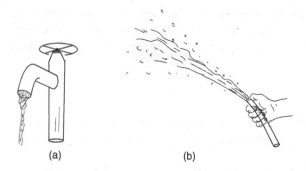

Figure 7.24 Faucet with valve (a) and squeezing a garden hose (b)

(a)   (b)

the capillary part of the venous parts. The hemodynamic resistance of the vessel is analogous to the discharge or flow coefficient of the valve. An FFR of 1.0 is generally accepted as normal. An FFR lower than 0.75–0.8 is generally considered to be associated with myocardial ischemia.

7.6.3 Faucet with Valve vs. Squeezing a Garden Hose

In internal flows, pressure and velocity fields are greatly affected by the exit flow passage. We are often asked why there is a marked difference between the flow of a faucet that is nearly closed with a valve and the flow of a hand-pressed garden hose (Figure 7.24) [11]. The answer to this question is that there is a big difference in the discharge coefficient between these two systems.

Let us suppose we have a large tank connected with a tube (Figure 7.25). If the tube is fully opened, the tube itself acts like a nozzle to the fluid in the tank. The fluid at rest is gradually accelerated toward the entrance of the tube so that the convective acceleration vs. pressure change is balanced, conserving the energy as in potential flow. The pressure in the tube is more or less the same as the atmospheric pressure and the velocity is $\sqrt{2gh}$, according to Bernoulli's principle (dotted lines).

Now, we can regulate the flow rate of the system with a valve by imposing the frictional head loss to the system (Figure 7.25a). As the valve is closed, the oncoming fluid tends to get stagnated, increasing the pressure to $\rho g h_f + p_\infty$ from the ambient pressure p_∞ and decreasing the velocity from $v_j = \sqrt{2gh}$ to $v_p = \sqrt{2g(h - h_f)}$. The changes of pressure and velocity in the pipe are indicated by solid lines in Figure 7.25a. Across the valve, the large pressure difference $\Delta p = \rho g h_f$ is dissipated, leading to a head loss. Since the downstream flow aft the valve is enclosed by the tube (not directly exposed to the ambient fluid), the downstream velocity must return to the upstream velocity. This is how the valve at the exit works.

In contrast, if we place a nozzle at the exit (in similar to squeezing a garden hose), the frictional head loss will increase the pressure to $\rho g h_f + p_\infty$ from the ambient pressure p_∞ and decrease the velocity of the oncoming fluid from $v_j = \sqrt{2gh}$ to $v_p = \sqrt{2g(h - h_f)}$ (Figure 7.25b). One difference is that the fluid past the nozzle is directly exposed to the ambient condition. Thus, the locally accelerated fluid at the nozzle end with velocity $v_n = v_p/\beta$ escapes while its pressure is fixed to the ambient condition. It is to be noted that the frictional head loss in the internal system directly imposes a resistance on the system. Thus, the upstream pressure and velocity are determined by this resistance condition. The exit velocity is determined by the downstream boundary condition; to be more specific, at what rate the kinetic energy of the flow aft the resistance is dissipated against friction.

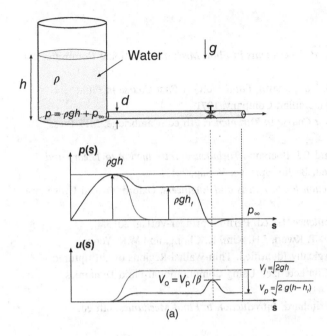

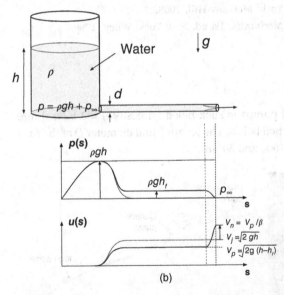

Figure 7.25 Reservoirs with valve (a) and nozzle exit (b); ideal flow (dotted lines) and frictionally resisted flow (solid lines); h_f: frictional head, v_p: velocity in the pipe, v_o: velocity past the orifice, v_n: velocity past the nozzle, v_j: jet velocity (ideal), $\beta = A_o/A_p$ or $A_n/A_p (> 1)$.

References

1 Stewart, R.W., "Turbulence," *Illustrated Experiments in Fluid Mechanics*, National Committee for Fluid Mechanics Films, 1980.

2 Sabersky, R.H., A.J. Acosta, and E.G. Hauptmann, *Fluid Flow—A First Course in Fluid Mechanics*, 2nd ed. New York: The Macmillan Company, 1971.

3 Tennekes, H., and J.L. Lumley, *A First Course in Turbulence*, 7th ed. Cambridge: MIT Press, 1981.

4 Nieuwstadt, F.T.M., J. Westerweel, and B.J. Boersma, *Turbulence: Introduction to Theory and Applications of Turbulent Flows*, 1st ed. Berlin: Springer-Verlag, 2018.

5 Davidson, P, *Turbulence: An Introduction for Scientists and Engineers*, Oxford: Oxford University Press, 2004.

6 Bailly, C., and G. Comte-Bellot, *Turbulence*, 1st ed. Berlin: Springer-Verlag, 2015.

7 Kim, J.H., H. Han, Y.J. Moon, S. Suh, T. Kwon, J.H. Kim, K. Chong, and W.K. Yoon, "Hemodynamic Features of Microsurgically Identified, Thin-Walled Regions of Unruptured Middle Cerebral Artery Aneurysms Characterized Using Computational Fluid Dynamics," *Neurosurgery*, 86(9378), 1–9, 2019.

8 Fox, R.W., A.T. McDonald, and P.J. Pritchard, *Introduction to Fluid Mechanics*, 6th ed. New York: Wiley, 2004.

9 Fay, J.A., *Introduction to Fluid Mechanics*, 1st ed. Cambridge: MIT Press, 1994.

10 White, F.M., *Fluid Mechanics*, 6th ed. New York: McGraw-Hill, 2008.

11 A.J. Smits, *A Physical Introduction to Fluid Mechanics*, 1st ed. New York: Wiley, 1999.

Problems

7.1 Mosquitos use cibarial and pharyngeal pumps to suck blood (viscosity $\mu = 3\ mPa.s$) into their body through food canals, as sketched below. The length L and diameter D of the food canal (or proboscis) are approximately 1600 and 20 μm.

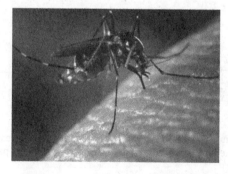 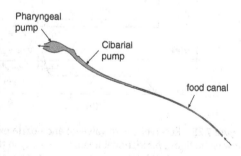

a) What forces should be considered in the equation of motion of the blood flow in the food canal?

b) Estimate the pressure difference across the food canal for sucking up the blood at an average volumetric flow rate of $Q = 1\ nL/s$.

c) Estimate the power required by the cibarial and pharyngeal pumps of the mosquito.

d) Estimate the mean velocity of the blood and hydrodynamic resistance in the food canal.

7.2 A person who sucks up honey from a jar with a straw consumes energy at a rate. Honey density is 360 kg/m^3 and viscosity is assumed to be 10 $Pa\ s$, although it ranges with moisture and temperature.

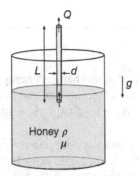

a) Estimate how much power a person consumes to suck up honey at a volumetric flow rate of $Q = 1\ mL/s$ with a straw of length 10 cm and diameter 3 mm.
b) Discuss the gravitational effect.
c) Compare the pressure difference Δp and power required for a person to suck up honey with those for a mosquito to suck up blood.

7.3 To see how much power a person uses while exhaling, a large can (diameter D) made of plexiglass is prepared with a tube connected to its side wall at the bottom. The top of the can has a small circular hole of diameter d_2 at the center. Assume no frictional effects along the tube wall.

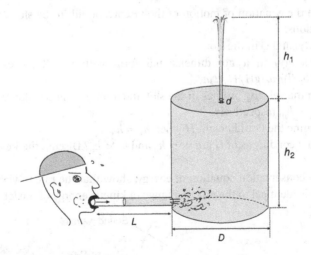

a) Explain how the person can measure the power while exhaling.
b) Express the power consumed while exhaling with the given parameters.

7.4 Two streams of an incompressible fluid are flowing concentrically as shown below, each occupying a flow area A, but the central stream's velocity is twice that of the outside stream. When they are mixed through a tube with constant area, the final velocity is uniform. Assume that the wall shear stress is zero.

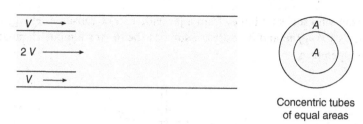

Concentric tubes
of equal areas

a) Find the rate of dissipation of mechanical energy due to the mixing process.
b) Find the rate of dissipation of mechanical energy for the following conditions: $V_i : V_o = \alpha : 1$ ($V_o = V$), and $A_i : A_o = 1 : \beta$ ($A_i + A_o = 2A$). Physically discuss the results.

7.5 A viscous liquid of density ρ and viscosity μ in the reservoir is driven by a belt moving at a constant speed of U_0. As soon as the liquid exits the slot of height h_0, its height decreases or increases to h_1 downstream, depending on the height H of the liquid in the reservoir. Note that the liquid height everywhere on the belt is very small compared with the belt length and the slot is of length l.

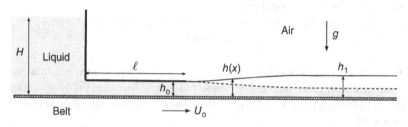

a) Derive the equation of motion of the viscous liquid in the slot and state the boundary conditions.
b) Find $u(y)$ and Q in the slot.
c) Express $u(y)$ in a non-dimensional form with the Reynolds number defined as $Re = (h_0/l)\rho\, h_0(gH/U_0)/\mu$.
d) Sketch the velocity profiles in the slot and after exiting the slot, when $h_1 > h_0$, $h_1 = h_0$, and $h_1 < h_0$, respectively.
e) Determine the liquid height H when $h_1 = h_0$.
f) On what conditions of H are $h_1 > h_0$ and $h_1 < h_0$? Discuss the Reynolds number effect.

7.6 Using the conservation equation of energy, show that the rate of viscous shear work done on the jet is identical to the rate of change of kinetic energy of the jet.

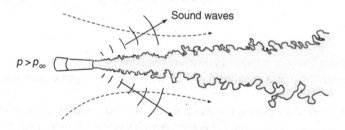

7.7 The figure below shows that the incoming boundary layer creates a so-called *passage vortex* on the suction side of the turbine blades.

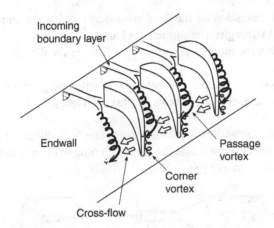

a) Explain why the passage vortex is formed at the corner of the suction surface of the blades.
b) Explain why this passage vortex degrades the turbine blade efficiency. (Hint: compare the surface pressure on the suction side of the blade with that of the blade without the passage vortex.)

7.8 The orifice flow meter (left) and flow nozzle (right) show different characteristics of the discharge coefficient C_d for different Reynolds numbers Re and different diameter ratios β $(= A_2/A_1)$.

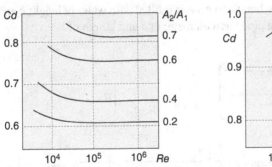
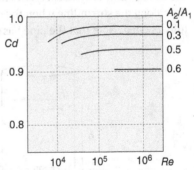

a) Explain why the orifice meters show that C_d decreases as β decreases and increases as Re decreases.
b) Explain why the flow nozzles show that C_d increases as β decreases and decreases as Re decreases.

7.9 An air cushion is used to as a damper by dissipating the kinetic energy of the road vehicle. Explain how the air cushion works and estimate how much energy can be dissipated.

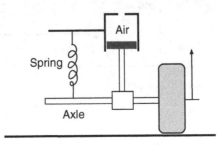

a) Derive an expression for the force exerted by the damper on the axle if the vertical displacement $x(t)$ is given as a function of time.

b) Explain why the damper always exerts a force in the direction opposite to the direction of motion.

c) Explain how the damper dissipates the kinetic energy of the vehicle and find how to design the damper to increase its function efficiency.

7.10 Consider a syringe with two different cases; one is pushing out the air inside the syringe through a narrow exit pipe, and the other is pulling in the ambient air outside, keeping the same flow rate in the exit pipe.

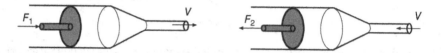

a) Sketch the streamlines for both cases and compare the pressure and velocity distributions along the streamline from the piston head to the jet (push) and to ambient air (pull).

b) Compare the forces applied to the piston for the same flow rate in the exit pipe.

7.11 To model the stenosis in the coronary arteries, a reservoir-pipe-valve system is considered as sketched below. The sketch shows the distributions of $p(s)$ and $u(s)$ of an ideal (or inviscid) flow along the streamline when the reservoir is filled with water of height h and the valve is fully-opened. Note that the pipe has a diameter of d and length of l.

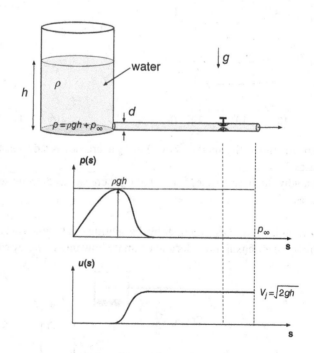

a) Sketch the distributions of $p(s)$ and $u(s)$ along the streamline for the water of heights h_1 ($> h$) and h_2 ($< h$).

b) Sketch the Δp vs. Q relationship for (a), where $\Delta p = p_b - p_\infty$ and p_b is the hydrostatic pressure at the bottom of the reservoir.

c) Sketch the distributions of $p(s)$ and $u(s)$ along the streamline for the Poiseuille flow.

d) Sketch the distributions of $p(s)$ and $u(s)$ along the streamline for 1/3 and 2/3-closed valves

e) Sketch the Δp_v versus Q relationship for (d), where Δp_v is the pressure difference across the valve.

f) Discuss the Δp versus Q relationship of (b) and that of (e).

7.12 (Problem 7.11 continues.) We have a manifold to discharge the water in the pipe (see below). Sketch the distributions of $p(s)$ and $u(s)$ along the streamline when the downstream valve is half-closed.

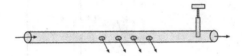

7.13 An analogy is made between a ducted-fan rotating at a certain rotational speed and a ducted-piston moving backward at a certain speed against the ambient pressure.

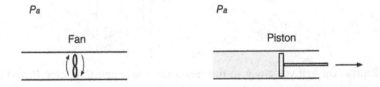

a) For both cases, sketch the streamlines and pressure distributions along the streamline.

b) What are the similarities and dissimilarities between the two cases?

c) If an orifice valve is mounted at the entrance and the valve opening is controlled, explain how the system resistance changes the pressure distribution and flow rate.

7.14 A blow-down type fan tester is shown below.

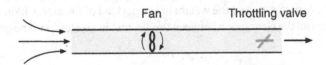

a) Sketch the pressure distribution along the streamline for low and high rotational speeds of the fan.

b) Sketch the pressure distribution along the streamline for two throttling valve positions: horizontal and inclined.

c) Physically explain why the flow rate decreases when increasing the inclination angle of the throttling valve.

d) For the inclined throttling valve, sketch the pressure distribution along the streamline for different rotational speeds of the fan.

e) On the fan performance curve, $|\Delta p|$ vs. Q, mark the points for the throttling valve of the same inclined angle. Explain the reason.

7.15 An air-cushion system is tested as sketched below. The system consists of a compressor, canvas-like skirt (diameter: D), and top-cover with an entrance hole (diameter: d), and the system is fixed to the ground.

For a given power input ($\dot{W}$) to the compressor, the clearance (h) between the air-cushion system and the flat plate is changed by applying a supporting force (F) to the plate. Note that air density is denoted by ρ.

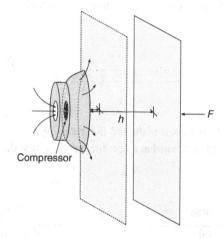

a) Express the exit velocity v_j at the clearance, the supporting force F, and the volumetric flow rate of the system Q, in terms of the given parameters, ρ, h, D, and $\dot{W}$.
b) Sketch the exit velocity v_j versus distance h.
c) Sketch the volumetric flow rate of the system Q versus distance h.
d) Sketch the supporting force F versus distance h.
e) Express the compressor performance curve, Δp versus Q, where Delta; $p = p_c - p_\infty$ and p_c denotes the pressure inside the canvas-like skirt.

7.16 The functionality of the air-cushion system can be illustrated by changing the power input ($\dot{W}$) to the compressor and the weight of the payload of the air-cushion system (F). Explain how to control the clearance h between the air-cushion system and the ground with $\dot{W}$ and F.

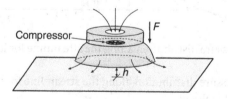

7.17 Air-floating tables are often used in processing and packaging industries. One good example is glass cutting. To design such an air floating table, air compressors are used to pressurize air to float the glass by creating jets through small circular holes. Discuss the differences between the air-floating table and the air-cushion system described in the previous problems.

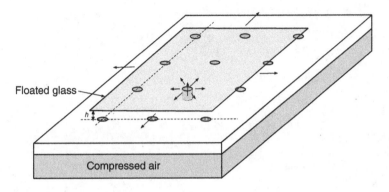

8

Compressible Flows

8.1 Compressibility of Fluids

8.1.1 Wave Propagation

The concept of compressibility is explained with a tube filled with a stationary compressible fluid of density ρ_0, pressure p_0, and temperature T_0. If the fluid is suddenly moved by a massless piston that accelerates from 0 to du with an infinitesimally small force dF, what will happen to the fluid? The fluid column in the tube is not moved as a whole, as a solid bar, or as an incompressible fluid would be. This is due to the compressibility of the fluid.

For the given force dF ($= p'A$), the volume of fluid next to the piston is locally compressed from V_0 to $V_0 + dV$ over a time interval dt (where V_0 is the volume subject to compression, and $dV < 0$ for compression) through a reversible and adiabatic[1] process (Figure 8.1). The compressed fluid absorbs the volume compression work done by the piston as elastic potential energy; this corresponds to internal energy in the conservation of thermal energy. Furthermore, the compressed fluid attains kinetic energy through the flow work done by the piston in mechanical energy conservation. Thus, the internal energy and kinetic energy of the compressed fluid in the volume of $V_0 + dV$ are locally increased in time, under the conservation principle of total energy. Once the fluid is locally compressed (or dilated), the compressed (or expanded) region will propagate via interactions of the fluid molecules.[2]

8.1.2 Volumetric Dilatation Rate

In compressible fluids, the fluctuations of physical quantities such as pressure, density, temperature, and velocities from an ambient condition (or a time-averaged value) are related to the volumetric dilatation rate $\dot{D}$:

$$\dot{D} = \frac{1}{V}\frac{DV}{Dt} = \nabla \cdot \vec{v} \tag{8.1}$$

where the volumetric dilatation rate represents the total rate of change of the volume per unit volume, i.e. a relative measure of volume change per unit time. For example, if $\dot{D} = -0.1$ (1/s), then 10% of the original volume has been shrunk per unit time. In other words, the volume has been compressed to be 90% of the original volume per unit time. Note that by the conservation law of mass, the volumetric dilatation rate $\dot{D}$ is directly related to the rate of change of the density per

---

1 The time interval is too short for heat to be transferred to the surroundings.
2 More discussion continues in Section 8.2 *The speed of sound*.

Introduction to Fluid Dynamics: Understanding Fundamental Physics, First Edition. Young J. Moon.
© 2022 John Wiley & Sons, Inc. Published 2022 by John Wiley & Sons, Inc.
Companion website: www.wiley.com/go/Moon/IntroductiontoFluidDynamics

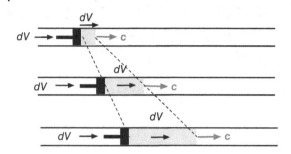

Figure 8.1 Sound wave propagating into a fluid at rest

unit density:

$$\dot{D} = -\frac{1}{\rho}\frac{D\rho}{Dt} \tag{8.2}$$

8.1.3 Isentropic and Isothermal Compressibilities

The volumetric dilatation rate $\dot{D}$ is also related to the isentropic compressibility β_S and isothermal compressibility β_T as follows:

$$\beta_S = -\frac{1}{V}\left(\frac{\partial V}{\partial p}\right)_S = -\left(\frac{\dot{D}}{dp/dt}\right)_S = \frac{1}{\rho}\left(\frac{\partial \rho}{\partial p}\right)_S \tag{8.3}$$

and

$$\beta_T = -\frac{1}{V}\left(\frac{\partial V}{\partial p}\right)_T = \frac{1}{\rho}\left(\frac{\partial \rho}{\partial p}\right)_T \tag{8.4}$$

where the specific heat ratio γ $(= c_p/c_v)$ is the ratio of these two compressibilities:

$$\frac{\beta_T}{\beta_S} = \gamma \tag{8.5}$$

Note that the concept of compressibility is opposite to that of bulk modulus K_d (or elastic stiffness).[3]

8.2 The Speed of Sound

8.2.1 Molecular Transport of Momentums

The interface between the compressed and uncompressed fluids represents a disturbance of pressure, density, temperature, etc., of which the strength is infinitesimally weak. This traveling disturbance will be heard as sound and the traveling speed, or propagation speed, of the wave is referred to as the speed of sound. To define the speed of sound, a concept of dilatational stiffness is defined analogous to the concept of frictional stiffness (Figure 8.2). The details are as follows:

8.2.1.1 Frictional Stiffness Per Unit Density

Newton's law of viscosity states that the dynamic viscosity (or viscosity) μ is defined by

$$\mu = \frac{\tau}{du/dy} = \frac{\text{shear stress}}{\text{shear strain rate}} \qquad [Pa \cdot s] \tag{8.6}$$

---

3 The bulk modulus of a substance is a measure of how resistant to compression that substance is. It is defined as the ratio of the infinitesimal pressure increase to the resulting relative decrease of the volume.

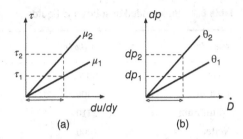

Figure 8.2 An analogy of stress vs. strain–rate relationship between the viscous effects and the compressibility effects; proportionality constants: frictional stiffness (viscosity, $\mu = \tau/2\epsilon$) vs. dilatational stiffness (dilatational resistivity, $\Theta = dp/\dot{D}$); τ: shear stress, dp: differential pressure, ϵ: shear strain rate, $\dot{D}$: volumetric dilatation rate

where μ/dt represents the measure of *frictional stiffness* K_f with a unit of *Pascal* (Figure 8.2a). Note that the viscosity is normalized by the related-time scale, since μ is proportional not only to the shear stress τ but also to the time scale dt associated with the shear strain; for example, the shear strain that occurs in 1 s is different from the shear strain that occurs in one day.

It is also to be noted that the shear strain of the fluid does not occur as a whole. If the plate is suddenly pulled, only the fluid in the vicinity of the plate is shear strained at a rate, and the created tangential momentum is continuously transported to the adjacent fluid via frictional intermolecular interactions. Therefore, the viscosity μ (to be more precise, frictional stiffness, $K_f = \mu/dt$) relates to the molecular diffusion speed v_d. Note also that, viscosity is a density-related quantity. Thus, it must be normalized by the density to represent the molecular diffusion speed as follows:

$$\frac{\mu/dt}{\rho} = v/dt = v_d^2 \quad [m^2/s] \tag{8.7}$$

where $v = \mu/\rho$ is called kinematic viscosity. The molecular diffusion speed of frictional effect is then expressed as follows:

$$v_d = \sqrt{\frac{\mu/dt}{\rho}} = \sqrt{\frac{K_f}{\rho}} \tag{8.8}$$

showing that it is proportional to the frictional stiffness μ/dt (or K_f) and inversely proportional to the density ρ. The more molecules in a given space of volume, the more interactions of molecules via frictions; thus, the heavier the fluid is, the longer the process time of frictional diffusion takes (or the lower the molecular diffusion speed becomes).

8.2.1.2 Dilatational Stiffness Per Unit Density

The propagation speed of a sound wave, or the speed of sound, can be deduced by the same analogy. We first define Θ (*dilatational resistivity*) as follows:

$$\Theta = -\frac{dp}{\dot{D}}\bigg|_s = -\frac{\text{normal stress}}{\text{volumetric dilatation rate}} \tag{8.9}$$

where Θ/dt represents the measure of *dilatational stiffness* (or bulk modulus, K_d), in analogy to $K_f = \mu/dt$ (Figure 8.2b). Here, the dilatational resistivity Θ is normalized again by the related-time scale, since Θ is proportional not only to the normal stress Δp but also to the time scale dt associated with the volumetric dilatation.

If the piston suddenly moves, the fluid in the vicinity of the piston is volumetrically dilated (or compressed) at a rate and the created normal momentum is continuously transported to the adjacent fluid via molecular interactions (Figure 8.1). Therefore, Θ (to be more precise, dilatational stiffness, $K_d = \Theta/dt$) relates to the speed of sound c. However, Θ is a density-related quantity, and thus, it must be normalized by the density to represent the speed of sound as follows:

$$\frac{\Theta/dt}{\rho} = \Theta'/dt = c^2 \tag{8.10}$$

Table 8.1 Bulk moduli of solids and liquids

Material	$\rho\ (kg/m^3)$	$K_d\ (N/m^2)$
Steel	8050	140×10^9
Cast Iron	7300	90×10^9
Aluminum	2710	70×10^9
Water	1000	2×10^9
Alcohol	789	1×10^9
Air	1.225	1.01×10^5
Human Lung	330–930	$1.5–9.8 \times 10^3$

where $\Theta' = \Theta/\rho$ is referred to as kinematic dilatational resistivity. The speed of sound is then expressed as follows:

$$c = \sqrt{\frac{\Theta/dt}{\rho}} = \sqrt{\frac{K_d}{\rho}} = \sqrt{\frac{1}{\rho\,\beta_S}} = \sqrt{\frac{\partial p}{\partial \rho}\Big|_S} \tag{8.11}$$

showing that the speed of sound is proportional to the dilatational stiffness, Θ/dt $(= K_d)$ (or inversely proportional to the isentropic compressibility, β_S) and inversely proportional to the density ρ. The more molecules that exist in a given space of volume, the longer the process time of compression becomes (or the lower the speed of sound becomes).

For an ideal gas, the density and pressure changes according to $p = c_1\,\rho^\gamma$ and the ratio of pressure to density differentials is a function of temperature:

$$\frac{\partial p}{\partial \rho}\Big|_S = \frac{\gamma\,p}{\rho} = \gamma R T \tag{8.12}$$

where R is the gas constant and shows that the speed of sound is only a function of temperature, i.e.

$$c = \sqrt{\gamma R T} \tag{8.13}$$

For example, the speed of sound in air at $T = 20°$ is

$$c_{air} = \sqrt{\gamma R T} = \sqrt{1.4 \times 287\ (J/kg\,°K) \times 293\ (°K)} = 343\ \ m/s$$

For solids and liquids, the speed of sound can be obtained using the bulk moduli (Table. 8.1):

$$c_{water} = \sqrt{K_d/\rho} = \sqrt{2 \times 10^9/1000} = 1414\ \ m/s \qquad \text{(for water)}$$

$$c_{steel} = \sqrt{K_d/\rho} = \sqrt{140 \times 10^9/8050} = 4170\ \ m/s \qquad \text{(for steel)}$$

8.2.2 Conservation Laws of Mass and Momentum in Moving Reference Frame

If a moving reference frame is used, the mass conservation reads as follows:

$$\rho\,c = (\rho + d\rho)(c - du) \tag{8.14}$$

where the compressed air in the state of $\rho + d\rho$ and $p + dp$ moves to the right with du. It then follows that

$$\rho\,du = c\,d\rho \tag{8.15}$$

The momentum equation in the moving reference frame is written as follows:

$$p - (p + dp) = \rho c \left((c - du) - c\right) \tag{8.16}$$

and it reads

$$dp = \rho c \, du \tag{8.17}$$

By combining Eq. (8.17) with Eq. (8.15), we obtain the following relation between dp and $d\rho$ in an isentropic process:

$$dp = c^2 \, d\rho \tag{8.18}$$

With the Gibbs equation and the perfect gas law, the speed of sound is determined as follows:

$$c = \sqrt{\left.\frac{dp}{d\rho}\right|_S} = \sqrt{\frac{\gamma \, p}{\rho}} = \sqrt{\gamma RT} \tag{8.19}$$

Note that the speed of sound is only a function of temperature T.

8.3 Fundamentals of Sound Waves

A wave propagates in elastic media such as fluid or solid via interaction of molecules. The wave speed is inversely proportional to the elasticity and density of the media. The more elastic (or compressible) the media, the lower the wave speed. The denser the media, the lower the wave speed.

8.3.1 Linearized Mass and Momentum Equations

If we define the instantaneous density, pressure, temperature, and velocity as follows:

$$\rho = \rho_0 + \rho', \quad p = p_0 + p', \quad T = T_0 + T', \quad u = 0 + u' \tag{8.20}$$

where ρ', p', T', and u' are their fluctuating quantities and substitute these variables to the mass conservation equation Eq. (8.2), it is expressed as follows:

$$\frac{\partial(\rho_0 + \rho')}{\partial t} + u' \frac{\partial(\rho_0 + \rho')}{\partial x} = -(\rho_0 + \rho) \frac{\partial u'}{\partial x} \tag{8.21}$$

and if we drop the high-order nonlinear terms, it is reduced to a linearized form:

$$\frac{1}{\rho_0} \frac{\partial \rho'}{\partial t} = -\frac{\partial u'}{\partial x} \tag{8.22}$$

which shows that $\partial \rho' / \partial t \sim \rho_0 \, \dot{D}$, i.e. the local rate of change of density fluctuation is proportional to the density ρ_0 and the volumetric dilatation rate $\dot{D}$.

Meanwhile, the compressed fluid is locally accelerated in time under the momentum principle, and the linearized momentum equation is written in a similar manner as follows:

$$\rho_0 \frac{\partial u'}{\partial t} = -\frac{\partial p'}{\partial x} \tag{8.23}$$

The velocity pick-up of fluid over dt will act as piston movement to the neighboring fluid at rest, and so on. This transient momentum pick-up process continues until it reaches the other end. Once the disturbance arrives, the entire compressed fluid will move at a constant speed of dV as sketched in Figure 8.1.

According to the Gibbs equation and the equation of state for a perfect gas, the pressure and temperature are increased as follows:

$$p' \sim (\rho')^\gamma \qquad \text{and} \qquad T' \sim (\rho')^{\gamma-1} \tag{8.24}$$

where $\gamma = c_p/c_v$ is the specific heat ratio of the fluid.

8.3.2 Wave Equation

If we differentiate Eqs. (8.22) and (8.23) with respect to t and x, respectively, and subtract one from the other, we obtain

$$\frac{\partial^2 \rho'}{\partial t^2} - \frac{\partial^2 p'}{\partial x^2} = 0 \tag{8.25}$$

If p is expanded about p_0 and ρ_0, it reads

$$p = p_0 + (\rho - \rho_0) \left.\frac{\partial p}{\partial \rho}\right|_{\rho_0} + \cdots \tag{8.26}$$

and by dropping the higher-order terms, we can obtain a linear relationship between p' and ρ' as follows:

$$p' = \rho' \left.\frac{\partial p}{\partial \rho}\right|_{\rho_0} = c^2 \rho' \tag{8.27}$$

With Eq. (8.27), Eq. (8.25) is written as a second-order wave equation of p':

$$\frac{\partial^2 p'}{\partial t^2} - c^2 \frac{\partial^2 p'}{\partial x^2} = 0 \tag{8.28}$$

where the waves travel at the speeds of $\pm c$. A general solution of the wave equation is

$$p'(x, t) = f(x - ct) + g(x + ct) \tag{8.29}$$

where $p'(x, t) = f(x - ct)$ is the solution of a pressure wave traveling in the direction of increasing x, and $p'(x, t) = g(x + ct)$ is the solution of the pressure wave traveling in the direction of decreasing x.

8.3.3 Acoustic Impedance

With the wave solution given by Eq. (8.29), we can obtain the density fluctuation ρ' or velocity fluctuation (or particle velocity) u'. For the right-traveling wave, $p'(x, t) = f(x - ct)$, a density wave is expressed as follows:

$$\rho'(x, t) = \frac{1}{c^2} p'(x, t) = \frac{1}{c^2} f(x - ct) \tag{8.30}$$

and the particle velocity can be obtained by the linearized continuity equation, Eq. (8.22).

With $\eta = x - ct$ and by using the chain rule, Eq. (8.22) is written as follows:

$$-\frac{1}{\rho_0 c} \frac{df}{d\eta} = -\frac{du'}{d\eta} \tag{8.31}$$

and can be integrated as follows:

$$\frac{1}{\rho_0 c} f = u' \tag{8.32}$$

It finally reads

$$p' = (\rho_0 c) u' \tag{8.33}$$

where $Z = \rho_0 c$ is called *acoustic impedance*. Similarly, the particle velocity for the left-traveling wave, $p'(x, t) = g(x + ct)$, reads

$$p' = -(\rho_0 c) u' \tag{8.34}$$

where p' and u' are off-phase by 180°.

8.3.4 Conservation of Energy in Sound Waves

A one-dimensional, total energy equation is written in conservative form as follows:

$$\frac{\partial(\rho \, e_t)}{\partial t} + \frac{\partial(\rho \, e_t \, u)}{\partial x} = -\frac{\partial(p \, u)}{\partial x} \tag{8.35}$$

where $e_t = c_v T + u^2/2$ is the specific total energy.

By expanding the pressure up to the second-order

$$p' = c^2 \rho' + \frac{\gamma - 1}{2\rho_0} c^2 \rho'^2 + \cdots \tag{8.36}$$

With the equation of state for perfect gas and $\gamma = c_p/c_v$, the internal energy of the fluid per unit volume can be expressed as follows:

$$\rho \, c_v \, T = \frac{p}{\gamma - 1}$$

$$= \frac{1}{\gamma - 1} \left(p_0 + c^2 \rho' + \frac{\gamma - 1}{2\rho_0} c^2 \rho'^2 + \cdots \right) \tag{8.37}$$

If we substitute the decomposed variables into Eq. (8.35) and combine it with Eqs. (8.36) and (8.37), it reads

$$\frac{\partial}{\partial t} \left(\frac{1}{\gamma - 1} \left(p_0 + c^2 \rho' + \frac{\gamma - 1}{2\rho_0} c^2 \rho'^2 \right) + \frac{1}{2} \rho_0 \, u'^2 \right)$$

$$+ \frac{\partial}{\partial x} \left(\frac{1}{\gamma - 1} \left(p_0 + c^2 \rho' + \frac{\gamma - 1}{2\rho_0} c^2 \rho'^2 \right) u' + \frac{1}{2} \rho_0 \, u'^3 \right)$$

$$= -\frac{\partial \, (p_0 + p')u'}{\partial x} \tag{8.38}$$

and if we keep up to the second-order and rearrange the terms, it is written as follows:

$$\frac{\partial}{\partial t} \left(\frac{c^2 \rho'}{\gamma - 1} + \frac{c^2 \rho'^2}{2\rho_0} + \frac{1}{2} \rho_0 \, u'^2 \right) + \frac{\partial}{\partial x} \left(\frac{p_0 \, u' + c^2 \rho' \, u'}{\gamma - 1} \right) = -\frac{\partial \, (p_0 + p')u'}{\partial x} \tag{8.39}$$

Now, the continuity equation in the conservative form reads

$$\frac{\partial \rho}{\partial t} + \frac{\partial(\rho \, u)}{\partial x} = 0 \tag{8.40}$$

Substituting the decomposed variables into Eq. (8.40) and multiplying by $c^2/(\gamma - 1)$ yields

$$\frac{\partial}{\partial t} \left(\frac{c^2 \rho'}{\gamma - 1} \right) + \frac{\partial}{\partial x} \left(\frac{\gamma \, p_0 \, u'}{\gamma - 1} + \frac{c^2 \rho' \, u'}{\gamma - 1} \right) = 0 \tag{8.41}$$

while retaining up to the second-order.

If we subtract Eq. (8.41) from (8.39), the final form of the energy equation for acoustically perturbed quantities is written as follows:

$$\frac{\partial}{\partial t} \left(\frac{1}{2} \frac{c^2}{\rho_0} \rho'^2 + \frac{1}{2} \rho_0 \, u'^2 \right) = -\frac{\partial(p'u')}{\partial x} \tag{8.42}$$

8.3.5 Physical Interpretations

8.3.5.1 Elastic Potential Energy

A volume compression (or expansion) work is done on the fluid changing volume against pressure force. As sketched in Figure 8.1, a sudden increase of velocity by an impulsive but weak motion of the piston locally compresses the fluid at rest so that the density is locally increased in time, according to the conservation law of mass:

$$\frac{1}{\rho_0} \frac{\partial \rho'}{\partial t} = -\frac{\partial u'}{\partial x} \tag{8.43}$$

Multiplying by p' $(= c^2 \rho')$ gives us the thermal energy equation:

$$\frac{c^2}{\rho_0} \frac{\partial \left(\rho'^2 / 2 \right)}{\partial t} = -p' \frac{\partial u'}{\partial x} \tag{8.44}$$

where the right-hand side is the rate of volume compression work done on the fluid per unit volume, i.e. a product of the excessive pressure and the volumetric dilatation rate of the fluid. Meanwhile, the left-hand side is the local rate of change of internal energy of the fluid per unit volume, which can also be interpreted as the local rate of change of *elastic potential energy* of the fluid per unit volume.

Now, $1/2 \, (\rho'c)^2/\rho_0$ needs to be proven to be the elastic potential energy density of the fluid that undergoes acoustic compression. For a fixed mass of gas, a volume compression work done on the gas can be expressed as $dW = - \int p' \, dV$, and for a small change of volume, $dV = -(V_0/\rho_0) \, d\rho$, due to mass conservation. Then, the volume compression work can be written with $p' = c^2 \rho'$ as follows:

$$dW = - \int p' \, dV = \frac{V_0 \, c^2}{\rho_0} \int \rho' \, d\rho' = \frac{V_0 \, c^2}{\rho_0} \frac{\rho'^2}{2} \tag{8.45}$$

which proves that the work done by the excess pressure in compressing the fluid per unit volume is $1/2 \, (\rho'c)^2/\rho_0$. If we express the volume compression work done on the gas, Eq. (8.45), as a rate and apply the divergence theorem, Eq. (8.45) becomes identical to Eq. (8.44).[4]

8.3.5.2 Kinetic Energy

In sound wave propagation, a balance of forces must be met between the inertia force with transient acceleration of the compressed fluid and the pressure force:

$$\rho_0 \frac{\partial u'}{\partial t} = -\frac{\partial p'}{\partial x} \tag{8.46}$$

Multiplying by u' yields the mechanical energy equation:

$$\frac{\partial}{\partial t} \left(\frac{1}{2} \rho_0 \, u'^2 \right) = -u' \frac{\partial p'}{\partial x} \tag{8.47}$$

where the right-hand side corresponds to the rate of flow work done on the fluid per unit volume and the left-hand side is the rate of increase of kinetic energy of the fluid per unit volume. By the volume compression work done on the fluid, the compressed fluid attains its momentum and moves with the same speed as the piston. This process continues in space and time as propagation of a compression wave (or sound wave). If the piston impulsively moves in the opposite direction, an expansion wave will propagate downstream, transferring the volume expansion work to the downstream fluid.

---

4 Dowling and Ffowcs Williams [1].

8.3.5.3 Total Energy

If Eq. (8.44) is added to Eq. (8.47), the total energy equation reads

$$\frac{\partial}{\partial t}\left(\frac{1}{2}\frac{c^2}{\rho_0}\rho'^2 + \frac{1}{2}\rho_0 u'^2\right) = -\frac{\partial(p'u')}{\partial x} \qquad (8.48)$$

where the term in the bracket on the left-hand side is the total energy density that consists of elastic potential energy density and kinetic energy density, whereas that on the right-hand side is the acoustic energy flux, known as *acoustic intensity*, by I $(= p'u')$.

8.4 Mach Waves

8.4.1 Mach Number

In compressible flows, the speed of a moving object or the speed of flow u is often compared to the reference velocity, i.e. the speed of sound c, and the ratio between two speeds is defined as follows:

$$M = \frac{u}{c} \qquad (8.49)$$

where M is called the Mach number. If $M < 1$, it is subsonic, if $M = 1$, sonic, and if $M > 1$, supersonic. If $0.7 < M < 1.2$, it is transonic, and if $M > 5$, hypersonic. Figure 8.3 shows the supersonic passenger airliner, Concorde, that reached its maximum speed of Mach 2.04 (2180 km/h) at cruising altitude.

8.4.2 Sound Source: A Point Object

When a point object moves through a fluid or if a fluid flows over it, a spherical sound wave is generated by omnidirectional compression. Although the point object is considered extremely small, it is still finite in volume such that it compresses the fluid in all directions with the same magnitude.[5]

8.4.2.1 Moving Mach Waves

If a point sound source, let us say, a small spherical object, is moving at a speed u in the stationary fluid, it continuously compresses the fluid and creates perturbations of density, pressure, and temperature, which will then propagate as sound waves. This process of compression is different from that of the piston movement in a tube (1D planar wave). This can be thought of as a small omnidirectional change of volume at an instant (3D spherical wave).

Figure 8.3 Supersonic commercial aircraft, Concorde; a British–French turbojet-powered supersonic passenger airliner that operated from 1976 until 2003; maximum speed: Mach 2.04 (2180 km/h at cruising altitude). Source: Undiscoveredcountry75/Pixabay

---

5 A spherical sound wave obeys the *inverse-square law*, i.e. $W = (4\pi r^2)I$, where W is the power of the sound source and I is the acoustic intensity. The latter is defined as the product of the acoustic pressure p and the particle velocity v, $I = p\,v$. This law indicates that the acoustic pressure is inversely proportional to the radius, $p \sim 1/r$, since p and v are in-phase.

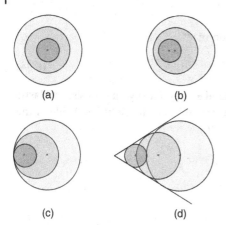

Figure 8.4 A point object moving through a stationary fluid at (a) $M = 0$, (b) $M = 0.5$, (c) $M = 1$, (d) $M = 2$

By compressibility, the disturbed (or compressed) fluid transports the mass, momentum, and energy to the surrounding fluid via molecular interactions. Thus, the interface between the perturbed and unperturbed fluids will propagate through the fluid; this interface is called sound wave. One important difference from the 1D planar wave is that the wave strength or perturbation of the spherical wave will decrease as $\sim 1/r$ under the energy conservation principle.

If the source is moving relative to the sound waves, there is an effect of frequency-shift created to an observer standing on the absolute reference frame. This is called the *Doppler effect* [2]. Figure 8.4 shows that a moving source creates the Doppler effect. When the speed of the moving source exceeds the speed of sound, it creates a Mach cone. In supersonic flows, $M > 1$, sound waves are accumulated at a certain angle, forming a Mach cone, where the Mach angle θ depends on the Mach number as follows:

$$\sin^{-1} \theta = \frac{c\,\Delta t}{u\,\Delta t} = \frac{1}{M} \tag{8.50}$$

8.4.2.2 Standing Mach Waves

It is important to note that the speed of sound is the relative velocity to the fluid into which the sound wave propagates.[6] Figure 8.5 shows a convection of sound with the flow. In this case, a sound wave is generated as the flow passes through the standing point object. Since the sound wave is created in the moving fluid, the distance of travel is uniquely determined by the flow speed and the speed of sound. If the flow speed is at $M = 1$, the left-running sound wave simply stands at the source (Figure 8.5b). If a supersonic flow passes a standing point object, sound waves form an envelope, as sketched in Figure 8.5c. This envelope, called the Mach wave, is the maximum reachable boundary of all sound waves. The higher the flow speed is, the smaller the angle of the Mach wave becomes. This Mach angle can be determined by the ratio between the flow speed and the speed of sound.

It is to be noted that Figure 8.5 is identical to Figure 8.4 if we use a coordinate (or Galilean) transformation from stationary to moving or moving to stationary, because this is the same physics described with different frames of reference.

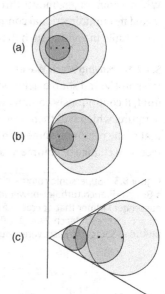

Figure 8.5 A fluid flowing over a standing point object at (a) $M = 0.5$, (b) $M = 1$, (c) $M = 2$; dots: time-tracking markers

------

6 John and Keith [3].

8.4.3 Zone of Silence

It is interesting to note that the Mach wave is the boundary where a component of the incoming supersonic flow velocity vector normal to the Mach wave becomes equal to the speed of sound. Then, the sound wave that emanated from the origin of the disturbance moving at supersonic speed does not move in the absolute reference frame; therefore, the Mach wave is standing in space, if the source of sound is standing. The area before the Mach wave is called the *zone of silence* and the area after that is called the *zone of action*.

The higher the incoming supersonic flow speed is, the more acute the angle becomes, since the component of the incoming supersonic flow velocity vector normal to the Mach wave becomes identical to the speed of sound at more acute angles; the lower the incoming supersonic flow speed is, the larger the Mach angle becomes, since its normal velocity component to the Mach wave becomes sonic with a larger angle. Note that if the incoming flow is sonic, the angle becomes 90° (i.e. the normal component becomes the incoming flow speed itself) and the Mach wave is a vertical line standing right at the point object.

8.5 Normal Shock Waves

8.5.1 Sound Source: A Planar Surface

If a piston is incrementally accelerated to push forward a fluid at rest in a tube, the adjacent fluid is locally compressed and moves with the same speed as the piston. The planar interface between the compressed and uncompressed regions (i.e. a planar sound wave) propagates through the fluid via molecular interactions; to be more specific, the molecules in the compressed region interact with the neighboring molecules, while transporting the momentum and energy (Figure 8.6a). The speed of sound is determined by the isentropic compressibility β_s (or bulk modulus K_d) and density of the fluid

$$c = \sqrt{1/(\rho\,\beta_s)} = \sqrt{K_d/\rho}$$

8.5.1.1 Sudden Compression

If a piston is accelerated to a finite speed, it produces a series of planar sound waves. At each incremental increase of the piston speed, the fluid next to the piston creates a sound wave in the *moving compressed* fluid. If we look at them in the absolute reference frame, the later-produced sound wave moves faster than the preceding wave. The wave speeds can be expressed as follows:

$$c_1 = c(T)$$
$$c_2 = c(T + dT) + dV$$
$$c_3 = c(T + 2\,dT) + 2\,dV$$
$$\cdots$$
$$c_n = c(T + (n-1)\,dT) + (n-1)\,dV \tag{8.51}$$

As time proceeds, the gradient of the speed of sound will steepen so that the sound waves will finally form a normal shock wave. It is, however, to be noted that the overtaking of the waves is prevented by the action of excessive normal viscous stresses that dissipate the energy. The speed of the shock wave V_s will become some intermediate value between the two speeds, i.e. c_1 and c_n, while accelerating the front-running wave and retarding the last-following wave (Figure 8.6b).

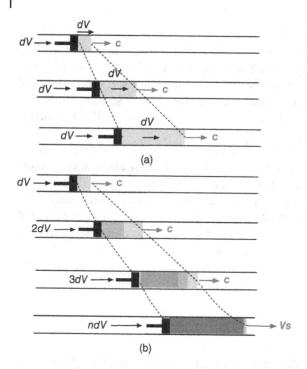

Figure 8.6 Sound wave (a) and normal shock wave (b) formed by compression

8.5.1.2 Irreversibilty

Across the normal shock wave, the excessive normal viscous stresses arise from rapid changes of volume in a short distance, i.e. $-\beta \nabla \cdot \vec{v}$, which is to be added to the thermodynamic pressure, i.e. $\sigma_{ij} = -(p + \beta \nabla \cdot \vec{v}) \delta_{ij}$ [4]. Note that β is the bulk viscosity defined as $\beta = \lambda + \frac{2}{3} \mu$, where λ is the second or dilatational viscosity and μ is the dynamic viscosity.

Irreversibility in the shock wave comes from heat generation produced by viscous dissipation of mechanical energy in an adiabatic process. From the Gibbs equation, an iso-enthalpy process, $dh = 0$ across the shock wave can be expressed as follows:

$$ds = d(\ln p^R) \tag{8.52}$$

In this case, the Stokes hypothesis is not valid except in monoatomic gases, and the mechanical pressure $\bar{p}$ is not the same as the thermodynamic pressure p:

$$\bar{p} - p = -\beta (\nabla \cdot \vec{v}) \neq 0 \tag{8.53}$$

where the bulk viscosity $\beta \neq 0$ (or $\lambda \neq -2/3 \ \mu$).

8.5.1.3 Sudden Expansion

If the piston is incrementally accelerated in the backward direction, a sound wave is generated by expanding the volume, which propagates downstream at the speed of sound (Figure 8.7a). If it is accelerated to a finite speed in the backward direction, a series of expansion waves will be created, moving at different speeds. In this case, the later-produced wave moves slower than the preceding wave in the absolute reference frame, because it propagates through an *expanded backward-moving fluid*, e.g. the wave speeds are

$$c_1 = c(T)$$
$$c_2 = c(T - dT) - dV$$

Figure 8.7 Sound wave (a) and a series of expansion waves (b) formed by expansion

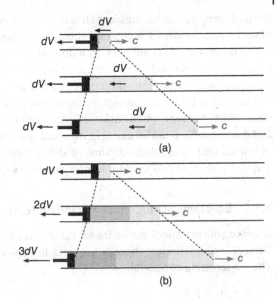

$$c_3 = c(T - 2\,dT) - 2\,dV$$

$$\cdots$$

$$c_n = c(T - (n-1)\,dT) - (n-1)\,dV \tag{8.54}$$

As time proceeds, the expansion waves will spread and never form a shock (Figure 8.7b).

8.5.2 Shock Tube

A shock tube is initially partitioned into two by a diaphragm that will rupture at a certain pressure ratio. The space on the left is filled with a compressed gas, while that on the right is often left for the ambient air (Figure 8.8). When the diaphragm breaks, the compressed gas suddenly moves

Figure 8.8 Characteristic lines of shock wave, contact surface, and expansion waves in shock tube

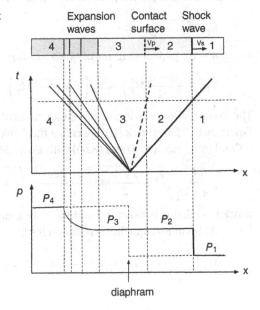

forward, accelerating the ambient air downstream. As a result, a normal shock formed by collapse of sound waves moves at a constant supersonic speed through the ambient air.

The normal shock is followed by a downstream-moving contact surface, an interface between the compressed gas and the ambient air. Note that this is a material surface, analogous to the piston head. Expansion waves are also created by the pressure difference between the two gases, which will propagate upstream through the compressed gas. In this case, the contact surface is the piston head which suddenly pulls backward the uncompressed gas. It is interesting to note that waves and contact surface can be traced in time and space by *characteristic lines*. If the shock tube is used as a wind tunnel, the duration time of steady flow can be estimated by the wave characteristics. The shock tube is often used for a short-duration supersonic wind tunnel.

8.5.3 Equations of Motion for a Standing Normal Shock Wave

A moving normal shock can be transformed into a stationary one by employing a moving reference frame. For the stationary shock wave, we define the variables of 1 and 2 before and after the shock. The conservation of mass shows

$$\rho_1 u_1 = \rho_2 u_2 \tag{8.55}$$

or in terms of Mach number,

$$\frac{p_1}{RT_1} M_1 \sqrt{\gamma RT_1} = \frac{p_2}{RT_2} M_2 \sqrt{\gamma RT_2} \tag{8.56}$$

With pressure being the only force acting on the control surface, the conservation of momentum shows

$$p_1 - p_2 = \rho_2 u_2^2 - \rho_1 u_1^2 \tag{8.57}$$

or in terms of Mach number,

$$p_1(1 + \gamma M_1^2) = p_2(1 + \gamma M_2^2) \tag{8.58}$$

For an adiabatic process, the conservation of energy also shows

$$h_1 + \frac{u_1^2}{2} = h_2 + \frac{u_2^2}{2} \tag{8.59}$$

where $h = e + p/\rho$, or in terms of Mach number,

$$T_1 \left(1 + \frac{\gamma - 1}{2} M_1^2\right) = T_2 \left(1 + \frac{\gamma - 1}{2} M_2^2\right) \tag{8.60}$$

The conservation equations of mass, momentum, and energy for the normal shock wave, Eqs. (8.56), (8.58), and (8.60), are called the Rankine–Hugoniot relations (or jump conditions).

Combining Eqs. (8.58) and (8.60) with Eq. (8.56) yields

$$\frac{M_1}{1 + \gamma M_1^2} \sqrt{1 + \frac{\gamma - 1}{2} M_1^2} = \frac{M_2}{1 + \gamma M_2^2} \sqrt{1 + \frac{\gamma - 1}{2} M_2^2} \tag{8.61}$$

which is a relation between M_1 and M_2. If we square both sides and let the left-hand side be L, then Eq. (8.61) becomes a quadratic equation for M_2^2:

$$M_2^4 \left(\frac{\gamma - 1}{2} - \gamma^2 L\right) + M_2^2(1 - 2\gamma L) - L = 0 \tag{8.62}$$

Figure 8.9 M_1 vs. M_2 across a normal shock wave

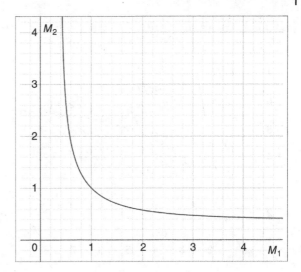

from which we obtain a solution for M_2^2:

$$M_2^2 = \left(M_1^2 + \frac{2}{(\gamma - 1)} \right) \Big/ \left(\frac{2\gamma}{(\gamma - 1)} M_1^2 - 1 \right) \tag{8.63}$$

Figure 8.9 shows a variation of M_2 vs. M_1, which has meaning only if $M_1 > 1$. If $M_1^2 \to (\gamma - 1)/2\gamma$ (or $M_1 \to 1/\sqrt{7}$, for $\gamma = 1.4$), then $M_2 \to \infty$. If $M_1 \to \infty$, then $M_2 \to 1/\sqrt{7}$. Note that M_1 is always greater than 1, since it is impossible for expansion shock to occur, as it violates the second law of thermodynamics.

From the energy equation, Eq. (8.60), we can obtain T_2/T_1 as a function of M_1:

$$\frac{T_2}{T_1} = \left(1 + \frac{\gamma - 1}{2} M_1^2 \right) \left(\frac{2\gamma}{\gamma - 1} M_1^2 - 1 \right) \Big/ \left(\frac{(\gamma + 1)^2}{2(\gamma - 1)} M_1^2 \right) \tag{8.64}$$

and from the momentum equation, Eq. (8.58), we can obtain p_2/p_1 as a function of M_1:

$$\frac{p_2}{p_1} = \frac{2\gamma M_1^2}{\gamma + 1} - \frac{\gamma - 1}{\gamma + 1} \tag{8.65}$$

and from the continuity equation, Eq. (8.56), we can obtain ρ_2/ρ_1 as a function of M_1:

$$\frac{\rho_2}{\rho_1} = \frac{(\gamma + 1)M_1^2}{(\gamma - 1)M_1^2 + 2} \tag{8.66}$$

Figure 8.10 shows variations of T_2/T_1, p_2/p_1, and ρ_2/ρ_1 vs. M_1 across a normal shock wave. It is to be noted that Figures 8.9 and 8.10 have meaning only if $M_1 > 1$. This can be proven by inserting Eqs. (8.64) and (8.65) into the Gibbs equation:

$$s_2 - s_1 = c_p \ln(T_2/T_1) - R \ln(p_2/p_1) \tag{8.67}$$

to obtain a variation of $(s_2 - s_1)/c_p$ vs. M_1 across the normal shock wave. Figure 8.11 clearly shows that the graph is only valid for $M_1 > 1$, since $(s_2 - s_1)/c_p$ is always positive by the second law of thermodynamics.

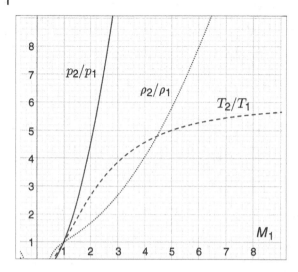

Figure 8.10 T_2/T_1, p_2/p_1, and ρ_2/ρ_1 vs. M_1 across a normal shock wave

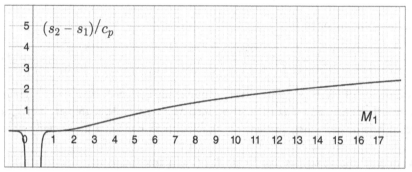

Figure 8.11 $(s_2 - s_1)/c_p$ vs. M_1 across a normal shock wave; only valid for $(s_2 - s_1)/c_p > 0$ by the second law of thermodynamics

8.6 Oblique Shock Waves

8.6.1 Sound Source: A Pointed Object

If a pointed object such as a wedge of an acute angle moves at a constant supersonic speed, or if a supersonic flow passes over such a pointed object, the stream must be deflected by the wedge angle through an oblique shock wave, of which the occurrence is the result of the collapse of multiple Mach waves [5].

8.6.1.1 Moving Oblique Shock Wave

If a pointed object, for instance, a wedge with an angle of less than 25°, moves at a constant supersonic speed in a fluid at rest, it hardly pushes the fluid forward. The wedge allows most of the fluid to pass over it; this is referred to as slip (or slip-backward) mode. Furthermore, the wedge apex can be viewed as a very small curved (concave) body made of a series of point objects, by which the fluid gradually changes its flow direction as it passes over the segmented surfaces.

Note that Mach waves are produced at each segmented surface but in different orientations. As illustrated in Figure 8.12, the orientation of the Mach wave depends on the moving direction of the point object or direction of the flow past the point object. If the velocity vector of a supersonic flow is tilted at an angle, then the Mach wave is tilted at the same angle. It is to be reminded that the

Figure 8.12 Mach waves in different orientations; a point object moving at $M = 2$ in two different directions (a); a fluid flowing over a point object at $M = 2$ in two different directions (b)

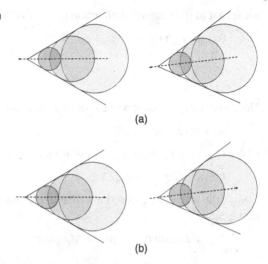

(a)

(b)

moving Mach waves produced from each segmented surface intersect each other but cannot pass through; instead, they collapse into an oblique shock wave. This is exactly the same physics of a normal shock produced by the piston suddenly accelerated in a tube (see Section 8.5.1).[7]

Now, the flow lifted up from the wedge apex will act as a wedge apex slightly displaced downstream to the flow above. As a result, another oblique shock wave is produced at a later time and space and this process continuously repeats. As a result, an oblique shock wave is attached to the apex of the wedge and moving with it.

8.6.1.2 Standing Oblique Shock Wave

If an incoming supersonic flow passes the wedge apex, it encounters a series of small segmented surfaces with successive changes of incremental angles, by which a series of standing Mach waves will be produced with steeper angles. It is, however, to be noted that the Mach waves cannot intersect each other due to irreversibility associated with viscous dissipation and heat transfer via excessive $\nabla \cdot \vec{v}$ (rapid changes of volume in very short distance); they will collapse into a single oblique shock wave.

Now, the incoming supersonic flow that has become tangent to the wedge surface at the apex plays like another wedge apex slightly displaced downstream to make the incoming flow above turn parallel to the wedge surface, producing again a small segment of oblique shock wave. Therefore, the whole supersonic incoming flow becomes tangent to the wedge surface through a standing oblique shock wave emanating from the wedge apex.

8.6.2 Equations of Motion for a Standing Oblique Shock Wave

8.6.2.1 Supersonic Compression by Turning

For a control volume taken over an oblique shock wave, the conservation equation of mass is

$$\rho_1 \, v_{1n} = \rho_2 \, v_{2n} \tag{8.68}$$

where the subscript n denotes the component normal to the oblique shock wave. The conservation equation of momentum in the direction normal to the oblique shock wave is written as follows:

$$p_1 - p_2 = \rho_2 \, v_{2n}^2 - \rho_1 \, v_{1n}^2 \tag{8.69}$$

---

7 Liepmann and Roshko [5].

while that in the tangential direction is

$$0 = (\rho_2 \, v_{2n}) \, v_{2t} - (\rho_1 \, v_{1n}) \, v_{1t} \tag{8.70}$$

or

$$v_{1t} = v_{2t} \tag{8.71}$$

The conservation equation of energy for the control volume reads

$$h_1 + v_1^2/2 = h_2 + v_2^2/2 \tag{8.72}$$

where $v^2 = v_n^2 + v_t^2$. Since $v_{1t} = v_{2t}$, Eq. (8.72) is written as follows:

$$h_1 + v_{1n}^2/2 = h_2 + v_{2n}^2/2 \tag{8.73}$$

Now, the wave angle and the wedge angle can be related to each other through the equations above. Since

$$M_{1n} = M_1 \, \sin\theta, \qquad M_{1t} = M_1 \, \cos\theta \tag{8.74}$$

$$M_{2n} = M_2 \, \sin(\theta - \delta), \qquad M_{2t} = M_2 \, \cos(\theta - \delta) \tag{8.75}$$

we can express the shock wave angle θ and deflection angle δ for a given M_1.

From Eq. (8.71),

$$M_{2t}/M_{1t} = \sqrt{T_1/T_2} \tag{8.76}$$

where T_1/T_2 is given from the normal shock condition, i.e. Eq. (8.64). Thus, it follows that

$$\frac{M_2 \, \cos(\theta - \delta)}{M_1 \, \cos\theta} = \sqrt{\left(\frac{(\gamma+1)^2}{2(\gamma-1)} \, M_1^2\right) \Big/ \left(1 + \frac{\gamma-1}{2}M_1^2\right)\left(\frac{2\gamma}{\gamma-1}M_1^2 - 1\right)} \tag{8.77}$$

where M_2 can be found from the normal shock condition, i.e. Eq. (8.63) as follows:

$$M_2 = \sqrt{\left(M_1^2 + \frac{2}{(\gamma-1)}\right) \Big/ \left(\frac{2\gamma}{(\gamma-1)} \, M_1^2 - 1\right)} \tag{8.78}$$

Combining Eqs. (8.77) and (8.78) yields a relation between θ and δ for a given M_1 and γ (Figure 8.13). From Eq. (8.77), we can also obtain a relation between M_2 vs. δ for a given M_1 and γ.

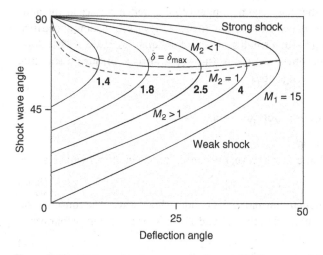

Figure 8.13 Oblique shock wave angle θ vs. deflection angle δ for a given M_1 and γ

8.7 Prandtl–Meyer Expansion Waves

8.7.1 Equations of Motion for a Prandtl–Meyer Expansion Fan

When a supersonic flow turns a convex corner, it expands and speeds up, forming a so-called Prandtl–Meyer fan, which is also referred to as supersonic expansion by turning. For a control volume taken over a Mach wave, the conservation equation of momentum in the direction tangent to the Mach wave shows

$$v\cos\mu = (v + dv)\cos(\mu + dv) \tag{8.79}$$

where v is the incoming flow speed, μ is the angle between v and v_t, and dv is the infinitesimally small turning angle of the convex corner. Since dv is a very small angle, Eq. (8.79) can be expressed as follows:

$$v\cos\mu = (v + dv)(\cos\mu - dv\sin\mu) \tag{8.80}$$

and can be simplified as follows:

$$dv/v = dv\tan\mu \tag{8.81}$$

Since $\sin\mu = 1/M$, it follows

$$\frac{dv}{v} = \frac{dv}{\sqrt{M^2 - 1}} \tag{8.82}$$

Now, differentiating $v = M\sqrt{\gamma R T}$ yields

$$\frac{dv}{v} = \frac{dM}{M} + \frac{1}{2}\frac{dT}{T} \tag{8.83}$$

and differentiating $T_t = T(1 + \frac{(\gamma-1)}{2}M^2)$ for adiabatic flow yields

$$\frac{dT}{T} + \frac{1}{2}\frac{(\gamma - 1)MdM}{1 + \frac{(\gamma-1)}{2}M^2} = 0 \tag{8.84}$$

If we combine Eqs. (8.83) and (8.84),

$$\frac{dv}{v} = \frac{dM}{M}\left(\frac{1}{1 + \frac{(\gamma-1)}{2}M^2}\right) \tag{8.85}$$

If we substitute Eq. (8.85) into Eq. (8.82), then it reads

$$dv = \frac{dM}{M}\left(\frac{\sqrt{M^2 - 1}}{1 + \frac{(\gamma-1)}{2}M^2}\right) \tag{8.86}$$

For a finite turning angle, Eq. (8.86) can be integrated to

$$\int_{v_{ref}}^{v} dv = \int_{M_{ref}}^{M} \frac{dM}{M}\left(\frac{\sqrt{M^2 - 1}}{1 + \frac{(\gamma-1)}{2}M^2}\right) \tag{8.87}$$

and if the reference state be $v = 0$ at $M = 1$, then it reads

$$v = \sqrt{\frac{\gamma + 1}{\gamma - 1}}\tan^{-1}\sqrt{\frac{\gamma - 1}{\gamma + 1}(M^2 - 1)} - \tan^{-1}\sqrt{M^2 - 1} \tag{8.88}$$

where v is the angle through which a sonic flow is expanded to become a supersonic flow of M. It is to be noted that the pressure, temperature, and density variations across the expansion waves can be found through the isentropic relations.

8.8 Bow Shock Waves

8.8.1 Sound Source: A Blunt Object

If a blunt object is placed in a fluid moving at a constant speed, the fluid approaching the body decelerates, changes direction (diverging away from the centerline), and passes over the object. By the Galilean transformation, the same physics can be explained as if a blunt object moves at a constant speed in a fluid at rest. The fluid in front of the body moves forward (i.e. push-forward mode) and diverges away from the centerline as it passes over the body (i.e. slip-backward mode), while drawing a curved trajectory (Figure 8.14). Note that the instantaneous streamlines and pathlines around a moving body are different; the former is a locus tangent to the velocity vectors at an instant, whereas the latter is the trajectory of a fluid particle traced in time.

In both cases, the compressed fluid around the body depends on its bluntness (or aspect ratio) and speed of the body or fluid. The more blunt the body is (or the slower the speed is), the more fluid there is pushed forward. By contrast, the more slender the body is (or the faster the speed is), the less fluid there is pushed forward. If the speed of the flow or body is supersonic, it exceeds the speed of sound so that the compressed area cannot be extended without limit. It is limited to the region that sound waves can reach.

8.8.2 Moving Bow Shock Wave

When a blunt body moves at a constant speed in a compressible fluid, it will compress the region ahead but not the entire fluid, as done by a piston accelerating in a tube.[8] The fluid in front of the blunt body is pushed-forward but soon diverged and slipped backward over the body. Due to the two modes of motion, i.e. a push-forward mode and a slip-backward mode, the compressed area cannot be extended without limit; it is finite, but its size depends on the moving speed of the body and its shape.

If a blunt body moves at a supersonic speed, the sound waves produced from the body will propagate upstream and collapse into a bow shock. The shape of the bow shock wave is determined by the balance of fluid motions: a push-forward mode and a slip-backward mode. Which mode is more dominant determines how far the bow shock can stand off from the blunt body. This dominance is also determined by the shape (or figure of which the blunt shape takes) and moving speed of the blunt object.

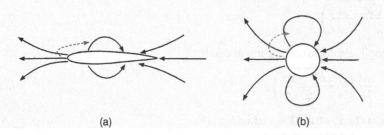

(a)   (b)

Figure 8.14 Fluid in two modes (in front of a moving body): dominance of slip-backward mode (slender body) (a), dominance of push-forward mode (blunt body) (b); pathlines (broken), instantaneous streamlines (solid)

---

8 In the tube, the shock will move faster than the piston since the fluid ahead of the piston cannot escape in the tube.

Obviously, if the slip-backward mode is more dominant (e.g. the faster the moving speed and the slender the body shape), the stand-off distance becomes shorter and the Mach angle becomes smaller. By contrast, if the push-forward mode is more dominant (e.g. the slower the moving speed and the more blunt the body shape), the distance becomes longer and the Mach angle becomes bigger. It is interesting to note that the wedge at a shallow angle is in full slip-backward mode, resulting in zero standing-off distance, whereas the piston moving in the tube is in full push-forward mode. In this case, the shock stand-off distance continuously increases over time since no flow can slip away from the body.

8.8.3 Standing Bow Shock Wave

As the supersonic flow approaches the blunt body, the flow at the centerline decelerates to a complete stop at the stagnation point of the body, while some other diverges away from the centerline and passes over the body. These two modes of flow motion (i.e. push-forward and slip-backward) will determine the shape, stand-off distance, and strength of the bow shock wave.

The bow shock (i.e. detached shock) formation is due to the fact that the decelerating fluid in front of the body produces sound waves that propagate upstream, as if a piston is accelerating upstream in a tube. These sound waves will then collapse into a normal shock because the later-produced ones move faster the preceding ones (i.e. piston acceleration effect in a relative sense). It is, however, to be noted that the shock wave moves upstream until it reaches a certain distance, where the supersonic inflow speed and the shock wave speed are equal and opposite in direction. The so-called *stand-off distance* is determined by how soon the normal shock speed becomes equal to that of the incoming supersonic flow.

For the same incoming flow speed, the more blunt the body, the more dominant the push-forward mode than the slip-backward mode. If then, the compression of the fluid will be stronger and the upstream-moving shock speed will be greater. As a result, the shock stand-off distance becomes longer and the Mach angle becomes bigger. By contrast, the slender the body, the more dominant the slip-backward mode. Thus, the shock strength will be weaker and the upstream-moving shock speed will be slower, which results in a shorter shock stand-off distance and smaller Mach angle.

It can also be said that for a given shape of body, the faster the incoming supersonic flow, the more dominant the slip-backward mode than the push-forward mode. Thus, the stand-off distance becomes shorter and the Mach angle becomes smaller (Figure 8.15). This slip-backward

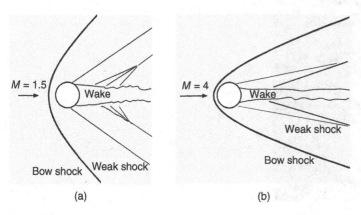

Figure 8.15 Bow shock in supersonic flow past a sphere; push-forward mode dominance at $M = 1.5$ (a), slip-backward mode dominance at $M = 4$ (b)

mode dominance works effectively for incoming Mach numbers less than 3. At Mach numbers higher than 3 or 4, however, this slip-backward mode dominance starts to break; the stand-off distance is nearly the same, although the Mach angle still becomes smaller [5]. Note that the overall structure of the bow shock wave is completed by the flow passing over the afterbody that produces the oblique shock wave, and the shock in between is determined by the flow in the junctional area.

8.9 Rocket Nozzle Flows

In rocket engines, the combusted gas rapidly expands in an isochoric process, greatly increasing the pressure and temperature. Due to the large pressure difference between the combustion chamber and the ambient air, the combusted gas is discharged through a convergent and divergent nozzle, called the C–D nozzle in short (Figure 8.16).[9] The flow in the rocket nozzle can be considered as steady, inviscid, quasi-one-dimensional, and isentropic (or reversible and adiabatic), since changes in cross-sectional area often dominates heat transfer and friction at the nozzle sidewall.

8.9.1 Equations of Motion in Isentropic Process

The conservation law of mass shows

$$\rho \, uA = (\rho + d\rho)(u + du)(A + dA) \tag{8.89}$$

Figure 8.16 Space X Merlin1c engine in Falcon 9 rocket; length: 2.92 *m*, dry weight: 630 *kg*. Source: SpaceX-Imagery/Pixabay

---

9 Rocket: Solid and Liquid Propellant Motors (Youtube).

and by dropping the higher-order terms, it can be expressed as follows:

$$\frac{d\rho}{\rho} + \frac{du}{u} + \frac{dA}{A} = 0 \tag{8.90}$$

Meanwhile, the conservation law of momentum shows

$$pA + (p + dp/2)dA - (p + dp)(A + dA) = \rho\, uA\, ((u + du) - u) \tag{8.91}$$

and it can be rearranged in a similar way as follows:

$$\frac{dp}{\rho} = -u\, du \tag{8.92}$$

which shows that the sign of pressure change is always opposite to that of velocity change.

In isentropic flow, Eqs. (8.90) and (8.92) can be combined with $dp = c^2 d\rho$ as follows:

$$(1 - M^2)\, dp = \frac{\rho\, u^2}{A}\, dA \tag{8.93}$$

If $M < 1$ (subsonic), $dA < 0$ in the C-nozzle will yield $dp < 0$ and the momentum equation Eq. (8.92) shows $du > 0$. On the other hand, $dA > 0$ in the D-nozzle yields $dp > 0$ and thus, $du < 0$ (Figure 8.17, left column).

If $M > 1$ (supersonic), $dA < 0$ in the C-nozzle will yield $dp > 0$ and thus, $du < 0$ (which will then act like a diffuser in subsonic flow). On the other hand, $dA > 0$ in the D-nozzle leads to $dp < 0$ and thus, $du > 0$ (which will then act like a nozzle in subsonic flow) (Figure 8.17, right column) [6].

8.9.2 Physical Interpretations

In high-speed rocket nozzle flows, an incremental change of the cross-sectional area in the streamwise direction produces not only a change in the flow speed but also a change in the fluid volume. Therefore, changes of pressure, density, and temperature are coupled through isentropic process, obeying the laws of conservation of mass, momentum, and energy. The coupling processes in subsonic and supersonic flows are described as follows:

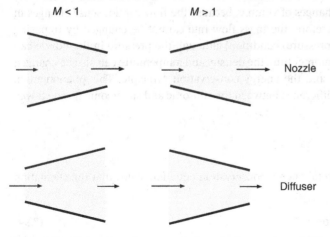

Figure 8.17 Subsonic and supersonic flows in the nozzle and diffuser; C-nozzle ($M < 1$) and D-nozzle ($M > 1$)

8.9.2.1 Subsonic Flows

In subsonic C-nozzle flow, the upstream fluid senses in advance the presence of the nozzle sidewall by sound waves traveling in the upstream and downstream directions. As a result, the incoming fluid is accelerated to compensate for the space decreased by the side wall. The pressure change is then determined by the momentum equation, while the density change should also be taken into account due to the volume change. In isentropic process, the density changes as $dp = c^2 d\rho$.[10] The temperature (or internal energy) also decreases due to the volume expansion work ($c_v \, dT = d\dot{w} = -p \, dv$) done at a rate on the fluid.

In subsonic D-nozzle flow, the fluid is decelerated to fill up the space increased by the sidewall. When the fluid is decelerated, the pressure increases through conservation of momentum. Due to the compressibility effect, density and temperature increase through volume contraction.

8.9.2.2 Supersonic Flows

In supersonic C and D-nozzle flows, an incremental change of the cross-sectional area is not sensed in advance because the signaling wave speed associated with geometric change is slower than the flow speed. As a result, the incoming supersonic flow successively encounters a sudden incremental change of volume, as if a piston in a tube is either running toward the flow (in the C-nozzle) or running away from the flow (in the D-nozzle).

In supersonic C-nozzle flow, the upstream fluid cannot sense in advance the presence of the nozzle sidewall. Due to sidewalls (i.e. blockage effects), the volume is abruptly compressed. Therefore, the fluid decelerates in a space where the pressure is increased by sudden compression of volume. In this case, a force balance is set between the inertial force of the decelerating fluid and the normal reaction force from downstream. Therefore, the instant compression of volume increases density and temperature through volume compression work done on the fluid (as in subsonic D-nozzle flow).

In supersonic D-nozzle flow, the anti-blockage effect, which is not sensed in advance, makes the fluid instantly expand its volume and accelerates to fill up the space where the pressure is lowered due to volume expansion. This fluid acceleration is the inertial force acting backwards, and this thrust force balances with the normal reaction force from upstream. The temperature is also decreased by the volume expansion work done on the fluid through an isentropic process.

[**Notes**] One important physics in supersonic C and D-nozzle flows is that the fluid successively changes its speed through sudden changes of volume, because the flow cannot sense changes in cross-sectional area in advance. Therefore, the mass flow rate cannot be changed by imposing different upstream or downstream (pressure) condition, although the pressure in the nozzle can be changed through conservation of momentum; the density and temperature can also be changed according to the isentropic relation and the energy conservation principle. This phenomenon, called *chocking*, is the most distinct difference between the subsonic and supersonic nozzle flows.

8.9.3 Stagnation Properties

8.9.3.1 Stagnation Temperature

In compressible adiabatic flow, the total energy conservation equation states that the stagnation enthalpy is conserved

$$h_t = h + \frac{u^2}{2} = e + \frac{p}{\rho} + \frac{u^2}{2} = \text{const.} \tag{8.94}$$

---

10 $dp = c^2 d\rho \;\rightarrow\; \frac{dp}{p} = \gamma \frac{d\rho}{\rho} \;\rightarrow\; p \sim \rho^\gamma$

Since $h_t - h = c_p (T_t - T)$, Eq. (8.94) is written as follows:

$$T_t = T + \frac{u^2}{2 c_p} = \text{const.} \tag{8.95}$$

and in terms of Mach number, it reads

$$T_t = T \left(1 + \frac{\gamma - 1}{2} M^2 \right) = \text{const.} \tag{8.96}$$

Note that this relation is valid as long as the process is adiabatic, regardless of whether it is reversible or irreversible.

8.9.3.2 Stagnation Pressure

In isentropic flow, the total energy conservation equation states that the stagnation pressure is conserved. The Gibbs equation shows

$$Tds = dh - dp/\rho = 0 \tag{8.97}$$

It then follows that

$$dh = dp/\rho \quad \text{or} \quad de = -p \, dv \tag{8.98}$$

where the latter shows that the internal energy change is due solely to the volume expansion (or compression) work done on the fluid at a rate, not due to the rate of viscous dissipation.

For a perfect gas, Eq. (8.98) can be expressed as follows:

$$c_p \, dT = dp/\rho \quad \rightarrow \quad d(\ln T) = d(\ln p^{(\gamma-1)/\gamma}) \tag{8.99}$$

If we integrate both side, then it reads

$$T = c_1 \, p^{(\gamma-1)/\gamma} \quad \text{or} \quad T_t/T = (p_t/p)^{(\gamma-1)/\gamma} \tag{8.100}$$

Combining Eq. (8.96) with Eq. (8.100) then yields

$$p_t = p \left(1 + \frac{\gamma - 1}{2} M^2 \right)^{\gamma/(\gamma-1)} = \text{const.} \tag{8.101}$$

8.9.3.3 Compressibility Effect

If we expand the bracket in Eq. (8.101) using the binomial expansion, it reads

$$p_t = p \left(1 + \frac{\gamma}{2} M^2 + \frac{\gamma}{8} M^4 + \frac{\gamma(2-\gamma)}{48} M^6 + \cdots \right) \tag{8.102}$$

where Eq. (8.102) is only valid if $(\gamma - 1/2) M^2 < 1$ (or $M < 1.0541$ for $\gamma = 1.4$). Since $\gamma \, p \, M^2/2 = \rho \, u^2/2$, Eq. (8.102) can be expressed as follows:

$$p_t = \underbrace{p + \frac{1}{2} \rho u^2}_{\text{(A)}} \underbrace{\left(1 + \frac{1}{4} M^2 + \frac{(2-\gamma)}{24} M^4 + \cdots \right)}_{\text{(B)}} \tag{8.103}$$

Equation (8.103) shows that the stagnation pressure in compressible flow is the sum of the stagnation pressure in incompressible flow Ⓐ and the correction factor for compressibility Ⓑ. The term Ⓐ is known as the Bernoulli equation for incompressible fluid, whereas Ⓑ represents the compressibility effect of the fluid. For example, the correction term for compressibility Ⓑ is approximately $1/400, 1/36, 1/16, 1/9, 1/4$ at $M = 1/10, 1/3, 1/2, 2/3, 1$. In contrast, the density changes are rather weak when the flow speed of a subsonic flow is less than $M = 0.3$. Otherwise, the density change plays its role in gas dynamics.

8.9.4 Mass Flow Rate in Isentropic Flows

At steady-state, the mass flow rate at a cross-sectional area of A can generally be expressed as follows:

$$\dot{m} = \rho\, A\, u = \frac{p}{RT}\, AM\sqrt{\gamma RT} \tag{8.104}$$

If the flow is isentropic, the stagnation temperature and pressure relations obtained in Eqs. (8.96) and (8.101) can be used. The mass flow then reads

$$\dot{m} = \frac{p_t}{RT_t}\, A\sqrt{\gamma}\, M\left(1 + \frac{\gamma-1}{2}M^2\right)^{(\gamma+1)/2(1-\gamma)} = \frac{p_t}{RT_t}\, A f(\gamma, M) \tag{8.105}$$

where $f(\gamma, M)$ is defined as follows:

$$f(\gamma, M) = \sqrt{\gamma}\, M\left(1 + \frac{\gamma-1}{2}M^2\right)^{(\gamma+1)/2(1-\gamma)} \tag{8.106}$$

Since p_t, T_t, and $\dot{m}$ are constant, we can find a relation for A/A^*, where $*$ indicates the sonic reference condition, i.e.

$$\frac{A}{A^*} = \frac{f(\gamma, 1)}{f(\gamma, M)} = g(\gamma, M) \tag{8.107}$$

where $g(\gamma, M)$ is defined as follows:

$$g(\gamma, M) = \left(\frac{\gamma+1}{2}\right)^{(\gamma+1)/2(1-\gamma)} / M\left(1 + \frac{\gamma-1}{2}M^2\right)^{(\gamma+1)/2(1-\gamma)} \tag{8.108}$$

Figure 8.18 shows the variation of A/A^* against M, indicating a possible shape of C–D nozzle, where the subsonic flow becomes sonic in the C-nozzle and supersonic afterward in the D-nozzle. Equation (8.105) can be rearranged as follows:

$$\frac{\dot{m}\, T_t\, R}{p_t\, A} = \sqrt{\gamma}\, M\left(1 + \frac{\gamma-1}{2}M^2\right)^{(\gamma+1)/2(1-\gamma)} \tag{8.109}$$

to show that the maximum flow rate occurs at $M = 1$ (Figure 8.19).

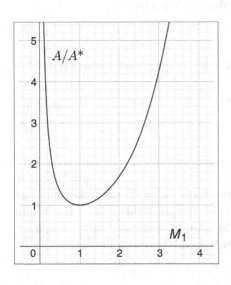

Figure 8.18 $A/A^* = g(\gamma, M)$ vs. M for $\gamma = 1.4$

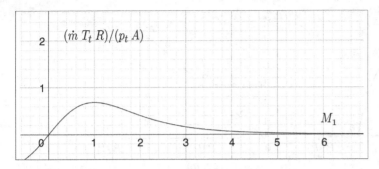

Figure 8.19 $(\dot{m}\,T_t\,R)/(p_t A)$ vs. M for a given A and $\gamma = 1.4$

8.9.5 Isentropic Flow in the C-nozzle

Let us consider a converging nozzle (or C-nozzle) attached to a high-pressure reservoir at $p = p_t$, $T = T_t$, and $u = 0$. At the outlet, a vacuum tank is used with a valve to control the discharge flow rate. As soon as the back pressure p_b is lowered from p_b, the disturbance propagates as sound waves that travel in both upstream and downstream directions, and the subsonic flow is accelerated through the C-nozzle, lowering the pressure as in Figure 8.20. The more the valve opens, the larger the mass flow rate becomes, thus lowering the pressure (a–d).

8.9.5.1 Chocking

This process continues until the flow speed becomes sonic at the exit (d), which we call *chocking*. For $p_b < p^*$, the mass flow rate cannot be increased because the sound wave produced by the action of lowering the back pressure cannot propagate upstream in the absolute frame of reference.

Figure 8.20 Subsonic isentropic flow and chocking (C-nozzle)

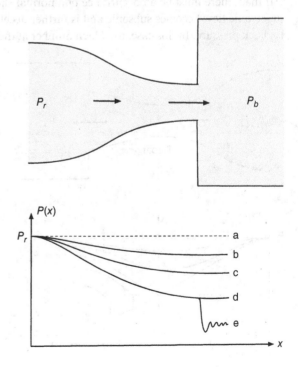

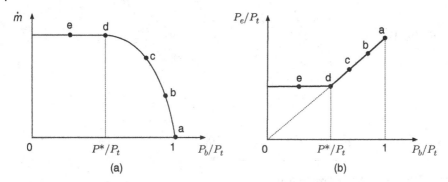

Figure 8.21 Mass flow rate $\dot{m}$ vs. back pressure ratio p_b/p_t (a) and exit pressure ratio p_e/p_t vs. back pressure ratio p_b/p_t (b) (C-nozzle)

What then happens is a sudden expansion of flow at the exit (e), producing the Prandtl–Meyer expansion waves. The mass flow rate and p_e/p_t are plotted against p_b/p_t in Figure 8.21.

8.9.6 The de Laval Nozzle: C–D Nozzles

8.9.6.1 Occurrence of a Normal Shock in the D-Nozzle

Once the flow is chocked or becomes sonic at the throat, mass flow rate in the present nozzle is fixed; it cannot be increased by lowering the back pressure. However, the momentum flux in the D-nozzle can be increased by lowering the back pressure from the critical pressure, p_b^*, although the mass flux of the system is fixed (Figure 8.22). It is to be noted that the flow in the D-nozzle cannot be supersonic all the way to the exit, since the back pressure is still higher than the pressure of the supersonic flow in the D-nozzle.

If then, there must be an occurrence of a normal shock wave in the D-nozzle. As a result, the supersonic flow becomes subsonic and is further decelerated to meet the downstream condition, i.e. back pressure. In this case, the Mach number at the exit can still be increased by lowering the

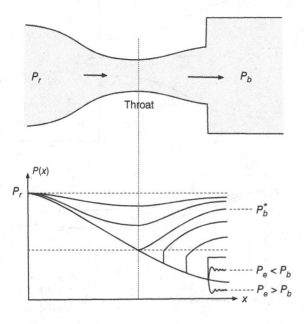

Figure 8.22 Supersonic flow in the de Laval nozzle (C–D nozzle); over-expanded $(p_e < p_b)$ and under-expanded $(p_e > p_b)$ supersonic jets

back pressure; it becomes subsonic or supersonic, depending on the back pressure ratio p_b/p_t. Note that the sonic speed flow at the throat can accelerate in the D-nozzle.

The normal shock wave in the D-nozzle results from the rapid steepening, upstream-propagating sound waves. Note that these sound waves are produced as the fluid decelerates in the plenum while lowering the back pressure. The normal shock stands at a point where the upstream flow speed (sonic or supersonic) is the same as the upstream-moving shock wave speed; therefore, the speed of the upstream-moving normal shock becomes zero in the absolute reference frame. Note that the normal shock starts to form at the throat where the flow becomes sonic, and as the back pressure becomes lower, its location moves toward the exit since the shock strength and its upstream-moving speed become greater.

8.9.6.2 Over-Expanded and Under-Expanded Supersonic Jets

Once the shock moves out of the exit, but the supersonic jet is still over-expanded – the local pressure of the supersonic flow is lower than the back pressure (i.e. $p_e < p_b$) – the exit flow jet must increase the pressure by forming an oblique shock around the exit edge; if then, the jet slightly converges toward the jet centerline. The oblique shocks cannot cross each other. Instead, they form a normal shock at the center of the jet. We call this the Mach disk. The jet continues to flow downstream by forming a series of shock waves and expansion waves, i.e. *shock cells*, to adjust to the ambient flow being entrained with the jet. If the back pressure is too low (i.e. $p_e > p_b$), the under-expanded supersonic jet is expanded at the exit edge by forming Prandtl–Meyer expansion waves; in this case, the jet flow diverges out slightly. The jet then continues to flow downstream with a series of expansion waves and oblique shock waves to adjust itself to the ambient fluid entrained by the jet [6, 7].

References

1 Dowling, A.P., and J.E. Ffowcs Williams, *Sound and Sources of Sound*, 1st ed. West Sussex: Ellis Horwood Limited, 1983.

2 Thompson, P.A., *Compressible-fluid Dynamics*, 1st ed. New York: McGraw-Hill, 1972.

3 John, J.E.A. and T.G. Keith, *Gas Dynamics*, 3rd ed. Pearson, 2006.

4 Lighthill, J., *Waves in Fluids*, 1st ed. Cambridge: Cambridge University Press, 1978.

5 Liepmann, H.W., and A. Roshko, *Elements of Gasdynamics*, 1st ed. New York: Wiley, 1957.

6 Shapiro, A.H., *The Dynamics and Thermodynamics of Compressible Fluid Flow*, Vol. 1, 1st ed. New York: Wiley, 1953.

7 Coles, D., "Channel Flow of a Compressible Fluid," *Illustrated Experiments in Fluid Mechanics*, National Committee for Fluid Mechanics Films, 1980.

8 NASA Goddard, *Insights on Comet Tails Are Blowing in the Solar Wind*, YouTube.

9 Shercliff, J.A., "Magetohydrodynamics," *Illustrated Experiments in Fluid Mechanics*, National Committee for Fluid Mechanics Films, 1980.

Problems

8.1 Discuss how the internal energy changes in the following:

　　a) Sound wave propagation in a tube.

　　b) Frictional intermolecular diffusion in the Rayleigh problem.

8.2 An analogy can be made between the viscosity μ defined as follows:

$$\mu = \frac{\tau}{du/dy} = \frac{shear\ \ stress}{shear\ \ strain\ rate}$$

and the dilatational resistivity Θ defined as follows:

$$\Theta = -\frac{dp}{\dot{D}}\bigg|_S = -\frac{normal\ stress}{volumetric\ dilatation\ rate}$$

a) Explain how the viscosity μ and dilatational resistivity Θ represent the rate of transfer of momentum through molecular interactions.
b) Explain why the latter can represent the speed of a longitudinal wave and why the former cannot.

8.3 Explain how the ratio between the isothermal compressibility and the isentropic compressibility equals the specific heat ratio γ

$$\frac{\beta_T}{\beta_S} = \frac{c_p}{c_v} = \gamma$$

8.4 The speed of sound is considered as a dilatational stiffness measure of a fluid. Physically explain why the speed of sound increases as the temperature increases in gases.

8.5 Explain why the speed of sound of water is approximately five times faster that of air, comparing both at sea-level.

8.6 Rocket nozzle flows can be considered as an isentropic process.
a) Derive the following isentropic flow equation:

$$p_t = p \left(1 + \frac{\gamma - 1}{2}M^2\right)^{\gamma/(\gamma-1)} = const.$$

b) Using the binomial expansion, derive the following:

$$p_t = p + \frac{1}{2}\rho u^2 \left(1 + \frac{1}{4}M^2 + \frac{(2-\gamma)}{24}M^4 + \cdots\right)$$

c) Plot the function and discuss the significance of the compressibility effect.

8.7 Sound Fixing and Ranging Channel (SOFAR) is an ocean channel that allows sound to carry great distances. Since this acts as an acoustic guide, it can be used as a communication channel for sea animals, submarines, etc. The existence of SOFAR is based on the fact that the speed of sound in the ocean varies with depth, due to the differences in salinity (salt content), temperature, and pressure variations.

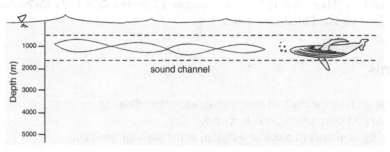

a) Sketch the temperature and pressure variations vs. depth.
b) Sketch the speed of sound vs. depth.
c) Explain why SOFAR can act as an acoustic guide.

8.8 Acoustical design of an orchestra hall is one of the most important areas in architectural acoustics. All aspects of wave reflection, diffraction, absorption, and refraction need to be considered for optimizing the acoustical performance. In engineering acoustics, scale resolution is very important. Explain the reason.

8.9 Let us consider a piston in a tube filled with a gas at rest.
a) Draw the characteristic lines of the shock and piston when the piston is rapidly accelerated to a subsonic speed on the $x - t$ diagram.
b) Draw the characteristic lines of the shock and piston when the piston is rapidly accelerated to a supersonic speed on the $x - t$ diagram.

8.10 A small diaphragm is actuated in a tube by a motor, which horizontally oscillates the diaphragm at a specific frequency f.

Dipole source

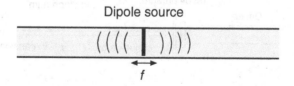

f

a) Assuming that the displacement of the diaphragm is very small, physically explain how the motion of the diaphragm will act as a dipole sound source.
b) Sketch the pressure and velocity distributions $p(x, t)$ and $u(x, t)$.

8.11 The supersonic flow over a wedge with an afterbody can be explained by the characteristics of the piston in the tube.

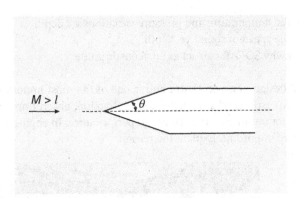

a) Describe the supersonic flow over a wedge with an afterbody.
b) Sketch a characteristics line of the piston path in (x, t) space.
c) Sketch a characteristics line of the shock path in (x, t) space.
d) Describe how the shock speed changes over time in the tube.
e) Sketch the characteristics lines of the Prandtl–Meyer expansion waves in (x, t) space.

8.12 The formation of a bow shock over a blunt object is described by two concept modes of flow motion: a push-forward mode and a slip-backward mode. Use the two modes to describe the following three limiting cases:
a) a supersonic flow over a wedge of an acute angle.
b) a supersonic flow in a tube with a fully closed downstream end wall.
c) a supersonic flow in a tube with a partially closed downstream end wall.

8.13 A supersonic-driven flow can be boosted to a hypersonic flow if we partition another low pressure gas (i.e. expansion tube) by a diaphragm located downstream of the shock tube. A high enthalpy flow can be produced in a short period time by an expansion tube, which can generate a hypersonic flow, as experienced in the re-entry of a rocket. Sketch the characteristic lines in an (x, t) space, identifying the first and second shock waves, contact surfaces, and expansion waves.

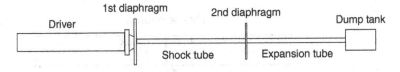

8.14 Let us consider a piston in a tube. If the piston is accelerated to move at a supersonic speed, the fluid ahead of the piston will be compressed while being confined between the piston and the moving shock.

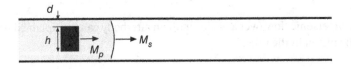

a) What will happen to the shock if $d/h = 0$?
b) What will happen to the shock if $d/h \to 1$?
c) What will happen to the shock if d/h increases from 0 to 1?

8.15 The formation of a standing normal shock in the de Laval nozzle can be explained by the formation of a bow shock wave standing in front of a blunt body at rest.
a) Explain how these two cases are comparable to each other.
b) Describe the correspondence between the two for changes in the incoming flow Mach number.
c) Describe the correspondence between the two for changes in the shock strength.

8.16 In the de Laval nozzle, the mass and momentum equations are written, respectively, as follows:

$$\frac{d\rho}{\rho} + \frac{du}{u} + \frac{dA}{A} = 0$$

and

$$\frac{dp}{\rho} + u \, du = 0$$

a) Express the momentum equation for isentropic flow in terms of Mach number.
b) By comparing the mass conservation equation with the isentropic momentum equation, explain why the pressure distribution is locally symmetric about the throat when M_t is less than 1. M_t is the Mach number at the throat [7].
c) Explain why the pressure distribution is not symmetric about the throat when the Mach number at the throat is sonic, i.e. $M_t = 1$.

8.17 In de Laval nozzle flow, mass flow rate can be controlled by changing the back pressure ratio p_b/p_t in the downstream plenum, where p_b and p_t are the plenum pressure and total pressure.
a) Sketch the mass flow rate in the nozzle against p_b/p_t.
b) Once the nozzle is chocked, the supersonic flow still accelerates in the D-nozzle by matching the pressure condition. Physically describe this mechanism of flow matching and discuss how this is different from the flowing process in a subsonic C-nozzle flow.
c) Physically explain why we cannot hear any flow sound from the nozzle when $M_t = 1$ at the throat.

8.18 Gun firing creates a spherical shock wave centered on the gunâ's muzzle, due to the explosive discharge of gases used to fire the bullet. The bullet also causes a conical bow shock wave as it travels supersonically out of the gun.

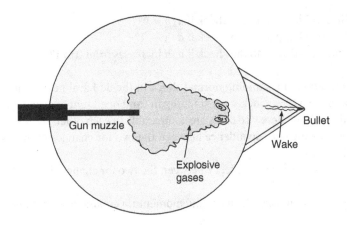

a) Describe how the flows and shocks have developed as shown in the figure.
b) Explain why the bullet moves faster than the spherical shock wave produced by the explosive discharge of gases used to fire the bullet.

8.19 The following picture shows Halley's comet. It shows two tails: the dust tail and gas (or iron) tail [8]. Identify which one is which and justify your answer. (Hint: indicate where the sun is.)

8.20 The photo below shows an interstellar bow shock wave from *Zeta Ophiuchi*, a star located in the constellation of Ophiuchus. This is the most famous bow shock of a massive star; image is from the Spitzer Space Telescope.

a) Explain how a bow shock wave can occur in space. (Hint: plasma physics)
b) Compare the governing equations between the hydrodynamics (HD) and the magneto-hydrodynamics (MHD) and discuss the related flow 0physics [9].
c) Discuss how the wave speeds can be defined in MHD.

Index

Introduction to Fluid Dynamics: Understanding Fundamental Physics, First Edition. Young J. Moon.
© 2022 John Wiley & Sons, Inc. Published 2022 by John Wiley & Sons, Inc.
Companion website: www.wiley.com/go/Moon/IntroductiontoFluidDynamics